ENERGY, POWER and
TRANSPORTATION
TECHNOLOGY

ENERGY, POWER and TRANSPORTATION TECHNOLOGY

Dr. Ralph C. Bohn
Dean of Continuing Education and
Professor, Division of Technology
San Jose State University

Dr. Angus J. MacDonald
Professor, Division of Technology
San Jose State University

Dr. James Fales
Associate Professor,
Industrial Technology Department
Purdue University

Dr. Vincent F. Kuetemeyer
Associate Professor,
Department of Industrial and
Technical Education
Louisiana State University

Bennett & McKnight
a division of
Glencoe Publishing Company
Encino, California

Send all inquiries to:
Glencoe Publishing Company
17337 Ventura Boulevard
Encino, California 91316

Printed in the United States of America

Library of Congress Card Catalog Number: 85-070090

ISBN 0-02-675400-2 (Student text)
ISBN 0-02-675420-7 (Student Activity Guide)
ISBN 0-02-675410-X (Instructor's Resource Guide)

  3 4 5   89 88 87 86

Cover and p. ii photos: Mobil Oil Corp.
                         Stemco Truck Products Co.
                         Colt Industries
                         NASA
                         Cathy Persin

Page i photo: Mobile Oil Corp.

Table of Contents photos: Quaker State
                           U.S. Bureau of Reclamation
                           Los Angeles Intl. Airport
                           Exxon Corp.
                           Cathy Persin

PREFACE

Energy, Power, and Transportation Technology provides a broad overview of both our energy sources and our use of controlled energy (power). One of our major power uses, transportation, is explored in detail. The transportation section in the text provides the material for an exploratory or first course in transportation.

The major emphases in the text are on sources of energy and systems for controlling power. Separate sections expand on this knowledge base to include transportation, heat engines, small engines, new power systems, and career development.

In addition, alternative energy sources are presented in considerable detail. While these sources provide only a fraction of our present energy, they will be extremely important in the future. Solar energy, in particular, will be a vital part of our society's future well-being.

The transportation material is presented in two sections. The first section surveys the operation of our country's transportation system, while the second section explains the five major transportation modes in detail.

We mustn't forget that energy use has many negative effects on the environment. To this end, the text has a separate chapter on the environmental impact of present and developing energy use systems. Conservation is stressed as a way of protecting the environment and our natural resources.

A dramatic photo essay opens the text and introduces the concepts of energy, power, and transportation. The photo essay will also help the reader understand the role of technology in bringing the world to its present state of development. One special emphasis is on the concept of a *systematic* way of doing things. This point in particular is stressed throughout the text.

The most important outcome of reading this text should be a greater understanding of the importance of energy, power, and transportation in our modern society. There are and will continue to be many controversies surrounding these areas. We hope that this book will enable the reader to make intelligent, informed decisions.

The Authors

TABLE OF CONTENTS

A Journey into Technology

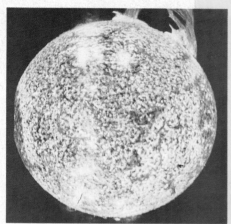

NASA

The sun — without it, nothing could exist on the earth. Our planet would be as lifeless as the moon. The sun is our greatest and most basic source of *energy*.

Our heat and light come from the sun. And the rains and the winds are results of its force. The sun makes plant life possible.

To capture the power of the wind . . .
This vertical-axis wind turbine is part of a research effort to find new and more effective ways of using wind power.

Plants have the special ability to capture and use the sun's energy directly. They use this **solar energy** to grow and to sustain themselves.

People use solar energy both directly and indirectly. For example, sunlight is a source of vitamin D. We absorb this important vitamin **directly** through our skin to help build our bones.

To capture the power of the waters . . .
Water trapped behind dams is released into the water turbines below them to produce electricity. Hydroelectric energy is the only major source of renewable energy now under control.

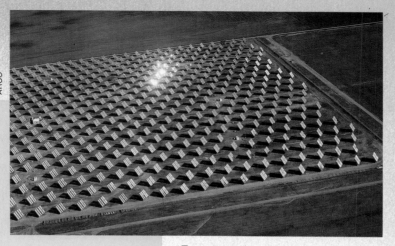

ARCO

*To capture the power of
the sun itself . . .*
Panels of solar cells track (follow) the
sun during the day, changing sunlight
into electricity. This solar power plant
can provide enough electricity for a
community of over 6,000 people.

People have also long used solar energy **indirectly** through their use of wind and water. Wind and flowing water are energy sources that result from the sun's influence. People once used windmills and waterwheels to operate equipment. Today, we use large wind turbines and hydroelectric (water power) plants to produce electricity to run machines. Scientists and engineers have also developed ways to change the sun's energy *directly* into electricity.

At Solar One, mirrors track the sun and focus sunlight on a central receiver. The focused heat changes water into steam. The steam, in turn, operates a turbine to create electricity.

Southern California Edison

During a large-scale power failure, this brilliant skyline would disappear into darkness.

How different the world would be if suddenly there were no electricity! There would be no TV or radio or telephone communications. Subways and elevators would stop. Machines in factories would be still and silent. Lights would go out, and even great cities would be dark.

Suppose it were winter and cold where you lived. Could you stay warm? Electric heating systems would be shut down. Natural gas furnaces would stop, too. Electric motors help force the gas through pipelines. You would be safe with a coal furnace for awhile. But where would the next supply of coal come from? Delivery trucks need fuel, too. And gas pumps operate on electricity. Our society today depends a great deal on electrical power.

Many people think that power is another word for electricity. But anytime we use an energy source to do work, we have *power*. There are many kinds of power, and our technology helps us to use it effectively.

Technology is simply the knowledge of doing things. It is knowing how to change the world to make life easier. We make these changes with tools (like a plow), with simple machines (like pulleys and levers), and with complex machines (like bulldozers and computers).

Technology is often very complicated. But it can be very simple. For example, thousands of years ago, farmers used pointed sticks to make holes in the ground. They would then drop in a few seeds, cover them with dirt, and wait for the plants to grow.

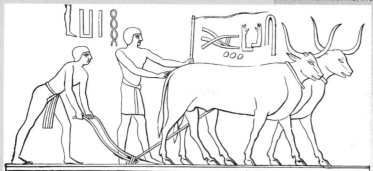

Yesterday . . .
Technology has developed through the ages. The plow is an example of early technology. Its use allowed farmers to grow more grain. Food surpluses brought about trading between countries and changed the course of history.

After many years, someone got the idea to make a curved stick with a handle on it. Using a harness, the stick could be connected to an ox. Then the ox would pull the stick to make a long trench or furrow in the ground. This was good news for the people who had to bend over so much to dig holes. Now they only had to guide the curved stick. This simple tool — the **plow** — is an example of *early technology*. Fuel transportation, electrical power generation, and robotics are examples of *modern technology*.

We can organize technological activities into systems. A **system** is an organized way of doing something. For example, suppose you want to take a ride in a car to get from the city to the country. Sounds like a simple activity, doesn't it? Well, let's look at just one thing involved in that activity — the fuel for the car.

Most cars run on gasoline. Gasoline is a highly refined fuel that comes from crude oil. The crude oil is taken from the ground and transported to an oil refinery. There it is processed into gasoline and other products. The gasoline must then be transported to the service station where you fill up the car's gas tank, start the engine, and take off for the country.

Today . . .
Technology is growing and becoming more complex. Robots have replaced many workers on assembly lines. Robots work faster and can do jobs that are dangerous for humans. But what about the people who have lost their jobs to robots? Are we prepared to deal wisely with the changes that result from advancing technology?

Tomorrow . . .
People used to think crude oil was worthless. But today's technology enables us to tap the energy it contains. We can process the oil into useable forms, such as gasoline. And tomorrow? . . . What new doors might technology open for us?

Do you see the system that is involved in producing and using fuel? First you need an energy source (like crude oil). This is the **system input.** Then you do something with the energy source (like refining it). All systems involve **change** or **processing.** More change goes on when you burn the refined energy source in an engine. What's the result, or *output*, of all this? The engine provides power, the car moves, and you get from the city to the country. Transportation is just one of the outputs that you can get from an energy-use system.

Transportation is a system in itself. The trucks that deliver gasoline to service stations are using the transportation system. If you ride a bus to school, you are using the system, too.

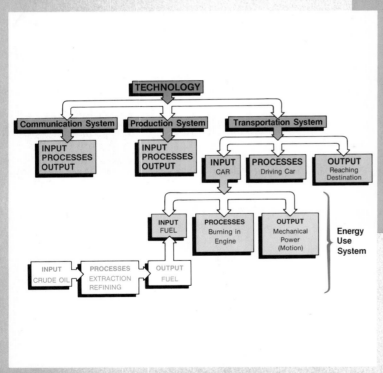

Modern technology is made up of many systems. The systems are all related to each other. For example, we develop *power* from energy with an energy-use system. The other systems — transportation, communication, and production — all need *power* to operate.

Energy use and transportation are two of the big systems that make up technology. Other important systems include communication and production (construction and manufacturing). These and other systems are all related to each other in the total system of technology.

It takes *power* to transport an energy source. This train uses diesel-electric engines to haul coal to generating plants. The engines use another energy source — diesel fuel — to develop the power to move the train.

Power enables communication between people. This solar-powered satellite relays television messages across the nation and around the world.

NASA

People like cars because they make traveling easy and convenient. However, as you can see, our technological advances can "backfire" on us and cause problems — like traffic jams and pollution.

Are you starting to see how wide-ranging technology can be? Remember, it can be as simple as a wooden plow. And it can be as complex as a communication satellite. But today's space shuttle astronauts are doing what the ancient Egyptians did — using technology. Just think of the multitude of tools and machines we have today. All of these things are technology.

The entire history of technology involves humans inventing ways to make life easier. Today we have become almost totally dependent on our technological systems. But that isn't the end of the story. These easier ways of living also change the ways we live!

Someone once said, "We shape our tools, and after that, our tools shape us." After Henry Ford invented the Model T, our world was never the same again. Cars made transportation easier and gave us new ways to spend our spare time. But they also poured pollution into the air, created a lot of noise, and were involved in the deaths of millions of people.

Yes, cars have negative effects. But most people would rather use a car than ride a bike or walk. We are very closely involved with our technology. We aren't happy about the problems caused, but we love the things technology can do for us.

Lawrence Livermore National Laboratory

Great advances in technology were made in the past. But *now* is an exciting time to live. Maybe in the future *you* will be the one to develop a new source of power!

Keep in mind two important ideas as you learn about technology:

• When you change one part of a system, you usually end up changing another part — whether you want to or not.

• You can't get something for nothing.

In very simple terms, if you want to enjoy the benefits of technology, you have to deal with its negative effects.

As responsible citizens, we must take a good look at technology. We must find out what we're gaining by it — and what we're losing. If we're not wise, we could lose the advantages that the human race has gained over thousands of years. But if we're thoughtful and careful, we have much to gain. We can find safe ways to keep our technological advances, and go even further in making the world a better place. Quite a journey!

And now it's time to begin. Tap into your own energy reserves! There are some powerful ideas in this book. Think of them as transportation devices that will carry you on your journey into the world of technology!

10 . . . 9 . . . 8 . . . 7 . . . 6 . . . 5 . . . 4 . . . 3 . . . 2 . . . We have ignition . . . We have lift-off . . . A mighty surge of power lifts the spacecraft into the heavens.

NASA

Energy and Power

Sources of energy are all around us. But before we can use this energy, we must learn to control it. Controlled energy is **power.** Power is used to move automobiles, airplanes, and all other transportation devices. Power is also used in industry and throughout our homes.

It is important to know how much our society depends on controlled energy. Where would we be without power? However, it is also important to know that our use of power can cause serious problems.

In this section, you'll learn about energy sources, and how we control energy to get power. Knowing how much energy or power we have is important, too. Therefore, you'll read about measurement systems. Finally, you'll get a look at some of the problems of power use — and how we might solve these problems.

Welcome to the world of power!

Sources of Energy

1

Most of our energy is produced by the sun. The sun gives us heat and light. We call this energy **radiant energy**. It warms the earth, sustains life, and produces our changing weather conditions.

David Falconer/Frazier Library

Fig. 1-1 **Wood is stored energy. Trees store the sun's energy in the form of wood. As the wood burns, this energy is released.**

The sun's energy is provided to us in many ways. Green plants capture and use radiant energy. They change sunlight, carbon dioxide, and water into food. Plants use this food for growth. The new growth is actually stored energy. Animals that eat plants take in and use this energy. The stored energy is released in them. Then it is used for survival and growth. As with plants, some energy goes into storage.

Energy is also released when plants are burned. Burning plants release the same energy they collected from the sun. See Fig. 1-1.

A great deal of the energy we use is energy that nature has placed in storage. We use coal, oil, and natural gas for much of our energy. All three of these energy sources were originally plants and animals that died millions of years ago.

Flowing water is another major energy source. Again, it is the sun that is behind this energy source. The sun's heat causes the water on our planet's surface to evaporate and form clouds. (**Evaporation** is the changing of a liquid into a gas or vapor.) The water then falls to earth as rain or snow. It collects in lakes or flows to the oceans. We can tap into this cycle of water flow. First we build dams to trap moving water. Then we release the water to generate electricity.

There are many other ways in which the sun provides us with sources of energy. However, if we are to maintain and improve our present way of life, we must learn to make better use of our basic energy sources. Sources of energy fall into three classifications. See Fig. 1-2. The three types of energy sources are the following:

- Exhaustible
- Renewable
- Inexhaustible

National Coal Association

California Redwood Association

Sun Company, Inc.

A. **Exhaustible**   *B.* **Renewable**   *C.* **Inexhaustible**

Fig. 1-2 **There are three basic types of energy sources.**

EXHAUSTIBLE ENERGY

Exhaustible energy sources are those that cannot be replaced once they are used. Coal, oil, natural gas, and uranium are all exhaustible energy sources. We use these four sources for over 90 percent of our present energy needs. Our major uses include transportation, heat, light, and industrial power. There are basically two classifications of exhaustible energy sources: fossil fuels and uranium.

Fossil Fuels

Fossil fuels are deposits of coal, petroleum (oil), and natural gas found underground. When we use these fuels, we are actually using the sun's energy that was stored in plants and animals millions of years ago.

Figure 1-3 shows how fossil fuels were formed. Heavy forest growths in swamplands were covered over by other plants and water. Fallen plants and dead animals were pushed down by

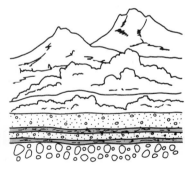

Fig. 1-3 **Fossil fuels originated from ancient plant and animal life that was buried in swamps and later subjected to high pressure and temperature: *A.* As plants and animals died, they fell into swamps, forming a layer of organic matter. *B.* More plants and animals died, forming a new top layer. High pressure and temperature produced *peat*, the first material in the formation of fossil fuels. *C.* Shifts in the earth and deposits of sand and gravel combined to raise the pressure and temperature. The end results were coal, oil, and natural gas.**

Fig. 1-4 **Miners use powerful machines to extract coal from underground.**

the weight of other plants, animals, and water. Eventually, shifts inside the earth pushed these decaying remains underground. There the tremendous weight of the earth pressed them together. Over millions of years, pressure and heat slowly changed the once-living matter into coal, oil, and natural gas.

We can release the energy stored in fossil fuels very easily. When we need energy, we simply burn the fuel. Fossil fuels release their energy in the form of heat. We can use this heat directly. We can also convert (change) it into other forms of energy, usually mechanical energy or electrical energy.

Unfortunately, our supplies of fossil fuels are decreasing. Because of the millions of years it takes to create fossil fuels, we cannot replace them. Therefore, we must start to develop other sources of energy. This way, we can rely less on fossil fuels.

Coal. Coal is a solid, combustible (burnable) substance that is brown to black in color. We obtain it through mining operations. See Fig. 1-4. Once it is mined, coal can be burned immediately as a fuel.

Coal is our most abundant (plentiful) fossil fuel. We have enough coal reserves (untapped supplies) to last 500 years at our present rate of use. However, our use of coal is expected to increase. Therefore, we may use up our coal supplies more quickly. Coal shortages may begin to occur in less than 200 years. This seems like a long time. However, it is really a very short time compared to the millions of years it took to make the coal.

At the present time, there is plenty of coal, and it is easy to use. However, coal has some disadvantages. Mining operations often deface the land. Aside from this, mining is a dangerous occupation. Coal also produces large amounts of air pollution when it is burned.

The problems involved in obtaining and using coal are solvable. However, the solutions are expensive. The costs of reclaiming land, protecting miners, and controlling pollution may increase the price of coal a great deal.

Coal used to be a major source of energy for home heating. However, it has been replaced during the last 30 to 40 years by oil and natural gas. These fuels are cleaner-burning (less polluting).

At one time, coal also fueled trains. The heat from burning coal produced steam from water. This steam powered the steam engine that drove the train. Today, diesel engines have replaced almost all steam engines. However, research is being done to improve the design and efficiency (effectiveness) of steam engines. One day we may see the return of coal-fueled steam trains. See Fig. 1-5.

Today, coal is mainly used to fuel electrical power plants. These power plants use three-quarters of all the coal mined.

American Coal Enterprises

Fig. 1-5 **Most people think of coal-fired steam trains as a thing of the past. However, coal may someday be popular again as a transportation fuel. This experimental train is powered by a highly efficient coal-fired steam engine.**

Fig. 1-6 **Most of our transportation fuels come from petroleum. Aviation fuel to operate jet planes and gasoline to power stock cars are both petroleum products.**

Oil (Petroleum). Almost all transportation devices, from cars to jet planes, are fueled by some form of oil. The most common petroleum products are gasoline and diesel fuel. These fuels can be transported easily. They also provide large amounts of power. This makes petroleum products the best energy sources for transportation. See Fig. 1-6.

Many homes and large buildings use oil as an energy source for heating. We also use oil to generate electricity. These uses, plus its use in transportation, have made oil the world's most important fuel.

There are three main sources of oil: oil pockets, oil shale, and tar sands.

Oil pockets underground hold a dark, usually thick liquid called **crude oil**. We obtain crude oil through drilling operations. Workers use large drills to reach the oil pockets. Pumps are used to bring the oil to the surface. See Fig. 1-7.

Oil shale is another source of oil. Oil shale is rock that contains a type of petroleum called **shale oil**. See Fig. 1-8. Over 80 percent of the oil reserves in the United States are in the form of oil shale.

It is hard to extract oil from oil shale. First, workers mine the shale like coal. Then the shale must be crushed and heated. The heat releases the oil from the rock.

Tar sands are a third source of oil. In tar sands, the oil is trapped in sand instead of shale. See Fig. 1-9. As with oil shale, heat is used to remove the oil.

After oil is obtained, it must be refined. **Refining** is the process of separating oil into several useful substances. See Fig. 1-10. This is done first by heating oil inside a tall tower. As it heats up, the oil changes into different vapors (gases). As the vapors rise in the tower, they condense (turn into liquids) at different levels. This part of the refining process is called **fractionating**.

Heavy vapors do not rise very high in the tower. They condense into fuels such as heating oil and diesel fuel. Light vapors rise higher

Fig. 1-7 **Pumps like these are used to bring crude oil to the surface.**

U.S. Department of Energy

Fig. 1-8 **The block at the left is a sample of oil shale. A variety of petroleum products can be obtained from this type of rock.**

and condense into products such as gasoline and kerosene.

The oil left from the fractionating process is further refined by a process called **cracking**. Cracking converts the less useful fractions into gasoline.

The refining process also includes the adding of chemical compounds to the different fuels. This improves the burning and performance abilities of the fuels.

American Petroleum Institute

Fig. 1-9 **Huge machines are used at this mining site in Alberta, Canada to extract tar sands at the rate of 100,000 tons per day.**

Fig. 1-10 **Refineries like this one in Texas convert crude oil into useful products such as gasoline, diesel fuel, and motor oil.**

Exxon

Oil has two major problems as an energy source. One of these is its limited supply. Our oil reserves can provide only about 1/30 the energy available from coal. We use much more oil than coal. Therefore, our reserves of oil will run out before our coal reserves do.

The other major problem with oil is that it creates pollution. Oil is not as polluting as coal. However, the pollution it produces seriously affects our environment. In the future, we will have to develop other fuels that can substitute for oil.

Natural Gas. When we speak of heating or cooking with "gas," we really mean *natural gas*. Natural gas is a mixture of several types of gases. These gases include ethane, propane, butane, and methane. The main gas is methane.

Natural gas is most commonly found in underground pools or with crude oil deposits. For many years, natural gas was obtained simply by drilling into oil deposits and collecting the gas. However, there is not enough gas in these kinds of deposits to meet the demand. Fortunately, scientists have found other reserves of natural gas. Two types of reserves — geopressure reserves and tight sand — are being explored.

Geopressure reserves consist of natural gas dissolved under high pressure in brine (salt water). This gas-filled brine is found in pools deep inside the earth. See Fig. 1-11. High pressure holds the gas in the brine. Drilling into a geopressure pool relieves this pressure. The natural gas then separates from the brine.

There are very large geopressure reserves. These reserves are hard to tap into because they are so far underground. Scientists also need to develop better ways to separate the gas from the brine. However, geopressure reserves may be our major natural gas source in the future.

Tight sand reserves consist of natural gas trapped in a type of hard, dense sandstone. This sandstone is found deep in the earth in the Rocky Mountain region. The natural gas is obtained by injecting a high-pressure fluid into the rock. This causes the sandstone to break and release the gas.

Once natural gas is collected from a reserve, it is transported through pipelines to **processing**

U.S. Department of Energy

Fig. 1-11 **The drilling rig shown here can drill to depths of 20,000 feet. It is being used to extract geopressured natural gas in Texas.**

plants. At these plants, impurities such as dust, water, and sulfur are removed from the gas. Specialized gases such as propane and butane are also separated out. The remaining gas, ready to use, flows through pipelines to homes and industry.

Natural gas is our cleanest fossil fuel. It produces the least amount of pollutants when burned. It is also cheaper than most other fuels. For these reasons, natural gas has become a major energy source for home and industry. We use it for heating, cooling, cooking, generating electricity, and many other uses.

Natural gas is harder to transport than oil. It is most often transported by pipeline. However, it can be transported by truck or ship if it is liquified. See Fig. 1-12. Natural gas is liquified by placing it under extremely high pressure and cooling it. The high pressure makes transportation of this **liquified natural gas (LNG)** potentially dangerous.

Synthetic Fuels. The word *synthetic* means *artificial* or *made by human beings*. Synthetic fuels, or **synfuels**, are liquid or gaseous fuels that are made from already existing solid fuels. Synfuels can be produced from coal, tar sands, and oil shale.

Coal is the most likely fuel to be converted into synfuels. Scientists are now experimenting with different processes for converting coal into liquid or gaseous fuels.

The process of liquifying coal it not new. It was developed in Germany in 1931. The process is basically very simple. Coal liquifies when hydrogen is added to it under great heat and pressure. The result is synthetic oil, or **synoil**.

Coal gasification was also first developed in Germany. Basically, the gasification process consists of adding steam and hydrogen to coal under high heat and pressure. The result is a low-grade gas called **syngas**. This gas must be purified and upgraded before it can be used. Both synoil and syngas projects are presently underway in the United States. See Fig. 1-13.

There are disadvantages to the production of synthetic fuels. Both gasification and liquefaction are very expensive. They also require large amounts of solid fuel. Therefore, synfuels are being produced mainly where there are limited supplies of oil or natural gas. For example, South Africa has a limited supply of petroleum. Processing plants in that country convert coal into 55,000 barrels of synoil each day.

Fig. 1-12 **Large tankers such as this one can carry enormous amounts of liquified natural gas.**

Mobil Oil Corp.

Fig. 1-13 Here a worker at a coal gasification plant is checking high-temperature units used in the gasification process.

Uranium

Like the fossil fuels, uranium is an exhaustible energy source. It is one kind of **nuclear fuel** used to obtain **nuclear energy**. In fact, uranium is the most commonly used nuclear fuel.

Uranium is a heavy substance that is found in many metal ores. There are two basic types of uranium, called **isotopes**. One of these

Fig. 1-14 The nuclear reactor is the heart of this nuclear power plant. The reactor uses nuclear fuel to produce nuclear energy. This energy is then converted into electrical power.

isotopes, uranium-235 (U-235), is rare. The other isotope, uranium-238 (U-238), is abundant. The rare isotope, U-235, provides the basic fuel for **nuclear reactors**. See Fig. 1-14.

We obtain uranium ore through mining. At processing plants, the uranium is separated from the ore. At other plants, the uranium is separated into U-235 and U-238.

To understand how uranium is used as a fuel, you should know something about the atomic structure of matter.

The Structure of Matter. The term *matter* refers to any substance, such as water, iron, or oxygen, that has weight and takes up space. All matter is made of tiny particles called **atoms**. There are over 100 different kinds of atoms. Each type of atom is called a **chemical element**.

Two or more atoms often combine to form a **molecule**. The smallest amount of any substance is either an atom or a molecule.

Atoms are made up of even smaller particles called **sub-atomic particles**. Most atoms have at least three types of sub-atomic particles: protons, neutrons, and electrons. **Protons** have a positive electric charge. They combine with **neutrons** (uncharged particles) to form a nucleus. The **nucleus** is the center of the atom. (The plural form of *nucleus* is *nuclei*.) **Electrons** are negatively charged particles that orbit around the nucleus. See Fig. 1-15.

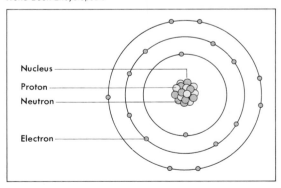

Nucleus
Proton
Neutron
Electron

Fig. 1-15 An atom resembles a miniature solar system, with the "planets" (electrons) orbiting around the "sun" (the nucleus).

Releasing Energy from the Atom. There is a great deal of energy locked in the nuclei of atoms. This energy is called **binding energy**. The larger the nucleus, the greater the binding energy.

Uranium is an atom with a large and heavy nucleus. Therefore, it has a large amount of binding energy and can be used as a nuclear fuel.

In nuclear reactors, uranium atoms are bombarded with neutrons. This causes the uranium nuclei to split apart. This splitting, or **nuclear fission**, releases energy and other neutrons. The released neutrons in turn split other nuclei. The result is a **chain reaction** that releases huge amounts of energy in the form of heat. See Fig. 1-16. Heat from the chain reaction is used to produce steam to drive turbine-powered electrical generators.*

Our Uranium Supplies. The United States' supply of uranium is limited. If this uranium is used only in present-type nuclear reactors, it will provide about twice the energy available from our total reserves of oil. (This amount is

*Chapter 23 provides more detailed information on nuclear energy and its uses.

about 5 percent of the energy available from our coal reserves.)

There is another type of reactor being developed, however. The **breeder reactor** uses uranium-238 as its fuel. This type of reactor converts U-238 into plutonium. **Plutonium** is another substance that can be used to power nuclear reactors. If the breeder reactor became widely used, the energy available from our uranium supplies would increase sharply. It would then be about three to four times the energy available from our coal reserves.

RENEWABLE ENERGY

Renewable sources of energy are those that can be used indefinitely if they are properly managed and maintained. Wood and plants are renewable sources of energy. See Fig. 1-17.

Renewable fuels are only renewable if they are carefully managed. For example, cutting too many trees without proper replanting could destroy our source of wood. In the years to come, renewable sources of energy will grow in importance. This is because we can use many of them to replace oil.

Fig. 1-16 **In nuclear fission, uranium nuclei are bombarded with neutrons. The neutrons split the nuclei and release both energy and other neutrons. The released neutrons then cause a chain reaction as they hit other uranium nuclei.**

World Book Encyclopedia

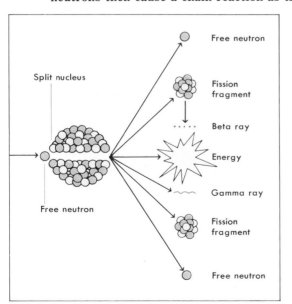

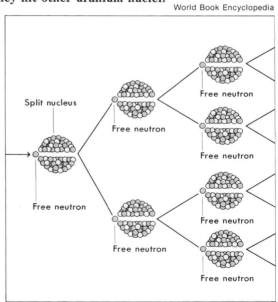

Fig. 1-17 **A forest represents a renewable source of energy.**

Gasohol

Gasohol is a mixture of nine-tenths unleaded gasoline and one-tenth **ethyl alcohol**. It is used as a substitute for gasoline in cars and trucks. By using gasohol, we can save 10 percent of the oil needed to produce gasoline.

The sources of ethyl alcohol include sweet sorghum, sugar beets, and grain. At distilleries, these crops are processed into alcohol. The alcohol is then mixed with gasoline.

The main by-product of gasohol processing is a mash that contains all the protein of the original product. This mash is not wasted. Farmers use it as a livestock food supplement.

Methanol

Methanol, or **methyl alcohol**, is a clean-burning liquid fuel. It can be made from natural gas and coal. These sources are nonrenewable. However, methanol can also be made from renewable sources. These sources include wood, plants, and waste products from homes, farms, and industry.

Like gasohol, methanol can be used as a transportation fuel. Methanol produces more energy than the ethyl alcohol in gasohol. Therefore, it does not have to be mixed with gasoline.

Methanol burns more slowly than gasoline. However, with changes in engine design, methanol can produce as much power as gasoline. Because it is slower-burning, it produces smoother engine performance.

Methanol has less energy per gallon than gasoline. As a result, a car that uses methanol needs a larger fuel tank. Methanol is already being used as a substitute for gasoline. See Fig. 1-18.

Bioconversion

Bioconversion is the process of obtaining energy from society's waste products. There are many sources of waste **biomass**: food product waste, animal wastes, paper, cardboard, and wood. All of these can either be burned or converted into fuels such as alcohol, oil, and methane.

Some cities burn solid trash, such as paper and wood products. They then use the heat energy to drive turbines to generate electricity. There have also been successful experiments in producing fuel from sewage, cannery processing wastes, and gardening and agricultural wastes. See Fig. 1-19.

California Energy Commission

Fig. 1-18 **In traffic, there will be no black smoke coming from this bus. It uses methanol for fuel.**

U.S. Department of Energy

Fig. 1-19 In this biomass gasifier, biomass is mixed with steam and oxygen at high temperatures and pressures to produce methane and liquid fuels.

Bioconversion is not a very efficient process. However, this energy source may add to other sources. It will also help us dispose of unwanted waste.

Wood

Wood is one of our oldest sources of energy. People have long used wood for cooking and heating. In pioneer days, wood was one of the major sources of energy. Coal and oil later replaced it.

Today, more and more people use wood as a fuel. Wood stoves again heat millions of American homes. See Fig. 1-20. In a few cases, wood also provides energy to generate electricity. Figure 1-21 shows a wood-fueled power plant.

As an energy source, wood has one major disadvantage. It is not a clean-burning fuel. It creates high levels of air pollution.

Researchers are studying other uses for wood. Wood can be converted to liquid and gaseous fuels easier than fossil fuels can. One of these wood-based fuels is methanol. As you know, methanol is already being used in some vehicles. Wood could become a major source of fuel for automobiles.

Russo Manufacturing Corp.

Fig. 1-20 Wood stoves have changed a great deal in the last 100 years. Today's wood stoves are safer and much more efficient.

Competition for the use of wood will increase in the future. Wood is a renewable resource. However, its use as a fuel will compete with its use in construction. Other competing industries include paper making and furniture production. Very large forests would be necessary to sustain all these uses.

Fig. 1-21 In Vermont, a power company uses wood chips to fuel two electrical power generation units.

Burlington Electric Co.

INEXHAUSTIBLE ENERGY

Inexhaustible energy sources are those that will always be available. It does not matter how much of them we use. These sources include the following:

- Solar energy
- Hydroelectric energy
- Wind
- Tides
- Ocean thermal energy
- Solar salt ponds
- Hydrogen
- Geothermal energy

The sources listed above are basically renewed by forces beyond our control. As mentioned earlier, most of our energy comes from the sun, in one way or another. The sun is a flaming ball of hydrogen and helium that will provide us with a constant supply of radiant energy for the next five billion years. For our purposes, then, the sun is an inexhaustible source of energy.

The sun also produces our weather. The weather, in turn, provides for hydroelectric power and wind power.

Another powerful force, the moon's gravity, creates another inexhaustible energy source — the tides. As you read on, you will find out just how we can take advantage of the tides, wind, solar salt ponds, and other "free" energy sources.

Solar Energy

Most of the energy sources you have read about so far are *indirect* sources of the sun's energy. However, we can also use the sun's energy directly. We can collect solar energy (heat and light) and put it to work.

Suppose that all the energy arriving from the sun each day could be collected and controlled. In this case, we would have all the energy we need. Unfortunately, our ways of collecting and controlling solar energy are inefficient and expensive. We also cannot collect solar energy at night or during cloudy weather. And what is more, it is hard to store solar energy.

However, we have made much progress in using solar energy in the past several years.

U.S. Department of Energy

Fig. 1-22   The solar panels on the roof of this home heat water. The water is then used to heat 1200 square feet of living space.

Pennsylvania Power & Light Co.

Fig. 1-23   Each circular photovoltaic cell collects solar energy and converts it into electricity. Together the cells form a *solar battery*.

Fig. 1-24 **The Hoover Dam, in the Black Canyon of the Colorado River, traps behind it a body of water 115 miles long and 589 feet deep.**

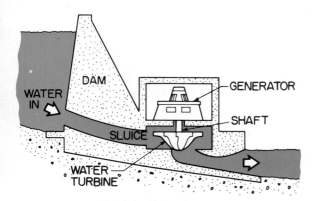

Fig. 1-25 **Flowing water drives the blades of the water turbine. The turbine rotates the generator shaft to produce electricity.**

If research and development continues, solar energy could become an important direct energy source.

Solar energy can provide both heat and electricity. **Solar panels** are devices that collect solar energy to heat water. The hot water is then used directly in homes or industry. It may also be used to heat houses, office buildings, or factories. See Fig. 1-22.

We can also control solar energy with mirrors. Mirrors concentrate the sun's rays on a small area. In this way, a great deal of heat is produced. This heat can be used to produce steam to operate electrical generators.

Photovoltaic cells are devices that convert sunlight directly into electricity. See Fig. 1-23. These cells may someday produce a large part of the electricity we use. At the present time, photovoltaic cells are expensive. Scientists are trying to reduce their cost with new designs.

Hydroelectric Energy

Hydroelectric energy has, for many years, supplied electricity for millions of people. Hydroelectric energy is simply electrical energy produced from flowing water.

Each year the sun evaporates 100,000 cubic miles (*416,550 cu. km*) of water. This is ten times the amount of water in the Great Lakes. All of it returns to the earth as rain or snow. Rain and snow sustain our streams, rivers, lakes, and oceans. We can build up run-off water in reservoirs by trapping it behind dams. See Fig. 1-24. We can then use much of this water to produce electrical power.

As water drains from a reservoir, it flows through a sluice (a special waterway or tunnel). A **water turbine** is at the end of the sluice. This turbine is a wheel that is driven by the flow of water. See Fig. 1-25. The rotating turbine drives a generator that produces electricity.

One-fifth of all the electricity produced in the United States is hydroelectric power. This large amount is less than 40 percent of the power available from flowing water. Norway, Switzerland, and Sweden use over 50 percent of their available water power.

Harvesting the Wind

At one time in history, oil was considered a worthless substance. People actually complained of the annoyance of oil seeping from their land. Then oil was discovered to be a fuel rich in energy. People who owned oil-bearing land became rich.

Much the same thing is now happening at places considered to be barren of natural resources. For example, some land is especially useful for capturing and using wind energy.

Altamont Pass, just east of San Francisco, is one of these places. The land is dry, hilly, and windy. A few years ago, the land was thought to be almost worthless. Today, the landowners are very fortunate. They hold the key to the renewable resource of wind.

Some people hold **mineral rights** to their property. They can sell or lease (rent) these rights to a mining company. The landowners at Altamont Pass hold the rights to put wind turbines on their property. They lease these rights to wind power companies. The power companies set up the turbines and sell the power generated to utility companies. In turn, the landowners receive monthly payments.

Pacific Gas & Electric

There are already 2000 wind turbines on Altamont Pass. And there is room for 10,000 more! When they operate at full capacity, the turbines now being used can generate electricity for over 150,000 people. The number of turbines and the amount of power generated are sure to grow. And all this from property that was once considered a wasteland!

At this time, we cannot store electricity easily or economically. Therefore, we must use it as soon as we produce it. At any given time, electricity production must equal the need for electricity. Workers at hydroelectric plants control electricity production by adjusting the amount of water passing through the sluice. The electricity is then transmitted to homes and industry. There it is put to work.

Electricity is one of the easiest forms of energy to transmit. Therefore, power generated at hydroelectric plants can be sent along power lines to cities many miles away. Mountainous areas serve as basins to store water needed to produce hydroelectric power. These areas are usually located long distances from the cities they serve.

Wind Energy

At one time, the wind was an important energy source. People built windmills at places where the wind blew much of the time. They used the wind energy to pump water, grind grain, and do other useful tasks.

However, the wind was also an *unpredictable* energy source. People could not control its availability. Therefore, when cheap fossil fuels became available in this country, most windmills disappeared. Today, fossil fuels are becoming more and more expensive. As a result, people are again becoming interested in the free energy of the wind. Wind energy is especially appealing because it generally does not create pollution, as fossil fuel use does.

Today's windmills, or **wind turbines**, are designed to generate electricity. The wind drives a propeller or turbine that is connected to a generator. As the turbine rotates, the generator produces electrical current.

Some experimental wind power plants are already in service. More are being planned. Wind power is growing rapidly. In California alone, over 4000 commercial wind turbines now provide electricity for over 75,000 homes. Wind could provide for nearly one-tenth of California's electrical needs by the year 2000.

Researchers are constantly working to improve the design and operation of wind turbines. An efficient wind turbine must produce electricity under many different wind conditions. It must also be strong enough to resist damage during storms. Finally, its energy production must cost less than fossil fuel energy production.

The output of a single wind turbine is very limited. Even the largest turbine can produce electricity for only about 5000 people. However, the output can be increased by grouping smaller turbines together on a **wind farm**.

There are two basic designs for wind generators. Some turbines are of **horizontal-axis** design. Traditional windmills also use this design. Figure 1-26 shows a **vertical-axis** type of wind generator. The "egg-beater" design allows the turbine to catch the wind from any direction.

Small wind turbines can provide for the electrical needs of individual users. See Fig. 1-27

Fig. 1-27 This home wind turbine generates electricity for a farmer's home and outbuildings.

Wind is also used to power sailboats. This use of wind is mainly limited to recreational sailing. However, experiments are underway to use large sails to help power ocean-going freighters. See Fig. 1-28.

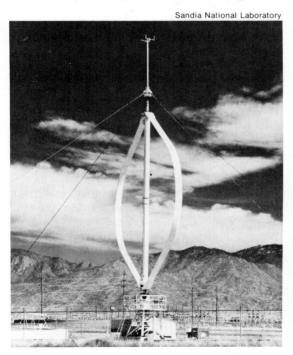

Fig. 1-26 This vertical-axis wind turbine helps provide electricity to homes and industry in the Albuquerque, New Mexico area.

Fig. 1-28 This Japanese experimental ship, the *Shin Aitoku Maru*, uses sails to capture wind energy. The sails can provide more than half of the power needed to drive the ship.

Wind energy does have some disadvantages. People living near wind farms have complained about the whining noise. Also, turbines placed on hillsides distract from the natural beauty of the hills. However, wind energy is generally pollution-free.

Energy from the Oceans

There are many possibilities for using the inexhaustible energy of our oceans. We can grow plants in the oceans. We can then use these plants to manufacture fuels such as alcohol or methanol. We can also use ocean currents and waves to drive electrical generators. The most promising sources of energy, however, are tides and ocean thermal energy.

Tides. Large bodies of water raise and lower in a regular sequence every day. These changes in water level are called *tides*. Tides are caused mainly by the pull of the moon's gravity. As the earth spins on its axis, the moon's gravity pulls on different parts of the earth. This causes water in the oceans to raise and lower. Raised water is called **high tide**. It is high for about six hours. Then the water lowers and becomes **low tide**. Low tide also lasts about six hours. Therefore, there is a change in the water level four times a day.

Tides are different in different parts of the world. They are generally lower near the equator and get higher toward the earth's poles. In some northern areas, the difference in water level between high tide and low tide may be as much as 40 feet (*12.2 m*).

We can use the difference in the height of tides to generate electricity. See Fig. 1-29. This method requires a bay area that can be closed with a dam. Water is kept in or out of the bay depending on the tide. When the tide is rising, the dam is closed. This keeps water out of the bay. When the tide is at its highest, the water is allowed to flow through the dam into the bay. The flowing water turns turbines to produce electricity.

When the tide starts to lower, the dam is again closed. This traps the high-tide water behind the dam in the bay. When the tide is at its lowest, the dam is opened. This allows the water to flow back to the ocean and turn the electrical turbines again.

Not many locations are suitable for tidal power generation. For this reason, tidal power plants will provide only a small part of our energy in the future.

Ocean Thermal Energy. Tropical oceans collect and store tremendous amounts of heat from the sun. It is possible to use this **thermal energy** to produce electrical energy. Ocean thermal energy may have the potential to produce more than 200 times the entire world's present usage of electricity.

Producing electricity from ocean heat is called **Ocean Thermal Energy Conversion (OTEC)**. This process is based on the fact that the sun heats the oceans unevenly. Surface water becomes hot. Water deep in the ocean remains cold. The temperature difference between the hot surface water and the nearly freezing deep water can be used to produce electricity. See Fig. 1-30.

French Embassy, Press & Information Div.

Fig. 1-29 **This tidal power plant across the Rance River in France can produce 10 megawatts of electrical power.**

Lockheed Missiles & Space Co. for
The Natural Energy Laboratory of Hawaii.

Fig. 1-30 **This Mini-OTEC research facility has successfully generated 50,000 watts of electricity. A large OTEC plant could produce enough electricity to provide for the needs of a city of 100,000 people.**

The OTEC system operates like a giant refrigerator. Basically, it involves a loop between the hot surface water and the cold bottom water. A liquid, such as ammonia, circulates in the loop. See Fig. 1-31.

Ammonia is a liquid at low temperatures. It becomes a gas at higher temperatures. When ammonia circulates between the hot and cold water, it changes from a liquid to a gas, and from a gas to a liquid. When the ammonia changes into a gas, it expands and produces pressure. This pressure turns a turbine generator to produce electricity.

After passing through the turbine, the gas is condensed by the cool bottom water. The liquified ammonia then circulates to an evaporator. There it changes back into a gas. The evaporating-and-condensing process continues as the ammonia circulates through the system.

27

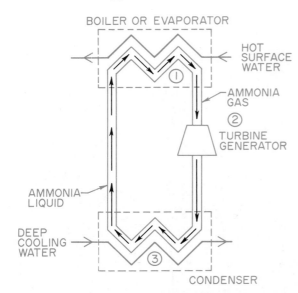
BOILER OR EVAPORATOR

HOT SURFACE WATER

AMMONIA GAS

①

② TURBINE GENERATOR

AMMONIA LIQUID

DEEP COOLING WATER

③

CONDENSER

Fig. 1-31 **This is a simplified diagram of an OTEC plant in operation. Heat from the surface water vaporizes the liquid ammonia (1). The ammonia gas produces enough pressure to drive the turbine generator (2). The deep cooling water then condenses the gas into a liquid (3) to repeat the process.**

The OTEC cycle is constant as long as the temperature difference between the surface and bottom water remains the same.

The United States is very well-located to take advantage of ocean thermal energy. We are next to areas of year-round warm surface water overlying cold deep water. Warm water from the Caribbean Sea flows northward into the Gulf of Mexico and around Florida. The average surface temperatures are 76-80° F (*20-22° C*), with the deep water at near-freezing temperatures. These same conditions exist around Hawaii and in most tropical ocean areas.

Ocean Thermal Energy Conversion has some strong advantages. It does not produce air pollution. It is also an inexhaustible source of energy. Furthermore, the construction of OTEC plants does not require any major technological developments. However, OTEC plants *do* require much special engineering consideration. Researchers are now working on this problem.

Ocean thermal conversion also has some potential environmental problems. OTEC plants will cool surface water and warm deep water. This may cause changes in the ocean environment. It may also affect the tropical climate. Scientists are now studying these environmental effects.

Energy from Solar Salt Ponds

Basically, solar salt ponds operate just the opposite of the OTEC process explained above. A solar pond is a shallow body of salt water that has a layer of less salty water near the surface. The water at the pond bottom contains more salt and is heavier than the top layer. The sun's rays pass through the top layer and heat the bottom of the pond. Because the bottom water layer is heavy with salt, it keeps the heat from passing up to the top layer. As a result, the temperature at the bottom of the pond can reach 250° F (*121° C*), while the surface stays fairly cool.

A loop containing a liquid such as ammonia is installed between the high and low temperature areas. The ammonia operates a turbine generator in the same way as in the OTEC system.

The energy potentially available from solar salt ponds is much smaller than the energy available from OTEC. However, solar ponds have fewer disadvantages. The electricity from solar ponds could help us meet our future energy needs.

Hydrogen

Hydrogen is one of the most common elements on earth. A molecule of water (H_2O) consists of two atoms of hydrogen bonded (joined) to an atom of oxygen. Water covers over two-thirds of the earth's surface. Therefore, in water we have an almost endless supply of hydrogen.

Pure hydrogen is very combustible. We can use it to power automobiles, operate turbine generators, and heat homes. One day, hydrogen may replace both gasoline and natural gas.

There are not many problems with *using* hydrogen. However, it is hard to *produce*

Fig. 1-32 **This machine produces hydrogen by the electrolysis of ordinary tap water. When installed in a garage, it can produce the fuel for a hydrogen-powered car.**

Consulate General of Iceland

Fig. 1-33 **Iceland has an abundance of easily tapped geothermal energy. Naturally occurring hot water can be piped in to heat buildings such as the greenhouse shown here.**

hydrogen. The bonds that hold water molecules together are very strong. These bonds must be broken to release the hydrogen. This process requires large amounts of energy.

There are several ways in which hydrogen can be released from water. The most promising method is called **electrolysis**.

In electrolysis, electricity is passed through water. This process separates the hydrogen from the oxygen atoms. Electrolysis requires electricity. See Fig. 1-32. So far, it costs more to produce hydrogen than the hydrogen is worth. However, advances in solar energy research may provide the answer. The conversion of solar energy into electricity is becoming more efficient. Someday solar energy may be used to power the electrolysis process.

Geothermal Energy

The term **geothermal** simply means **earth heat**. This heat comes from molten rock, called **magma**, miles beneath the earth's surface. The heat from magma is enough to raise the temperature of the earth 14° F with each foot of depth *(25° C per meter)*.

Usually, the heat from magma is trapped below the surface. However, sometimes magma erupts from the earth in a volcano. Other times

it heats water to produce geysers or hot springs. Geothermal energy has great possibilities as an energy source. See Fig. 1-33.

The greatest use of geothermal energy is for the generation of electricity. There are several places in the U.S. that are good for electrical generation through geothermal energy. By the year 2020, the power from geothermal energy could equal that of hydroelectric energy.

The Geysers is an area north of San Francisco where magma is only about five miles below the surface. The heat produces steam that boils up out of the ground. The first geothermal power plant was built at The Geysers in 1960. There are now 17 power plants in the area. Together, they produce enough electricity to power a city of over a million people.

The sources of usable geothermal energy fall into four classes:

- Hot, dry rock fields
- Dry steam fields
- Hot water fields
- Fields of lesser heat content

Pacific Gas & Electric

Fig. 1-34 The Geysers is located north of San Francisco. This facility uses dry steam trapped underground to produce electrical power.

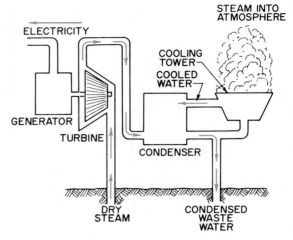

Fig. 1-35 This diagram shows how a dry-steam powered geothermal plant operates.

Where **hot, dry rocks** are close to the surface, water is injected to produce steam. The steam is then piped to the surface. There it is used to drive electrical generators. This kind of system is only in the experimental stage.

Dry steam fields are the most easily tapped form of geothermal energy. The Geysers (Fig. 1-34) uses a dry steam field for heat. Figure 1-35 shows a drawing of the power system in operation. Dry steam from the earth passes through pipes to a turbine. The turbine drives an electrical generator. The steam then passes to a condenser. There it condenses into water and is sent to a cooling tower. Some water is evaporated, some is used to cool more incoming steam, and some is returned to the earth.

Hot water fields contain water at temperatures ranging from 350 to 700° F *(177-371° C)*. At sea level, water boils at 212° F *(100° C)*. However, underground water can reach much higher temperatures. This is because the pressure on the water keeps it from changing to steam. As the hot water comes to the surface, the pressure reduces suddenly and the water "flashes." That is, some of the water changes immediately into steam. It can then be used to drive turbines.

Fields of lesser heat content also contain hot water. Its temperature is lower, though. The heat range is from 120 to 300° F *(49-199° C)*. This water is not hot enough to create usable steam. However, it can be used to boil another liquid that has a lower boiling point, such as ammonia. When the liquid is heated, it changes into a high-pressure gas. It can then be used to drive turbine generators.

THE USE AND AVAILABILITY OF ENERGY

Over 90 percent of all the energy we use in the United States comes from three fuels: oil, coal, and natural gas. Figure 1-36 shows the United States' usage of these fuels as compared to their use worldwide. Note that the only important energy sources other than fossil fuels are hydroelectric energy and nuclear energy. The remaining sources, including solar energy, account for less than one-half percent of current energy usage.

As you learn more about the use of energy in the United States and the rest of the world, you will need to know something about our energy reserves. You should also know how much of these reserves we are using.

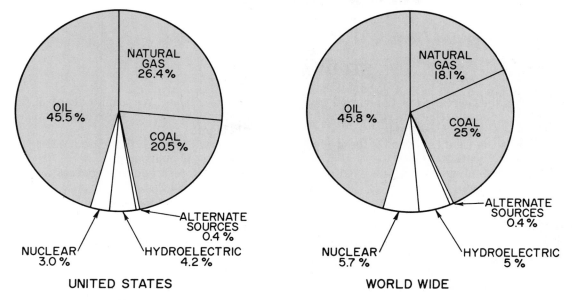

Fig. 1-36 These pie diagrams compare the present energy usage in the U.S. with that in the entire world. The fossil fuels are shown in color. Note that about 45% of the total energy, both in the U.S. and worldwide, is from oil.

Figure 1-37 shows a comparison of the U.S. and worldwide exhaustible energy reserves. Figure 1-36 shows which sources we are presently using the most of. As you can see, we presently rely very much on oil. You can also see that our reserves of oil are very low. At our present rate of usage, the world's known supplies of exhaustible energy will last only about 70 years.

If we continue to use exhaustible energy sources at the present rate, most people living today will see major fuel shortages happening throughout the world. Some fuels, such as natural gas and oil, may all but disappear. To avoid serious energy shortages, we will have to develop inexhaustible or renewable supplies of energy throughout the world.

Fig. 1-37 These pie diagrams compare U.S. and worldwide total reserves of exhaustible energy. Notice that oil — our most-used energy source — is in short supply.

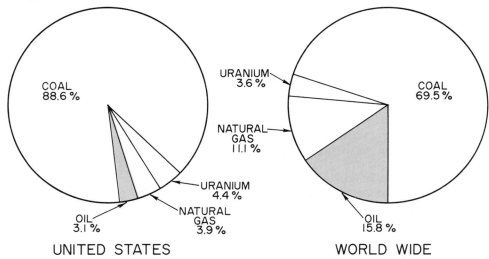

STUDY QUESTIONS

1. What is the source of most of our energy?
2. What are exhaustible energy sources?
3. Name the three fossil fuels.
4. What is the cleanest-burning fossil fuel?
5. What are synthetic fuels?
6. What kind of energy is obtained from nuclear fuels?
7. What process releases energy from uranium?
8. What are renewable energy sources?
9. Name two transportation fuels that can be obtained from renewable energy sources.
10. What are inexhaustible energy sources?
11. True or false: Our ways of collecting and controlling the sun's energy are efficient and inexpensive.
12. True or false: We cannot control the amount of energy produced at a hydroelectric plant.
13. Give two reasons that wind energy is becoming popular again.
14. What two ocean energy sources are the most promising?
15. Which is harder to do — use hydrogen or produce hydrogen?
16. What is the source of geothermal energy?
17. What is the most commonly used fuel in the world today?
18. Why should we be concerned about our supplies of fossil fuels?

ACTIVITIES

1. Conduct research to find out what sources of energy are being used in your community. To gather information for a written report, you could meet with the following groups and ask the questions listed.
   - **Gas company** — What is the source of natural gas? What proportion of natural gas is used in home heating as compared to oil, coal, and electricity? Are any alternative sources (sources other than fossil fuels) used to generate methane? Are there plans to use alternative sources?
   - **Electric company** — How is electricity for the community generated? If the power comes from different sources (such as coal-fired plants, nuclear reactors, and hydroelectric plants), what percentage of power is delivered by each source? Are any alternative sources being used or considered for the future?
   - **Local government officials** — Are there plans to use alternative energy sources for the needs of the community?

2. Conduct a study of the use and availability of wood stoves in your community. Local dealers would be a good source of information. Discuss stove improvements with them and find out what progress has been made to reduce pollution or increase stove efficiency. Find out what wood costs in comparison to other sources of heat. Your final report could include information on both the advantages and disadvantages of wood as a fuel.

3. Select an alternative energy source that you are interested in and prepare a class presentation. Your report should include the following:

   - A description of the source
   - Its potential for use at the community, national, and international level
   - Potential problems with the use of the source

The Control of Energy

Energy is the capacity to do work, or the capacity to produce motion, heat, or light. When we turn on an electric motor, drive a car, light a stove, or switch on a light, we are using energy.

To use energy, we must be able to *control* it. We can trace history through human efforts and successes in controlling energy. The first people were hunters. They had no sources of energy except their own muscles. Muscle power, then, was the first controlled source of energy.

Eventually, people learned to make better use of their strength by using tools. For example, they discovered that wooden clubs were more effective in killing wild animals than bare hands were. The club made a person's strength more effective. See Fig. 2-1.

Later, people learned to grow their own food. They found that grain had a better chance of growing when the soil was turned over. To do this difficult task, they developed a special tool, the plow. At first, people pulled or pushed plows. Their physical strength limited the amount of work they could do.

Later, people used animals to pull plows. Still later, animals provided the power to grind grain into flour. People could then control a form of energy greater than their own. See Fig. 2-2.

Centuries later, people used wind to power sailing ships. Wind also turned windmills to pump water. Waterfalls provided the energy to turn waterwheels on grain mills. See Fig. 2-3. Wind and water were powerful energy sources. However, they were not dependable. People found it difficult to regulate these sources. They also had to use these sources at the places where they occurred in nature.

Fig. 2-1 Early humans used clubs to multiply their muscle power.

Fig. 2-2 Using animals as power sources greatly increased the amount of work that people could do.

Michael LaForest

Fig. 2-3 Years ago, the waterwheel on this mill provided the power to grind grain.

The energy and power situation changed with the development of the steam engine. Steam engines were first used about 300 years ago. They had many advantages over earlier sources of power. For the first time, people could fully control the *placement* of a power source. They could move the steam engine to where its power was needed. See Fig. 2-4. They could also control the *amount* of power the engine produced. Finally, they could control the *duration* of the power (how long it lasted).

Since the steam engine, people have developed many new methods of controlling energy. We control energy for manufacturing, construction, communications, and transportation. See Fig. 2-5. Our ability to control energy grows each year.

Fig. 2-4 The first steam engines were used to pump water from mines. These engines could be moved to where their power was needed.

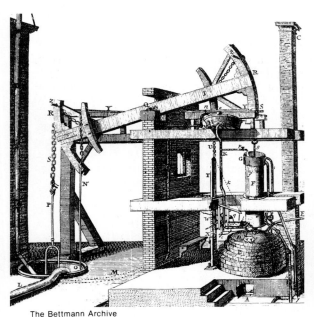

The Bettmann Archive

Metropolitan Area Transit Authority

Fig. 2-5 These modern trains use diesel-electric power.

Forms of Energy

As mentioned earlier, energy is the capacity to produce motion, heat, or light. These three products are also known as **mechanical energy**, **thermal energy**, and **light energy**. Energy also has three other forms: **chemical energy**, **electrical energy**, and **nuclear energy**.

The six forms of energy listed above are all related to each other. We can convert any of these forms of energy into any of the other forms. See Fig. 2-6. For example, wood is a

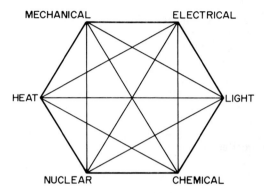

Fig. 2-6 **All forms of energy are related. Each form can be converted into any other form.**

Fig. 2-7 **Mechanical energy is simply the energy of motion. The mechanical energy of this moving hammer is used to drive a nail.**

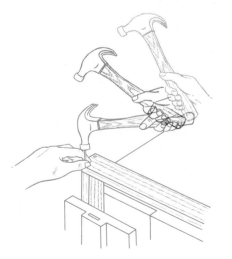

form of chemical energy. When we burn wood, we change its chemical energy into heat energy and light energy.

Not all energy conversions are as easy as burning wood. For example, an automobile is a fairly complex tool or instrument used for energy conversion. It converts the chemical energy in gasoline or diesel fuel into mechanical energy, the energy of motion. As you will learn later, the automobile uses several different control systems to bring about this conversion and to regulate the mechanical energy.

Mechanical Energy

Mechanical energy is the most familiar form of energy because it is the energy involved in motion. Every moving object has mechanical energy — whether it is a carpenter's hammer driving a nail, a leaf falling from a tree, or a rocket flying in space. See Fig. 2-7.

When we put mechanical energy to work in such devices as a hammer or an automobile, it is useful. However, mechanical energy can also be destructive. If you do not properly control the mechanical energy of a hammer, you could smash your finger. Uncontrolled automobiles can kill people or cause property damage.

Heat (Thermal Energy)

Heat, or thermal energy, also involves motion. However, the motion in heat energy is generally not visible. (An exception might be the shimmering "heat waves" you can sometimes see above a road on a hot day.) Usually, though, we can only feel or see the *effects* of heat. See Fig. 2-8. And what we are seeing is the result of atoms or molecules in motion.

Atoms and molecules are always in motion. The amount of heat energy given off by a substance depends on the speed and number of atoms or molecules in motion. The faster the atoms or molecules move, the greater the heat energy they give off. See Fig. 2-9. Also, the more atoms or molecules there are in motion, the faster they move and the more heat they produce.

Fig. 2-8 The movement of a hot-air balloon is a visible effect of thermal energy. The "fuel" for the balloon is the heated air. As the hot air rises, so does the balloon.

ICE   FIRE

Fig. 2-9 When matter is heated, its atoms or molecules increase in speed. The higher the speed, the more thermal energy the matter has.

Heat energy is an important form of energy. The sun's heat energy sustains all life on earth. Without it, all life on our planet would die.

It is interesting to note that the word *fuel* comes from words related to *fire*. We can release heat energy from fossil fuels by burning them. We get heat energy from nuclear fuels through the fission process, which produces a kind of nuclear fire. Whatever the source of our "fire," we can put the resulting heat energy to work. Heat energy cooks our food and warms our homes. We use it to generate electricity. It powers our cars, planes, ships, and even hot-air balloons.

Light Energy

In Chapter 1, you learned that the sun produces radiant energy (heat and light). Light energy is the part of radiant energy that we can see. However, even though we can see it, light energy is difficult to understand. That is, it is hard to see how light does work.

But just look around. Where would we be without the light from our sun? One basic fact of life is that light is necessary for the growth of plants. Plants convert sunlight, carbon dioxide, water, and nutrients into food. This process, called **photosynthesis**, is the most important use of light energy. Without light energy, we would not have plant or animal growth. We also would not have the fossil fuels that were originally plants and animals millions of years ago.

Every time we eat something or burn a fossil fuel, we are indirectly tapping into the light energy that came from the sun. One *direct* way to use light energy is to use photovoltaic cells. As you recall, photovoltaic cells convert light directly into electricity.

Chemical Energy

Chemical energy is the energy of all living things. Think again about photosynthesis. Photosynthesis is an example of a *chemical change*. See Fig. 2-10. Plants and animals use the result of photosynthesis — sugar — as a source of energy. Since sugar is a chemical, it is a type of chemical energy. We can define chemical energy as *energy produced by chemical changes*.

Many of the energy sources described in Chapter 1 are forms of chemical energy. Think about the fossil fuels — coal, oil, and natural

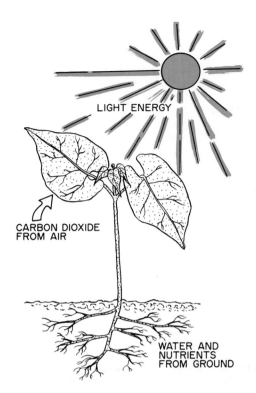

LIGHT ENERGY

CARBON DIOXIDE
FROM AIR

WATER AND
NUTRIENTS
FROM GROUND

Fig. 2-10 **Green plants are food-making factories. They combine light, water, nutrients, and carbon dioxide to form a simple sugar. This reaction is an example of a chemical change. The sugar is a form of stored chemical energy.**

gas. They are all chemicals or combinations of chemicals. All of them were produced by chemical changes. When we need energy, we burn certain amounts of these fuels. This burning converts the chemical energy into a certain amount of heat energy. For example, a gallon of gasoline burned in an automobile engine provides the heat energy to drive a car a certain distance.

Chemical energy can also be converted into other forms of energy. For example, a flashlight battery is a device for converting chemical energy into electrical energy. The chemicals in a battery are arranged so that an electric current will result when the ends of the battery are connected. When you turn on a flashlight, you make this connection. The battery's chem-

ical energy changes to electrical energy and lights the bulb.

Some chemicals contain a great deal of energy that we can release all at once. These chemicals are called *explosives*. We can convert the chemical energy of explosives into several forms of energy. For example, when dynamite explodes, its chemical energy changes into heat, light, and motion (wind and pressure). Explosive chemicals are used extensively in mining and road construction. We control the power of explosives to uncover mineral deposits and make underground passages. See Fig. 2-11.

Electrical Energy

Electrical energy is the motion of tiny invisible particles of matter called **electrons**. Electrons are too small to see, even with the most powerful microscope. Therefore, we study electrical energy by watching the changes it produces.

Electrical energy has a limitless number of uses. We can convert it into light energy with electric lamps. Electric motors change electrical energy into mechanical energy. Electric stoves and heaters change it into thermal energy. We also use electrical energy for radio, television, and telephone communications. See Fig. 2-12.

Electrical energy is also easy to transport from place to place. This characteristic allows us to use electrical energy almost anywhere.

Fig. 2-11 **An explosion is the rapid conversion of chemical energy into heat, light, and motion. Here dynamite is being exploded to clear an area for construction work.**

NICOR, Inc.

Knight-Ridder Newspapers

Fig. 2-12 **Electrical energy is the "life-blood" of all kinds of communication systems.**

Nuclear Energy

Nuclear energy is developed by changing matter into energy. Hydrogen and uranium are two kinds of matter used to produce nuclear energy. In a nuclear reaction, part of each hydrogen or uranium atom is changed into energy. This conversion lowers the original amount of matter. In a nuclear reaction, a very small amount of matter is converted into a tremendous amount of energy. See Fig. 2-13.

A conversion from energy back into matter is possible but very difficult. This process involves bringing large amounts of energy together in a tiny space. It can be accomplished only under carefully controlled laboratory conditions. These conditions have been produced on only a few occasions.

The practical uses of nuclear energy are somewhat limited. The release of nuclear energy involves high operating temperatures. There is also the possibility of radiation leaks. Therefore, complex equipment and extensive safeguards are needed to safely convert nuclear energy into other energy forms.

The most common conversion takes place in nuclear power plants, which convert nuclear energy into heat energy. This heat is then used to drive steam turbines. The steam turbines operate electrical generators.

Nuclear energy also has other uses. In medicine, nuclear energy is used in the form of radiation therapy to treat cancer. The United States Navy also uses nuclear energy to power submarines and some large ships. See Fig. 2-14. A submarine powered by nuclear energy can operate for several years without refueling.

One other use of nuclear energy, of course, is in nuclear weapons. For more detailed information on nuclear energy, see Chapter 23.

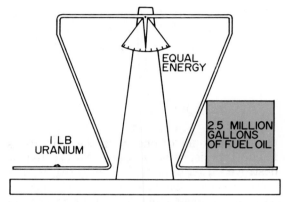

Fig. 2-13 **We can release enormous amounts of energy from the nuclei of uranium atoms. From one pound of uranium, we can get energy equal to 2.5 million gallons of fuel oil.**

Fig. 2-14 **This submarine is powered by nuclear energy. The nuclear energy is converted to heat, which is then used to drive turbine generators.**

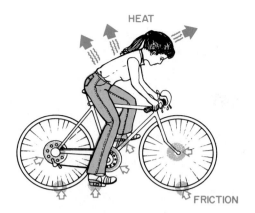

HEAT

FRICTION

Fig. 2-15 Part of the effort this girl is exerting is going into the motion of the bicycle. However, some of her energy is lost as heat through her skin. Energy is also lost as friction at the bike's moving parts.

CONSERVATION OF ENERGY

The **law of conservation of energy** states that the amount of energy in the universe is fixed. That is, energy cannot be created or destroyed. Therefore, when we say that energy is being *used*, we do not mean that it is being used up. We mean that it is being changed from one form to another. For example, when we exercise, we convert chemical energy (food) into mechanical energy (motion) and thermal energy (heat).

The control and use of energy consists of converting energy from one form into another. There is no energy destroyed during this process. However, some energy *is* converted into unwanted forms. In fact, it is impossible to convert one form of energy into another without wasting some energy.

To see how energy is lost during conversion, let's look at a simple energy conversion system — a person riding a bicycle. When people ride bikes, their bodies convert chemical energy into mechanical energy to work the pedals. However, some of the chemical energy also changes into heat. This heat is given off through the rider's skin. It cannot be used to move the bicycle. See Fig. 2-15. Heat is also generated between the moving parts of the bike and between the bike and the air. This **friction** reduces the usable energy further.

In many energy conversions, there is more energy lost than there is applied to accomplishing a task. These types of conversions are **inefficient**. For example, automobile engines waste more than two-thirds of the total energy used to drive them.

Some conversions, however, are highly **efficient**. The conversion of electrical energy into mechanical energy with electric motors is one example. Only about 10 percent of the electrical energy is wasted.

In Section II of this book, you will learn about the major energy control (power) systems that have been developed over many years. Every **energy control system** has three parts:
1. The original source of energy
2. All the conversions the energy goes through, including the moving of energy from one place to another (**transmission**)
3. The eventual use of the energy

One example of an energy control system is the development, transmission, and use of electrical energy. See Fig. 2-16. The original source

Fig. 2-16 A complete energy system.

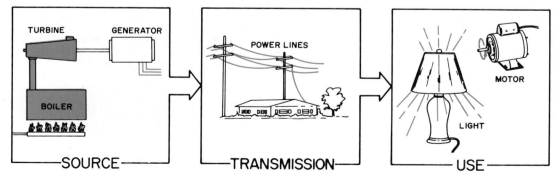

TURBINE   GENERATOR

BOILER

SOURCE

POWER LINES

TRANSMISSION

MOTOR

LIGHT

USE

of energy is the fuel used at the generating plant (for example, coal). From this original source, the energy goes through the following conversions:

- Burning coal (chemical energy) is changed to heat energy.
- Heat energy is applied to water to produce steam (mechanical energy).
- Mechanical energy is used to drive a turbine generator, which produces electricity (electrical energy).

The electricity is then transmitted along power lines to where it is needed. There the electricity is used to operate motors, provide light and heat, and operate radios and television sets.

POTENTIAL AND KINETIC ENERGY

All forms of energy can be put into two categories: kinetic energy and potential energy. See Fig. 2-17.

Kinetic energy is energy in motion. The radiant energy that travels from the sun to the earth is a kind of kinetic energy. Radiant energy produces wind and rain, which are other forms of kinetic energy. See Figs. 2-18 and 2-19. Rushing water, ocean breezes, and winter storms are all forms of kinetic energy.

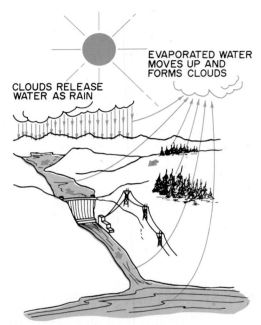

Fig. 2-18 In the *water cycle*, the sun evaporates water from the earth. The water forms into clouds, then falls to the earth as rain. Rain is kinetic energy produced by the sun.

Fig. 2-19 As the sun warms the air at the earth's surface, it rises and produces wind. Wind, therefore, is kinetic energy produced by the sun.

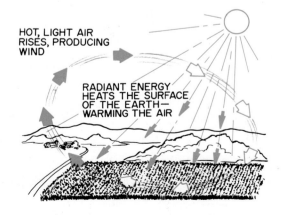

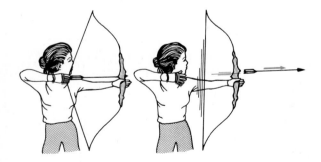

Fig. 2-17 The stretched bowstring has a certain amount of potential energy. When the archer releases the string, the potential energy changes into kinetic energy and propels the arrow.

Mechanical energy and electrical energy are also forms of kinetic energy. A moving car and the electrical current that powers a motor are both examples of kinetic energy. See Fig. 2-20. Whenever we use energy, it is in the form of kinetic energy.

Potential energy is any form of *stored* energy. In Chapter 1, you learned about one good example of potential energy — the water stored behind a dam at a hydroelectric plant. When energy is needed, the water is released and flows through a turbine generator. The generator produces electricity. Therefore, the overall conversion is from potential energy (water) into kinetic energy (electricity).

The fuel in a nuclear reactor is another form of potential energy. Uranium has a tremendous amount of stored energy. When energy is needed, engineers at the nuclear power plant start the fission process. This process produces heat, which is converted into mechanical energy, then electrical energy. The overall conversion is from potential energy (uranium) into kinetic energy (electricity).

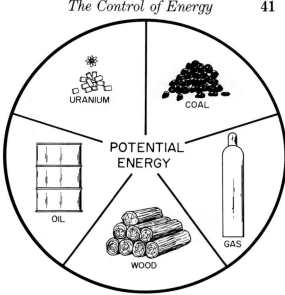

Fig. 2-21 **Fossil fuels and uranium are forms of potential energy.**

Fig. 2-20 **Mechanical energy and electrical energy are both forms of kinetic energy. (The electric motor has an electrical input and a mechanical output.)**

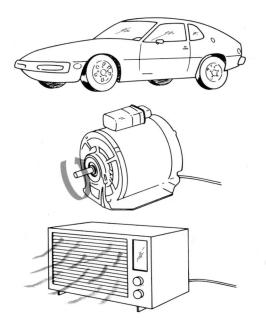

Most of the energy under our control is in the form of potential energy. See Fig. 2-21. Wood, coal, oil, and natural gas are all forms of potential energy. When we need energy, we convert these sources into forms of kinetic energy. We use generating plants and various types of engines to make the energy conversions. See Fig. 2-22.

Potential energy refers to all types of available or motionless energy. A raised hammer has potential energy. When dropped, the hammer can drive in a nail. Food is potential energy. Our bodies convert its chemical energy into kinetic energy as needed. A boulder at the top of a hill also has potential energy. This energy is released when the boulder rolls down the hill.

Storing Energy

Kinetic energy and potential energy are both important in energy control systems. We can change potential energy into kinetic energy, and we can change kinetic energy into potential energy.

Chrysler Corp.

Fig. 2-22 An automobile engine converts the potential energy of gasoline or diesel fuel into kinetic energy (motion).

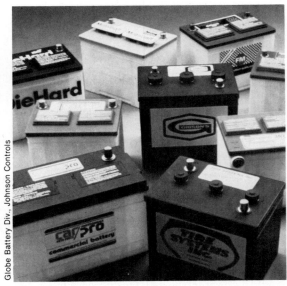

Globe Battery Div., Johnson Controls

Fig. 2-23 Storage batteries can convert kinetic energy (electricity) into potential energy (chemical compounds).

Changing potential energy into kinetic energy is fairly simple. For example, it is easy to set fire to wood and release its potential energy in the form of heat and light, which are both kinetic energy. However, it is much harder to change kinetic energy into potential energy. In the case of the wood, the original kinetic energy was the sun's radiant energy. This energy was put into storage by a tree. The storage process lasted many years. Although the release of energy can be a sudden process, the storage of energy is usually a long and slow process.

We can use storage batteries to convert small amounts of electrical energy into chemical energy for storage. See Fig. 2-23. This method, however, is inefficient and costly. Also, storage batteries hold small amounts of energy for their size.

Another way of changing kinetic energy into potential energy involves the use of machines. We can use machines to pump water into raised reservoirs. Like the storage battery method, storage by machines is not efficient for storing large amounts of energy.

However, batteries and pumped water *are* being used with some success in different parts of the United States. For example, one small town collects solar energy and stores it in batteries in the form of electricity. See Fig. 2-24. The sun generates electricity through photovoltaic cells during the day. The townspeople use some electricity right away. They store the surplus electricity in batteries to use when the sun goes down. The batteries also provide electricity for cloudy days.

There are problems if the sun does not shine for a few days. The batteries run down, and the community is without electricity. Therefore, this kind of energy system works best in regions that have sunshine almost every day.

In another community, the power company uses hydroelectric power produced at night to operate pumps that raise water to a reservoir. See Fig. 2-25. The company releases this water during the day to provide additional electricity. This system is based on the fact that energy needs are lower at night than they are during the day.

American Electric Power

Fig. 2-24 **At an Indian village in Arizona, photoelectric cells are used to produce electricity to charge the storage batteries shown here. The batteries then provide electrical power during the night.**

Fig. 2-25 **At night, water is pumped uphill to the reservoir behind this dam at Smith Mountain, Virginia. During the day, the water is released to generate electricity.**

The Real McCoy

Elijah McCoy lived in Michigan in the late 1800s. As he was growing up, he became interested in mechanical things. He learned all he could about mechanics. He even went to Scotland to study mechanical engineering!

After completing his studies, Elijah came back to America. He got a job oiling the big steam engines for a railroad. McCoy noticed right away that there was a major problem with the lubrication method used. The engine's moving parts could not be oiled while the engine was running. The engine had to be shut down completely first. Therefore, the oiling process took a lot of time. It also cost the railroad a lot of money.

McCoy thought the problem over. Could there be a way to oil the moving parts *automatically*, without shutting the engine down? He used his knowledge of mechanics to design a system to do just that!

First McCoy built an experimental model of his lubrication system. It worked! In 1872 he patented (registered) his design. McCoy's system could work on almost any machine that had moving parts. The system worked so well that people started asking for "the real McCoy" instead of other systems. Soon automatic lubrication began to be used on all kinds of machines.

Elijah McCoy received many patents during his long and productive career. Each patent had something to do with lubrication. In his later years, McCoy spent much of his time working with young people. He told them how important it was to get a good education. McCoy died at the age of 86.

TRANSPORTING ENERGY

We can transport (move) energy as either potential energy or kinetic energy. Potential energy, in the form of fuels, is easy to transport. The fuel is simply moved from one location to another. For example, gasoline is refined from oil and then stored in tanks. From the refinery, trucks transport the gasoline to gas stations, where it is stored again. We then pump the gasoline into the gas tanks of our cars. Cars transport the gasoline and change its potential energy into kinetic energy as needed.

We can also transport kinetic energy. Electricity travels along power lines long distances from generating plants.

We can transfer mechanical energy for short distances with belts, gears, and shafts. We can also transfer motion by using **pneumatics** (moving air) and **hydraulics** (moving liquids).

STUDY QUESTIONS

1. Define *energy*.
2. What three advantages did steam engines have over earlier sources of power?
3. Name the six forms of energy.
4. What do mechanical energy and heat energy have in common?
5. How is heat energy released from fossil fuels?
6. What is the most important use of light energy?
7. Define *chemical energy*.
8. Give three examples of how electrical energy can be converted into other forms of energy.
9. How is nuclear energy developed?
10. State the law of conservation of energy.
11. True or false: We can often convert one form of energy into another without wasting any energy.
12. Name the three parts of an energy control system.
13. What is kinetic energy?
14. What is potential energy?
15. True or false: It is fairly simple to change kinetic energy into potential energy.
16. Does energy have to be in a certain form (potential or kinetic) in order to be transported?

ACTIVITIES

1. Prepare a simple model or set of drawings showing the water cycle. The cycle should include (1) evaporation from the ocean, (2) clouds, (3) rain, (4) collection of water for power generation, drinking, and irrigation, and (5) distribution of water and electrical power. Present your project to the class, then set it up as a display in the classroom.

2. Select a special problem in the control of energy, such as the storage of electricity in batteries. Prepare a class report on your subject. Two other good topics are (1) other efforts to change kinetic energy into potential energy and (2) the conversion of light energy into other forms of energy.

Measuring Energy and Power

In Chapters 1 and 2, we studied energy sources and energy control. We learned that we can control many forms of energy, including heat, light, and motion. Our next step is to learn how to *measure* energy. When we measure energy, we can find out how much work we can accomplish with the energy.

Our measurement of controlled energy is **power**. In technical terms, power is *energy per unit of time*. See Fig. 3-1. As you know, energy is the capacity to do work. Power is work accomplished in a given period of time. For example, you must use a certain amount of energy to climb a flight of stairs. Your weight and the height of the stairs determine the amount of energy needed. This energy is the same whether you walk or run up the stairs. However, the amount of *power* used is different. The faster you climb the stairs, the more power you must use. See Fig. 3-2.

The measurement of power is important in determining the effectiveness of any energy

Fig. 3-1 **Power is energy or work per unit of time. The engine of this rocket produces enough power to lift a payload of 38,000 pounds at a speed of 17,000 miles per hour.**

NASA

20 SECONDS   IO SECONDS

Fig. 3-2 **Climbing stairs at different speeds requires different amounts of power. Both boys use the same amount of *energy* to climb the stairs. However, the boy on the right must use twice as much *power* as the other boy.**

45

control system. Controlled energy, in fact, is often identified as power. Later in this book, you will learn about power control and transmission devices. As you study these devices, it will be important to know how both energy and power are measured. This chapter will introduce you to the measuring systems and units used to make these measurements.

Fig. 3-3 **Basic units of measurement for the metric and customary systems of measurement.**

Measurement	Metric Units	Customary Units
Length	millimeters centimeters meters kilometers	inches feet yards miles
Weight	grams kilograms metric tons	ounces pounds tons
Volume	milliliters liters cubic centimeters	ounces cups pints quarts gallons
Area	square centimeters square meters	square inches square feet square yards
Temperature	degrees Celsius	degrees Fahrenheit
Speed	kilometers per hour	miles per hour
Force	newtons	pounds
Torque	newton-meters	pound-feet
Pressure	pascals	pounds per sq. inch inches of mercury
Energy Mechanical Heat Electrical	joules joules calories joules	foot-pounds British thermal units joules
Power Mechanical Heat Electrical	watts watts watts	horsepower BTUs per second watts

MEASURING SYSTEMS

There are two measuring systems presently used in the United States: the customary system and the metric system. The **customary system** is the measuring system we have traditionally used in the United States. This system is based on units such as the inch, the pound, the quart, and degrees Fahrenheit. Customary units used to measure power include horsepower, the foot-pound, pounds per square inch, and the BTU. See Fig. 3-3.

The **metric system** is the measuring system used by most of the industrialized countries of the world. This system is formally called the **International System of Units**, abbreviated **SI**. The SI metric system is based on units such as the meter, the liter, the gram, and degrees Celsius. Metric units used to measure power include the newton, the watt, the joule, and the pascal. See Fig. 3-3 again.

The metric system is easy to learn. It does not have as many base units as the customary system. There are 50 units of measurement in the customary system. There are only seven in the metric system. For example, in the customary system, length can be measured according to four units: the inch, the foot, the yard, and the mile. The metric system has only one unit for length: the **meter**. This is possible because the metric system uses *prefixes* with its base units. We use these prefixes to indicate amounts smaller or larger than the base unit. The most common prefixes are *milli-* (1/1000), *centi-* (1/100), and *kilo-* (1000). See Fig. 3-4.

Fig. 3-4 **By using one of three prefixes, you can change the value of a metric base unit.**

Prefix	+	Base Unit	=	Metric term	Value
milli- (m)	+	meter (m) liter (l) gram (g)	=	millimeter (mm) milliliter (ml) milligram (mg)	1/1000th of base unit
centi- (c)	+	meter liter gram	=	centimeter (cm) centiliter (cl) centigram (cg)	1/100th of base unit
kilo- (k)	+	meter liter gram	=	kilometer (km) kiloliter (kl) kilogram (kg)	1000 base units

The metric system is also easy to use because it is a decimal system. In a **decimal system**, all of the units and prefixes are multiples of the number 10. This means that all metric measurements can be easily multiplied and divided by 10. All we have to do is move a decimal point to the left or right.

Both the customary system and the metric system are used in the United States. Most people still use the customary system in their daily measurements. However, the metric system is often used in industry.

Measurement Conversions

When you work with energy and power, you will find that both the customary system and the metric system are used. It is important to be able to convert a measurement from one system into a measurement in the other system.

It is easy to convert customary measurements to metric measurements. We do this by multiplying the customary measurement by the **metric equivalent** of the customary unit used. Figure 3-5 shows the metric equivalents for the

Fig. 3-5 **Conversion table: customary to metric.**

To change a customary measurement to a metric measurement, multiply the customary measurement by the metric equivalent of the basic customary unit used.
Example: Change 127 pounds to kilograms.
    127 lbs. x .4536 kg = 57.61 kg

Measurement	Customary Unit	Metric Equivalent
Length	1.000 inch (in.) 1.000 foot (ft.) 1.000 yard (yd.)	2.540 centimeters (cm) 0.3048 meter (m) 0.9144 meter (m)
Distance	1.000 mile (mi.)	1.609 kilometers (km)
Area	1.000 square inch (sq. in.)	6.452 square centimeters (sq. cm)
Volume	1.000 cubic inch (cu. in.)	16.387 cubic centimeters (cu. cm)
Mass (Weight)	1.000 ounce (oz.) 1.000 pound (lb.)	28.349 grams (g) 0.4536 kilogram (kg)
Force	1.000 pound (lb.)	4.448 newtons (N)
Torque	1.000 pound-foot (lb.-ft.)	1.356 newton-meters (N-m)

Pressure	1.000 pound per square inch (psi)
	
	1.000 inch of mercury (at 60° F)
		6895 pascals (Pa) or
		  6.895 kilopascals (kPa)
		3377 pascals (Pa) or
		  3.377 kilopascals (kPa)

Energy
  Mechanical
  Heat
 	
  Electrical
	
	1.000 foot-pound (ft.-lb.)
	1.000 British thermal unit (BTU)
	
	1.000 joule (J)
		
		1.356 joules (J)
		1054 joules (J),
		  252 small calories (cal), or
		  0.2520 large calorie (kcal)
		1.000 joule (J)

Power
  Mechanical
  Heat
 	
  Electrical
	
	1.000 horsepower (hp)
	1.000 British thermal unit
	  per second (BTU/s)
	1.000 watt (W)
		
		746 watts (W)
		1054 watts (W)
		
		1.000 watt (W)

Temperature	To change degrees Fahrenheit (°F) to degrees Celsius (°C), use the formula:
	    °C = 5/9 (°F-32)
	Example: Change 68° F into degrees Celsius.
	    °C = 5/9 (68-32)
	    °C = 5/9 (36)
	    °C = 20	

basic customary units. We call these equivalents **conversion factors.** To convert 10 inches into centimeters, for example, simply multiply 10 by 2.540 (the metric equivalent for 1 inch). The answer is 25.40 centimeters.

Changing a metric measurement to a customary measurement is simple, too. We do this by multiplying the metric measurement by the customary equivalent of the metric unit used. See Fig. 3-6. For example, to find out how many

pounds there are in 10 kilograms, multiply 10 by 2.205 (the customary equivalent for 1 kilogram). The answer is 22.05 pounds.

Notice that equivalents are often not whole numbers. The conversion of a whole customary number usually produces a decimal metric number. For example, 20 feet is equal to 6.096 meters. This is one reason that people feel uncomfortable with the metric system.

Fig. 3-6 **Conversion table: metric to customary.**

To change a metric measurement to a customary measurement, multiply the metric measurement by the customary equivalent of the basic metric unit used.
Example: Change 175 kilometers to miles.
175 km x .6214 mi. = 108.7 mi.

Measurement	Metric Unit	Customary Equivalent
Length	1.000 millimeter (mm) 1.000 centimeter (cm) 1.000 meter (m) 1.000 kilometer (km)	0.04 inch (in.) 0.3937 inch (in.) 3.281 feet (ft.) or 1.094 yards (yd.) 0.6214 mile (mi.)
Area	1.000 square centimeter (sq. cm)	0.1550 square inch (sq. in.)
Volume	1.000 cubic centimeter (cu. cm)	0.06102 cubic inch (cu. in.)
Mass (Weight)	1.000 gram (g) 1.000 kilogram (kg)	0.03527 ounce (oz.) 2.205 pounds (lbs.)
Force	1.000 newton (N)	0.2248 pound (lb.)
Torque	1.000 newton-meter (N-m)	0.7376 pound-foot (lb-ft.)

Pressure	1.000 pascal (Pa)
	
	1.000 kilopascal (kPa)
		0.0001450 pound per square inch
		  (psi) or .0002961 inch of
		  mercury (in. of Hg)
		0.1450 pound per square inch (psi)
		  or .2961 inch of mercury
		  (in. of Hg)

Energy
  Mechanical
  Heat
 	
  Electrical
	
	1.000 joule (J)
	1.000 joule (J)
	
	1.000 small calorie (cal)
	1.000 large calorie (kcal)
	1.000 joule (J)
		
		0.7376 foot-pound (ft.-lb.)
		0.0009485 British thermal unit-
		  (BTU)
		0.003968 British thermal unit (BTU)
		3.968 British thermal units (BTUs)
		1.000 joule (J)

Power
  Mechanical
  Heat
 	
  Electrical
	
	1.000 watt (W)
	1.000 watt (W)
	
	1.000 watt (W)
		
		0.001341 horsepower (hp)
		0.009485 British thermal unit
		  per second (BTU/s)
		1.000 watt (W)

Temperature	To change degrees Celsius (°C) to degrees Fahrenheit (°F), use the formula:
	  °F = 9/5 °C + 32
	Example: Change 100° C to degrees Fahrenheit.
	  °F = 9/5 (100) + 32
	  °F = 180 + 32
	  °F = 212	

Metrics, American Style

Have you ever wondered what the numbers and letters on an automobile tire mean? This strange series of numbers and letters is a code that identifies the type of tire. But what about the numbers themselves? Are they metric or customary? Well, actually they're a combination of the two. Let's take a look at the code shown on the tire at the right:

First of all, the **P** means that the tire is meant to be used on a passenger car. The **195** means that the **cross-sectional width** of the tire is about 195 millimeters.

The **75** is a percentage. It represents the relation of the **cross-sectional height** of the tire to its cross-sectional width. In this case, the sectional height is 3/4, or 75 percent, of the sectional width. The lower the number, the wider the tire will appear to be. For example, a tire rated 50 appears wider than a tire rated 75.

The **R** means that the tire is a radial tire. And the **14** is the diameter — in inches — of the wheel rim onto which the tire will fit.

The code that we have just deciphered is an example of a tire-sizing system called the **P-metric system**. This system is one example of how industry is using both customary measurements and metric measurements in producing and identifying its products.

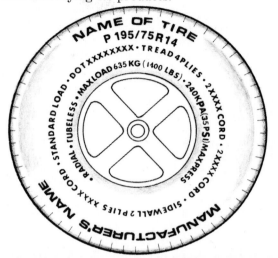

However, someday the United States may switch completely over to the metric system. People will begin to think in terms of metric numbers, just as they now think in terms of customary numbers. Someday we may think of a very tall person as being over two meters tall, just as we now think of a tall person as being over six feet tall. (Two meters is just over six feet, six inches.)

Metric/customary conversions do not affect our measurements of time. The traditional measurements of time — the second, minute, hour, day, and year — are universal (used everywhere).

MEASURING ENERGY AND POWER

As people learned to control energy, they developed ways to measure energy and power. Through measurement, we can find out how much energy we need to perform a particular task. We can also tell when we have developed enough energy to perform the task. Accurate measurement allows us to convert the correct amount of potential energy into kinetic energy. For example, we can figure out the electrical needs of a city by measuring the amount of power used by consumers. This tells us how much fuel must be converted into electrical energy at the power plant serving the city.

We commonly use a number of measurement terms to determine how much energy we have or need. In this section, you will learn how to use the following terms in calculations:

- Energy
- Work
- Power
- Force
- Torque
- Pressure
- Heat

Energy and Work

Energy has already been defined as the capacity to do work. **Work**, in turn, can be defined as *useful motion, or motion that results in something useful being done.* The motion produced by energy may be the movement of a bicycle or a rocket, or the lifting of a weight. It may be the movement of electrons in a wire (electricity). It may be the movement of atoms and molecules (heat). "Useful motion" may also involve the *stopping* of something that is already moving. For example, the driver of a car applies the brakes to reduce the motion of the vehicle.

There is no work if nothing is accomplished. Imagine a boy trying to lift a 1000-pound barrel of nails. If he cannot move it at all, he does not perform any work. However, he *does* exert energy. See Fig. 3-7. Sensitive measuring instruments will show that the boy gives off heat energy as he tries to lift the barrel. The instruments will also show a very slight increase in the temperature of the barrel. (This is the amount of heat that is transferred to the barrel by the boy's touch.)

When we think in terms of measuring something, our definition of work must become more specific. Perhaps the best definition is that work is *a measurement of mechanical energy.* In the customary system, work (or mechanical energy) is measured in **foot-pounds (ft.-lbs.).** One foot-pound of work is equal to the lifting of 1 pound a distance of 1 foot. See Fig. 3-8. The mathematical formula is:

Work = Weight (in pounds) × Distance (in feet)

Example problem:

How much work does a 120-pound boy accomplish when he climbs a 20-foot flight of stairs?

Work = Weight × Distance
Work = 120 lbs. × 20 ft.
Work = 2400 ft.-lbs.

Power

Work is not a complete measure of energy. Suppose a man were trying to move a 1000-pound barrel of nails up 20 feet to the second floor of a building. Now, he can accomplish this task in two different ways. He could operate a crane to pick up the barrel and lift it the 20 feet in 20 seconds. Or, he could carry 100 pounds of nails up a ladder and make 10 round trips. The second method will get the nails to the second floor as well as the first method. But the second method might take 20 minutes. In each method, the amount of *work* done is the same. However, there is a big difference in the *time* it takes to do the work. See Fig. 3-9.

It is obvious that the crane can do work much faster than the man. Since the measurement of work does not show this, we have to use another term — power. Power is defined as *energy per second or energy per unit of time.* However, when we commonly talk about power, we mean *work per unit of time.*

Fig. 3-7 The boy is using energy in his attempt to lift the barrel. However, since he cannot lift the barrel, he is not accomplishing any work.

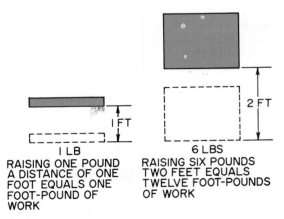

RAISING ONE POUND A DISTANCE OF ONE FOOT EQUALS ONE FOOT-POUND OF WORK

RAISING SIX POUNDS TWO FEET EQUALS TWELVE FOOT-POUNDS OF WORK

Fig. 3-8 Work is a measurement of accomplishment and is recorded in foot-pounds.

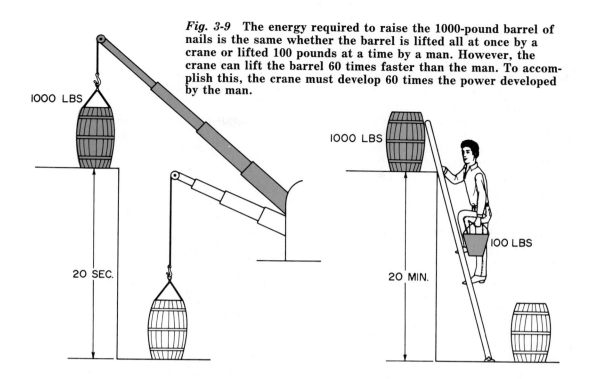

Fig. 3-9 The energy required to raise the 1000-pound barrel of nails is the same whether the barrel is lifted all at once by a crane or lifted 100 pounds at a time by a man. However, the crane can lift the barrel 60 times faster than the man. To accomplish this, the crane must develop 60 times the power developed by the man.

1000 LBS

20 SEC.

1000 LBS

100 LBS

20 MIN.

Power, therefore, is a measurement of work accomplished in a given period of time. To increase power, we must do more work in a given period of time. We can also increase power by accomplishing a given amount of work in a shorter period of time.

The crane in Fig. 3-9 accomplished the same amount of work as the man. However, it accomplished the work in 1/60 of the time. Therefore, in the 20 seconds it took the crane to lift the nails, the crane produced 60 times the power of the man.

Horsepower (hp) is the most common measurement of power. It is based on the amount of work that a horse can do in one minute. See Fig. 3-10. When this standard was set, the amount of work done by the horse was multiplied by one and one-half. This makes 1 horsepower somewhat above the effort of a strong horse. Therefore, an engine with an output of 1 horsepower can normally do more work than a horse.

One horsepower is equal to the energy needed to lift 33,000 pounds 1 foot in 1 minute. This is the same as the energy needed to lift 550

pounds 1 foot in 1 second. In mathematical equation form:

$$1 \text{ hp} = \frac{33,000 \text{ foot-pounds}}{\text{minute}}$$

or

$$1 \text{ hp} = \frac{550 \text{ foot-pounds}}{\text{second}}$$

Fig. 3-10 The unit of *horsepower* is based on the amount of work that a horse can do in one minute.

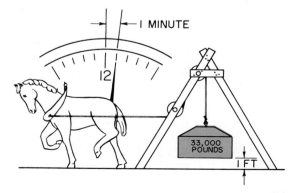

1 MINUTE

12

33,000 POUNDS

1 FT

To calculate horsepower, we usually divide the foot-pounds of work by the time (in seconds) multiplied by 550. The formula is:

$$hp = \frac{\text{Weight (lbs.)} \times \text{Distance (ft.)}}{\text{Time (secs.)} \times 550}$$

Example problem:
How much horsepower does a 165-pound man develop in climbing a 20-foot flight of stairs in 12 seconds?

$$hp = \frac{165 \text{ lbs.} \times 20 \text{ ft.}}{12 \text{ secs.} \times 550}$$

$$hp = 1/2 \text{ or } 0.5$$

Force

Up to this point, we have used *weight* in calculating both work and horsepower. We can usually substitute *force* for weight in both calculations.

Force (F) is any push or pull on an object. The earth's gravity is a force that pulls down on every object on earth. When we speak of the "weight" of an object, we are really talking about how much gravity is pulling on the object. When we lift anything, we must exert a force equal to the pull of gravity on the object. Therefore, **weight** is considered to be force applied in a vertical (up and down) direction. We use pounds to measure this kind of force. See Fig. 3-11.

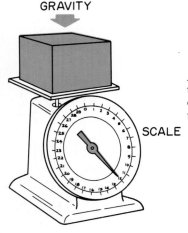

GRAVITY

SCALE

Fig. 3-11 Force is the weight of an object when measured vertically.

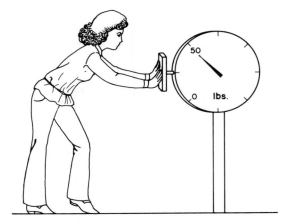

Fig. 3-12 Force is also the measurement of a push in directions other than vertical. Here the girl is exerting a horizontal push of 50 pounds.

Force, however, also applies to a push on an object in any direction. See Fig. 3-12. Force applied in directions other than vertical is also measured in terms of weight — ounces, pounds, and tons.

The term *weight* is used instead of force only when measuring work in a vertical direction, as in simple lifting. In our calculations of work and horsepower, we will substitute force for weight. Therefore, our formula for work will be:

Work = Force × Distance

Example problem:
A man pushes a 200-pound weight a distance of 10 feet along the floor. He must exert a force of 55 pounds to slide the weight. See Fig. 3-13.

How much work does the man accomplish?

Work = Force × Distance
Work = 55 lbs. × 10 ft.
Work = 550 ft.-lbs.

Substituting force for weight also affects the formula for horsepower. (This is true for all situations except simple lifting. In lifting situations, force and weight are the same.) In equation form:

$$hp = \frac{\text{Force} \times \text{Distance}}{\text{Time (secs.)} \times 550}$$

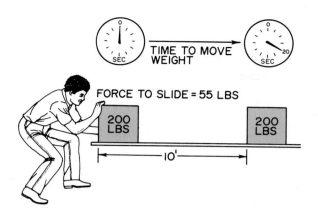

FORCE TO SLIDE = 55 LBS

Fig. 3-13 The term *force* should be used in all calculations of work except for those involving simple lifting.

Example problem:
   What is the torque if a force of 40 pounds is applied to a radius of 2 feet?

   Torque = Force × Radius
   Torque = 40 lbs. × 2 ft.
   Torque = 80 pound-feet

   The units used to measure torque (pound-feet) and work (foot-pounds) may cause you to confuse torque with work. These terms are not the

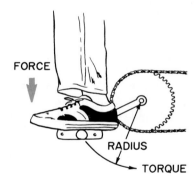

Fig. 3-14 Twisting open a jar requires torque (twisting force).

   Continuing with the example from Fig. 3-13, how much horsepower does the man produce if it takes him 20 seconds to slide the weight along the floor?

$$\text{hp} = \frac{\text{Force} \times \text{Distance}}{\text{Time (secs.)} \times 550}$$

$$\text{hp} = \frac{55 \text{ lbs.} \times 10 \text{ ft.}}{20 \text{ secs.} \times 550}$$

$$\text{hp} = 1/20 \text{ or } 0.05$$

Torque

   Force can also be measured in terms of **torque**. Torque is turning or twisting effort. See Fig. 3-14. We use torque whenever we turn a steering wheel or tighten a bolt. Torque is the force applied to push the pedal of a bicycle around in a circle. It is also the force that turns the wheels on a car.

   Notice that all of our examples of torque involve circular motion. We can define torque as *a force applied to a radius*. See Fig. 3-15. In the customary system, torque is measured in **pound-feet (lb.-ft)**. We calculate torque by multiplying the force applied by the distance from the center of the object being turned:

   Torque = Force (lbs.) × Radius (ft.)

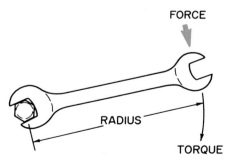

Fig. 3-15 Torque is force applied to a radius. (A radius is one-half the diameter of a circle.)

same. Torque is a certain kind of force only. It does not automatically accomplish anything. Work, on the other hand, always involves accomplishment.

Work is force applied through a distance. The *foot* part of *foot-pounds* describes this distance. The measurement of torque also involves distance. The *feet* in *pound-feet* describes this distance. However, the distance involved in pound-feet is only the distance from the force to the center of the object being turned. Torque accomplishes work only when it moves *through* a distance. The wrench in Fig. 3-15 applies torque to the bolt head whether it actually moves the bolt or not.

Pressure

Pressure is another measurement of force. Pressure is determined by the area over which a force is applied. Therefore, pressure is *force per unit of area*. For example, Fig. 3-16 shows a 100-pound weight with a base of 10 inches by 10 inches. The total area equals the length times the width.

Area = Length × Width
Area = 10 in. × 10 in.
Area = 100 sq. in.

The pressure is equal to the total force (weight) divided by the total area.

$$\text{Pressure} = \frac{\text{Force}}{\text{Area}}$$

$$\text{Pressure} = \frac{100 \text{ lbs.}}{100 \text{ sq. in.}}$$

$$\text{Pressure} = \frac{1 \text{ lb.}}{\text{sq. in.}} \text{ or } 1 \text{ lb./sq. in.}$$

This means that the weight is "spread out" over the whole area. Each square inch of surface area supports only 1 pound. Note the units of the answer. **Pounds per square inch (psi)** is the most common unit for measuring pressure. (We use another unit — **inches of mercury** — to measure atmospheric pressure.)

The example in Fig. 3-16 helps to explain pressure. However, we do not commonly measure the force of solid objects in units of pressure. Units of pressure are normally used to measure the force exerted by **fluids** (gases or liquids).

A confined fluid under pressure exerts equal force on all enclosing surfaces. The air in the balloon in Fig. 3-17 is pushing with a small

Fig. 3-16 **Pressure is the amount of force applied to a certain unit of area. The 100-pound block shown here applies 1 pound of force to each square inch of supporting area.**

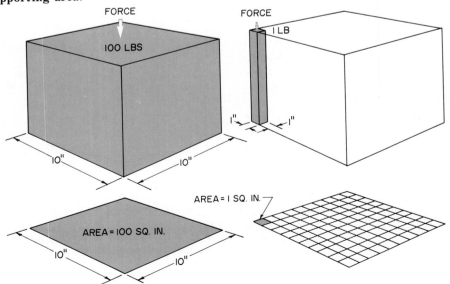

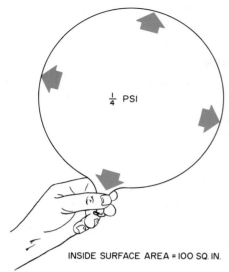

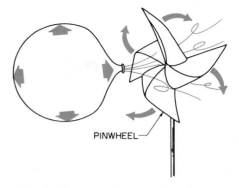

Fig. 3-18 Releasing air from the balloon releases pressure. The force of this pressure can be used to produce motion.

Fig. 3-17 **An inflated balloon contains air under pressure. When the stem of the balloon is held closed, the pressure is evenly balanced on the entire inside surface.**

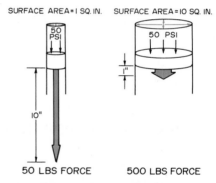

Fig. 3-19 **Here the same amount of *pressure* (force per unit of area) is being applied to the top of each piston. However, there is 10 times as much *force* exerted on the large piston as there is on the small piston.**

pressure in all directions. The pressure is only 1/4 pound per square inch. However, it presses evenly over the entire inside surface area.

The fact that the balloon contains force can be shown by releasing the air. The escaping air can rotate a pinwheel or blow out a match. See Fig. 3-18.

We can calculate the total force produced by a certain amount of pressure. To do this, we simply multiply the pressure (force per unit area) by the total area. In the case of our balloon, the inside surface area is 100 square inches. See Fig. 3-17 again.

Force = Pressure × Area
Force = 1/4 psi × 100 sq. in.
Force = 25 lbs.

The difference between pressure and force is important. You must remember that pressure is a special measurement of force. It is force per unit of area. However, the *total* amount of force

depends on the total amount of area. For example, Fig. 3-19 shows two pistons of different sizes. The small piston has an area of only 1 square inch. The large piston has an area of 10 square inches. Now, the pressure above the piston in each cylinder is 50 pounds per square inch. However, this pressure produces different amounts of force.

Small Cylinder:
  Force = Pressure × Area
  Force = 50 psi × 1 sq. in.
  Force = 50 lbs.
Large Cylinder:
  Force = 50 psi × 10 sq. in.
  Force = 500 lbs.

The large piston produces 10 times the force of the small piston. However, it will move only 1/10 the distance of the small piston.

Heat

Heat energy is measured in **British thermal units (BTUs).** As you recall from Chapter 2, the word **thermal** refers to heat. The BTU was originally used as a measurement by the British. One BTU is the heat needed to raise the temperature of 1 pound **(0.45 kg)** of water 1 degree Fahrenheit **(0.56° C).** See Fig. 3-20.

In itself, the BTU is a measurement of energy, not power. This is because power is energy per unit of time. However, we can and do measure heat energy in **BTUs per hour.**

MEASURING ENGINE POWER

We can measure the power produced by an engine with a device or machine called a **dynamometer.** A dynamometer shows how much work the engine can do in a given period of time.

There are three basic kinds of dynamometers: the prony brake, the hydraulic, and the electric.

Clayton Manufacturing Co.

Fig. 3-21 This car is being tested on a hydraulic dynamometer. The rear wheels drive a pump that pumps water to a turbine. The turbine absorbs the engine's power.

The **prony brake dynamometer** is a mechanical device that is rarely used today. The **hydraulic dynamometer** is the type most commonly used to measure automobile engine power. It uses water to transmit the power produced by the engine. The water turns a turbine that provides a way to measure the power output. See Fig. 3-21. The **electric dynamometer** is the most accurate type of dynamometer. It is used in engine research and development.

DEVELOPING POWER FROM ENERGY

There are three common forms of power in use: electrical, mechanical, and fluid. Most people easily recognize two of these forms. **Electrical power** provides light and operates motors. **Mechanical power** moves cars, trains, and airplanes.

Fluid power is equally important but not as familiar. It is the use of fluids (gases or liquids) to produce motion. Fluid power operates hydraulic brakes in cars and air brakes in

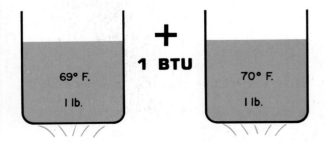

Fig. 3-20 It takes 1 British thermal unit of heat to raise the temperature of 1 pound of water 1 degree Fahrenheit.

trucks. Fluid power also operates the moving parts of most heavy construction equipment, such as road graders and earth movers.

Like energy, power can be changed from one form to another. Electrical power can operate an electric motor to produce mechanical power.

The same motor can drive a hydraulic pump to produce fluid power. Another example of power conversion is the mechanical power that turns a generator to produce electrical power. Each form of power can easily be changed into either of the other forms.

STUDY QUESTIONS

1. What can we find out by measuring energy?
2. Define *power*.
3. What two measuring systems are used in the United States?
4. Name four of the basic units used in the customary system.
5. Name four basic units used in the metric system.
6. Give the meaning of the following prefixes: *milli, centi,* and *kilo*.
7. How do you convert a customary measurement into a metric measurement?
8. Define *work*.
9. True or false: In the use of energy, there is work even if nothing is accomplished.
10. What is one foot-pound of work equal to?
11. Give two ways in which power can be increased.
12. What is one horsepower equal to?
13. Define *force*.
14. True or false: When we measure work, it does not matter whether we use the term *force* or the term *weight*.
15. What is torque?
16. What is the difference between work and torque?
17. Define *pressure*.
18. What is the most common unit used in measuring pressure?
19. What does the total amount of force produced by a certain pressure depend on?
20. How much is one British thermal unit?
21. What do we use to measure engine power?
22. In your own words, describe the difference between energy and power.
23. Name the three common forms of power.

ACTIVITIES

1. Identify the places where the metric system is used in your community. Find out if there are plans to convert to metrics. This study can include library research, factory observations, and meetings with community leaders.
2. Solve the following problems:
   A. Change 3 feet into meters.
   B. Change 2000 pounds into kilograms.
   C. How many feet are in 1000 meters?
   D. How many miles are in 100 kilometers?
   E. How much work does a 120-pound boy do in climbing to the top of the Washington Monument — a climb of 550 feet?
   F. If the climb in Problem E takes 20 minutes, what is the average horsepower produced by the boy?
   G. A 3000-pound car is pushed 500 feet by three people. A force of 60 pounds is required to move the car. How much work was accomplished? How much work did each person do if each did his or her share?
   H. A mechanic uses an 18-inch pipe wrench to loosen a 1/2-inch pipe fitting. What torque is applied to the fitting if a force of 50 pounds is needed?

Energy, Power, and the Environment

In the United States, we enjoy a very high standard of living. We live better than kings did in the Middle Ages. In the winter, we heat our homes with electricity, fuel oil, or natural gas. In the summer, we cool our homes with electric fans or air-conditioners. We cook our food with electricity or natural gas. Electricity powers our stereos and televisions. Gasoline and diesel fuel power cars, trucks, and buses. It is easy to see that our standard of living depends on large supplies of energy.

We must pay a price for the benefits of power. Energy use has had destructive effects on our environment. (**Environment** refers to our natural surroundings.) Automobile engines give off harmful gases that combine to form smog. Sulfur gases from coal-fired power plants combine with water to form acid rain. Waste heat from power plants pollutes lakes and rivers. Hydroelectric power production disrupts natural biological systems. Energy use even affects the delicate temperature balance of the earth. These problems are all harmful **side effects** of energy use.

Fig. 4-1 **Air pollution results mainly from the burning of fossil fuels. This Los Angeles scene shows the pollution created by car exhaust.**
South Coast Air Quality Management District

Early in this century, people thought that pollution from energy was not important. Tall smokestacks poured smoke into the city air. Factories dumped industrial wastes into the rivers. There was little pollution control of any kind. People used energy without thinking of the environment. They assumed that the atmosphere and the earth would somehow *automatically* dispose of the pollution.

This assumption was partly correct. Natural processes — photosynthesis, rain, and wind — make many side effects harmless. However, we produce pollution faster than natural processes can deal with it. As a result, pollution is damaging the environment. It reduces the quality of life for everyone.

Almost every source of controlled energy produces pollution. Some forms of pollution are worse than others. Pollution from fossil fuels is among the worst. It also causes the most damage to our environment. Fossil fuel combustion pollutes the air with harmful gases. See Fig. 4-1. Nuclear energy is less harmful, if it is controlled with care. However, its waste products are extremely dangerous. Also, a nuclear accident could cause extensive and long-lasting damage to our environment.

Of the current major energy sources, hydroelectric energy is the least harmful. Solar energy and wind energy, as they are developed, may be even less harmful. However, these new energy sources and power systems could also have side effects. Scientists are already concerned about how these side effects may change the environment.

Pollution is defined as *any undesirable change in the air, land, or water that harmfully affects living things.* There are many sources of pollution. Industrial wastes, trash, fertilizer run-off, sewage, and pesticides are some of them. These sources account for much of the land and water pollution. Energy use accounts for a large share of the world's air pollution.

POLLUTION FROM FOSSIL FUELS

Most of our energy comes from the burning of fossil fuels: coal, oil, and natural gas. As you know, natural gas is a relatively clean-burning fuel. It produces the least amount of pollution of the three fossil fuels. Petroleum products and coal, however, produce many harmful gases when they are burned. The following sections describe the air and water pollution that result from the use of fossil fuels.

Air Pollution

Fossil fuels are the world's main source of air pollution. We cannot see many of the harmful gases produced by fossil fuel combustion. However, their effects on the environment are very real. See Fig. 4-2. This section will discuss the following types of air pollution:

- Smog
- Carbon monoxide
- Particulates
- Earth warming

Smog. Automobiles are a major source of air pollution. They produce the gases that form smog. The word *smog* originally meant a mixture of smoke and fog. It is actually a brownish mass of air. Smog may be our main air pollution problem. It is present in most large industrialized cities throughout the world.

Smog is the result of a very complex chemical reaction. The main "ingredients" are hydrocarbons, nitrogen oxides, and sunlight. **Hydro-**

National Coal Association

Fig. 4-2 Although they are invisible, harmful gases such as sulfur dioxide and nitrogen dioxide are being released into the atmosphere by this coal-burning power plant.

carbons are chemical compounds made of hydrogen and carbon. Gasoline and diesel fuel both consist of hydrocarbons. Automobile engines do not totally burn these hydrocarbons. The engines emit (give off) many partially or completely unburned hydrocarbons into the air. These hydrocarbons are one type of automotive **emission**.

Nitrogen oxides are combinations of nitrogen and oxygen. The heat of combustion in automobile engines produces nitrogen oxides. Power plants also produce them.

Smog forms when sunlight strikes hydrocarbons and nitrogen oxides. Together they form a dense, brownish-yellow air mass. Smog makes breathing difficult, irritates the eyes, and can injure the lungs. It is especially harmful to people who are old, weak, or ill.

Smog is also harmful to plants. It often shrinks and discolors leaves. See Fig. 4-3. Smog

Fig. 4-3 The grapes at the left grew on vines that were exposed to air pollution. The grapes at the right grew in clean air.

also fades and weakens rubber, fabric, and cellulose.

Smog eventually blows away. Sometimes, however, special weather conditions hold the smog over a city for days. This makes the harmful effects even worse.

Carbon Monoxide. Automobile engines produce one other harmful gas: carbon monoxide. Carbon monoxide is colorless, odorless, and poisonous. It is the result of incomplete combustion of fuel inside engines.

When carbon monoxide is inhaled, it reduces the oxygen-carrying capacity of the blood. Even small amounts are harmful to humans and animals. Large amounts of carbon monoxide can cause death.

Particulates. Fossil fuel combustion usually produces many tiny particles of matter. These particles are called *particulates*. They include dust, smoke, ash, and other materials.

Some particulates serve a good purpose. For example, rain drops form when water vapor collects around particles.

Most particulates, however, are harmful. Some metallic particles, such as lead, are poisonous. Some types of gasoline contain small amounts of lead. Lead helps gasoline burn smoothly in an engine. However, when gasoline is burned, lead particles are released into the air. Federal standards call for a reduction of lead in gasoline. Eventually, no gasoline will contain lead.

Earth Warming. Another product of fossil fuel combustion is **carbon dioxide**. We usually think of carbon dioxide as a harmless gas. In fact, plants need carbon dioxide to live. They convert carbon dioxide and water into sugar and oxygen. However, fossil fuel combustion produces a *great* deal of carbon dioxide. The amount is about twice as much as plants can use. As a result, the excess carbon dioxide builds up in the air.

Excess carbon dioxide is warming the earth. It acts in the same way as the glass in a greenhouse. Figure 4-4 explains this **greenhouse effect.** The sun's radiant energy consists of short-wavelength radiation. It can easily pass through the greenhouse glass. The radiation strikes solid objects, such as plants and soil. The solid objects then release the radiant energy as heat energy.

Heat energy has a longer wavelength than radiant energy. Its longer wavelength keeps it from passing through the glass. As a result, the heat energy is trapped inside the greenhouse.

Carbon dioxide has a similar greenhouse effect on the earth. Like glass, carbon dioxide permits radiant energy to travel to the earth. However, the carbon dioxide makes it hard for the heat energy to escape. The result is a gradual warming of the earth.

Scientists have estimated the results of the greenhouse effect. This effect may already have increased the earth's temperature about 0.5° F.

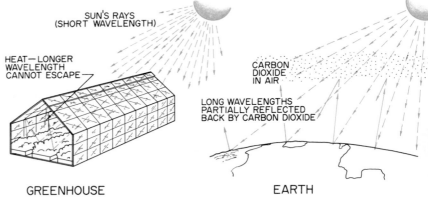

Fig. 4-4 **The Greenhouse Effect: The excessive carbon dioxide produced by burning fossil fuels acts like greenhouse glass. It allows short-wavelength radiant energy to pass through, but traps long-wavelength heat energy.**

SUN'S RAYS (SHORT WAVELENGTH)

HEAT—LONGER WAVELENGTH CANNOT ESCAPE

CARBON DIOXIDE IN AIR

LONG WAVELENGTHS PARTIALLY REFLECTED BACK BY CARBON DIOXIDE

GREENHOUSE

EARTH

This sounds like a small change. However, small temperature changes can have major effects on climate.

Suppose that the global temperature increased 4° F (*2° C*). This would produce a 9° F (*5° C*) increase at the earth's poles. There could be a major melting of ice. The level of the oceans would rise. Many of our coastal regions would be flooded.

We must be careful about predicting earth temperatures, however. The greenhouse effect is only one factor in long-range temperature changes. The earth is now in a very warm period. Scientists believe it may become cooler for reasons not fully understood. We are only beginning to learn why the earth's temperature changes. We know that carbon dioxide will continue to have a warming effect. However, this warming could be offset by an unknown cooling effect.

Water Pollution

The use of fossil fuels also affects lakes, streams, and oceans. Most water pollution results from the dumping of industrial wastes and improper sewage disposal. However, energy use accounts for some serious water pollution problems. Two of these problems are acid rain and thermal pollution.

Acid Rain. Acid rain results basically from the burning of coal and oil. These fuels may be burned to produce heat directly or to generate electrical power.

Fig. 4-5 **Brown trout fry appear healthy at pH 5.5 (a low acidity). However, they die at pH 4.5 (high acidity).**

SNSF Project Norway

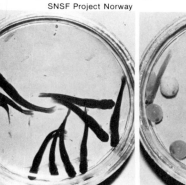
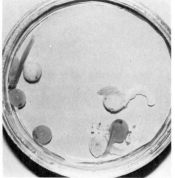

Coal and oil usually contain sulfur. Combustion changes the sulfur into **sulfur oxides** and releases them into the air. The sulfur oxides mix with water vapor in the air. The result is sulfuric acid. When it rains, *acid rain* falls to the earth.

Acid rain is very harmful to all forms of life. It can collect in lakes and rivers. Fish and other animal life may die from it. See Fig. 4-5. As the acidity increases, fewer fish survive.

There has probably been acid rain ever since the Industrial Revolution.* However, people are only beginning to see acid rain as a problem. In 1959, Norwegians noticed fewer fish in Norwegian and Swedish lakes. They connected the decrease in fish population to acid rain. In 1969, they traced the problem to industry in Europe.

Acid rain is a growing problem. Hundreds of lakes in Norway and Sweden no longer support fish. Many lakes in the eastern United States no longer have fish. Lakes in the Rocky Mountains and in the South are becoming more acidic.

The problem of acid rain crosses national borders. Sulfur oxides from the eastern United States produce acid rain in Europe. Sulfur oxides from China and Japan cause acid rain in our western states. All industrialized nations must cooperate to solve this problem.

Thermal Pollution. Power plants produce another form of pollution besides sulfur oxides. They release waste heat into rivers, lakes, and the oceans. This form of pollution is called *thermal pollution.*

Thermal pollution mainly comes from power plants that use steam turbines. These plants need water to condense the steam after it is used. Less than 50 percent of the heat generated by fuel combustion produces power. The water absorbs the waste heat.

Rivers, lakes, and oceans are natural "sinks" for waste heat. That is why power plants are often located near them. Waste heat from the plants is released into the water. This raises the water's temperature. Cool-water fish cannot survive in it. Warm water also promotes the

---

*The first Industrial Revolution was in England and lasted from 1750 to 1830. The industrial revolution in America took place in the second half of the nineteenth century.

Fig. 4-6 The cooling towers at this Ohio nuclear power plant cool water by evaporation at the rate of 600,000 gallons per minute.

growth of algae. The algae reduces the amount of oxygen in the water. The lack of oxygen then kills plants and animals in the water.

Some power plants use **cooling towers** to release waste heat. See Fig. 4-6. The water cools by evaporation as it passes through the towers.

Cooling towers work best in dry climates. Water evaporates more quickly in these areas. However, even the dry-climate cooling tower system has a disadvantage. It raises the humidity of the area. The increased humidity can reduce the quality of life.

POLLUTION FROM ALTERNATIVE SOURCES OF ENERGY

Alternative sources of energy are sources other than fossil fuels. Most alternative sources of energy are much less polluting than fossil fuels. However, each source must be thought about carefully. A few sources can cause serious pollution problems. Others may have little effect on the environment. Still others may have problems that we have not even identified yet. In this section, you will learn about the problems possible with the following five sources:

- Wood
- Geothermal energy
- Hydroelectric energy
- Nuclear energy
- Solar energy

Wood

The cost of gas and oil for heating has increased. Many people have turned to burning wood for home heating. Unfortunately, wood is not a clean-burning fuel. When it is burned, it releases two kinds of polluting materials. These are unburned chemical compounds and burned particles.

Part of the wood put into a stove is not completely burned. It ends up as **unburned chemical compounds** released into the air. Scientists have found that some of these compounds cause cancer in animals. One compound is a cancer-causing substance that is also found in cigarette smoke.

Burning wood also produces many small **burned particles.** Soot is one kind of burned particle. People can inhale these particles deep into their lungs. This makes them more likely to get respiratory (breathing-related) diseases.

Wood is even more polluting than fossil fuels. Wood burning can cause a large share of particulate pollution. It also produces a huge amount of harmful chemical compounds. It gives off over 200 times the amount of pollution produced by burning coal.

Designers are working to solve the problems of burning wood. One wood-burning stove design allows wood to be burned twice. The second burning removes some of the harmful unburned chemicals. See Fig. 4-7.

Fig. 4-7 This modern wood stove has a special catalytic combustion device. This device ignites and burns exhaust smoke before it reaches the chimney.

Geothermal Energy

Generating electricity from geothermal energy can also produce pollution. Most underground deposits of heat contain hydrogen sulfide gas. If this gas is released, it can combine with oxygen and moisture in the air. The result is acid rain, as described earlier.

In some parts of the world, hydrogen sulfide already goes into the air. For example, geysers at Yellowstone National Park release hydrogen sulfide. Therefore, the release of *some* sulfur is a natural process. However, geothermal plants can increase the amount of gaseous sulfur in the air. More plants may be built. It will then become more important to prevent the buildup of sulfur in the air.

Hydroelectric Energy

Hydroelectric power plants create large lakes where there were only rivers. These power plants provide two major benefits. They provide near pollution-free power, and they provide water for farming and recreation.

However, hydroelectric plants can harm the environment. The damming of a river changes natural biological systems. Animals using the river may not be able to live with these changes. For example, salmon cannot swim over a dam to get upstream. Also, hydroelectric plants are the direct cause of death for many fish. The fish can be destroyed by the rapidly turning turbine blades. They can also die from sudden pressure changes caused by passing through the hydroelectric installation.

Sometimes we can correct the problems caused by dams. For example, we can construct "fish ladders" that allow salmon to swim upstream. Also, a new type of hydroelectric system is being developed that does not harm fish. A **hydro-pneumatic system** works by using water to compress air. The air then drives turbine generators (*see boxed story*). In this system, the fish do not pass through the turbines. They also have time to adjust to pressure changes.

With each new hydroelectric plant that is built, we must carefully consider the possible harmful effects. The environmental changes must not be too severe.

Hydro-Pneumatic Power: A New Way to Tap the Energy of Moving Water

Using flowing water to do work isn't a new idea. People used waterwheels to power grain mills for years. Today, the most common use of flowing water is in hydroelectric plants. However, a new power-generating system being developed in Oregon may change all that.

In a hydro-pneumatic system, the moving water is directed into large cylinders filled with air. As the water rises in the cylinders, it compresses the air and sends the air to a device called a **receiver**. (You'll find out more about receivers and pneumatics in Chapter 6.)

From your reading of Chapter 1, you know that hydroelectric plants use water turbines to produce electricity. A hydro-pneumatic system also uses turbines. The difference is that the turbines are **air turbines**. The compressed air from the receiver turns the turbines, and they produce electricity.

Hydro-pneumatic power has many advantages over hydroelectric power. For one thing, it has a low **environmental impact**. (This means that it doesn't disturb the environment much.) Hydro-pneumatic power is also more efficient, because it can work year-round and can be used with gently sloping streams.

Another important point is that hydro-pneumatic systems may provide cheap electricity for many developing countries around the world. Since we all live on the same planet, this should be good news for everyone. Our world is an enormous system in itself. And whatever is done to better one part of the system benefits all other parts!

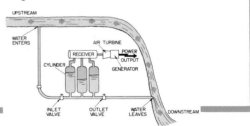

Fig. 4-8 **Thirty tanks like the ones shown under construction here are used to store highly radioactive waste from the Savannah River Plant in South Carolina.**

Fig. 4-9 **One proposed system to capture solar energy in outer space would reflect sunlight onto collection sites on earth.**

Fig. 4-10 **What looks like a small muffler on the exhaust system of this car is actually a catalytic converter. Material inside the converter helps eliminate harmful exhaust emissions.**

Nuclear Energy

Nuclear power plants produce almost no air pollution. However, they do produce the same thermal pollution as fossil fuel plants.

The main pollution from nuclear power is **radioactive waste**. This waste consists mainly of new substances created during nuclear fission. Most waste materials give off both heat and **radioactivity** (sub-atomic particles and high-energy rays). They must be handled with extreme care. Some nuclear wastes can be dangerous for thousands of years after their formation.

Many nuclear wastes are very hot liquids. They give off large amounts of heat and radioactivity. At the present time, most wastes are stored in above-ground containers. See Fig. 4-8. They will be kept there until a safe disposal method is found.

There are many possibilities for handling nuclear waste storage. The best method may be one that handles waste in three stages. First, liquid wastes would be stored for a few years. The wastes would cool over this time. They could then be converted into solid wastes. Next, the solid wastes would be sealed in containers. Finally, the wastes would be stored in special areas underground.

Nuclear accidents in nuclear plants can cause very severe pollution. Some small accidents can be corrected while the plant is operating. Major accidents can pollute the surrounding air, water, and land. As existing plants become older, the possibility of accidents increases. However, strict safety precautions can prevent accidents.

Solar Energy

As it is used today, solar energy does not cause pollution. Therefore, people are trying out solar energy for many different uses. In the future, however, solar energy may present some new pollution problems.

In the future, we may collect solar energy in space. Satellites would "beam" the energy to earth with microwaves or mirrors. See Fig. 4-9. There are some possible pollution problems with this system.

One problem is earth warming. The satellite system would send energy to the earth that normally would have missed the earth. This could increase the temperature of the earth.

Using microwaves to "beam" down the energy could also be dangerous. Scientists know that concentrations of microwaves can be harmful to living things.

Transmitting solar energy with mirrors could also be a problem. Mirrors would provide sunlight 24 hours a day to certain places on earth. Scientists know little about the environmental effects of constant sunlight.

Other Sources

Chapter 1 presented other possible sources of energy. These sources include wind, ocean thermal energy, tides, and others. Scientists must study each source carefully to determine its environmental effects. No sources are completely free of problems. However, some sources are better than others. Obviously, we should work to develop the sources that are more pollution-free.

PRESERVING THE ENVIRONMENT

Energy use causes environmental problems all over the world. Sulfur oxides produced on one continent can produce acid rain on another. Carbon dioxide in the air warms the entire earth. The smog from major industrial cities pours into air that circulates around the world. No country can escape the effects of pollution.

We can control pollution from energy use in two ways. First, we can develop the more pollution-free energy sources. Second, we can reduce the amount of pollution at the source.

Developing Alternative Energy

Much research and development has been done on alternative energy. Scientists have discovered many new energy sources. Some of these sources have added to our present energy supplies. Some have also reduced pollution.

Scientists will probably discover even better ways to obtain energy. They will also find better ways to control pollution. Energy is all around us at all times. The problem is to collect and control it without disturbing the environment.

Reducing Pollution

Pollution can be controlled and reduced at the source. Car makers have made much progress in controlling automobile pollution. Control devices have reduced hydrocarbons, nitrogen oxides, and carbon monoxide. See Fig. 4-10.

Sulfur oxides from coal-burning power plants can also be reduced. Sulfur can be removed from the coal before burning. This is an expensive process. However, it prevents the formation of acid rain.

Companies can also control particulate pollution. They can use filters to trap particulates in smokestacks. These filters allow only exhaust gases to pass.

We have the ability to reduce most forms of pollution. *Cost* is the major reason that pollution is not controlled better. Adding filters to smokestacks and adding pollution-control devices to cars is expensive. Removing sulfur from coal increases the cost of electrical power production. These costs are passed on to consumers. We must decide whether a clean environment is worth the cost of pollution control.

CONSERVATION

We can also preserve our environment by reducing our energy use. This is called **energy conservation**. We have made good progress in conserving energy. However, much more progress is possible.

Energy is one of our most important resources. We should not waste it. By conserving our energy supplies, we can accomplish two things. First, we can make our present energy supplies last longer. Second, we can reduce pollution. Since using energy usually creates pollution, using *less* energy will create *less* pollution.

There are two general possibilities for energy consumption in the future. We can either continue our present rate of energy use, or we can try to conserve energy.

Our energy consumption will increase a great deal if we do not conserve. It could almost triple during the next 20 years. If this happens, there will be much more pollution — even with the best pollution controls. The air would be more harmful to health. More lakes would be polluted by acid rain. The amount of carbon dioxide in the air would greatly increase.

However, we *could* reduce our use of energy by conserving. An all-out conservation effort would result in a much smaller increase in energy use. We must plan conservation in three areas:

- Home energy use
- Business, industry, and government
- Transportation

Fig 4-11 This house has many special energy-saving features. In addition, the solar panels on the roof can generate up to half of the electrical power needed by the residents.

Home Energy Savings

About 20 percent of all the energy consumed is for heating, cooking, lighting, and other home uses. In general, American homes waste energy. On the other hand, Swedish homes conserve energy. Many of these homes near the Arctic Circle use only a little more energy than homes in San Francisco. The difference is in insulation, special windows, elimination of air leaks, and thermostats that reduce the home temperature at night.

Energy-saving homes have been built in many parts of the United States. Conservation features used in the home in Fig. 4-11, for example, include the following:

- **Insulation in the ceiling, walls, and under the floor.** Some energy-saving homes are constructed with extra room in the exterior walls for additional insulation.
- **Double- or triple-pane windows.** These windows cut down on heat loss through glass. Further savings can be obtained by using windows framed with wood instead of metal, and adding storm windows.
- **Weatherstripping of all doors.** This step prevents air movement past the doors. Many homes also have insulation gaskets placed on the electrical outlets located on outside walls. Considerable heat can be lost at these outlets.
- **An efficient heating system.** For example, a hot-water heating system is generally more efficient than a forced-air system.
- **An insulated hot-water heater and insulated hot-water lines.** The water in the heater can also be kept at a temperature of 120° F (*49° C*). Many hot-water heaters are kept at 140° F (*60° C*) or higher. A lower water temperature saves considerable energy.

Energy-saving homes may also use solar energy for heating water or living areas. (More information on solar heating is given in Chapter 22.) Most of the conservation features listed above can be installed by the individual home owner. Each of us can contribute to the nation's overall energy conservation. See Fig. 4-12.

We can also conserve energy in the home by using energy-saving appliances. For example, microwave ovens use less energy than ordinary gas or electric stoves. See Fig. 4-13. And some efficient frost-free refrigerators use less than half as much electricity as other models.

Fig. 4-12 Home owners can reduce heat loss a great deal by installing insulation.

Admiral

Fig. 4-13 A microwave oven can provide savings in energy of over 25% of the energy used in traditional ovens.

Fluorescent lights are another energy-saving feature. They provide more light, with less electric current, than regular light bulbs. The list of conservation methods for the home is almost endless. If we use these methods properly, we can reduce our home energy consumption by 50 percent or more. And what is more, we can save money without any loss of comfort!

Business, Industry, and Government

The greatest savings of energy must come from the largest users — business, industry, and government. Commercial and government uses account for over half of all the energy consumed.

Industry requires great amounts of heat. This heat is used to refine ore into metals, make glass and ceramics, and produce electrical power for millions of machines. See Fig. 4-14.

Conservation in business, industry, and government can be carried out in two main areas:

Fig. 4-14 Steelmaking is just one of many industrial uses of great heat.

American Foundrymen's Society

Chevrolet Motor Division

Fig. 4-15 Lightweight materials and an efficient engine make this car a good one for fuel economy.

- **Planned conservation** — for example, using waste heat to heat buildings and water, using more efficient furnaces, controlling temperatures carefully, and shutting down furnaces when they are not needed.
- **Worker conservation** — efforts by employees to conserve energy as they work. For instance, workers can shut off truck engines while they are waiting to pick up loads. They can also keep doors to the outside shut to save on heat in the winter and air conditioning in the summer. If employees develop a "conservation attitude," energy waste can be reduced or eliminated.

Transportation

Over one-quarter of all energy is used for transportation. This is an important area of conservation since oil provides most of the energy for transportation. Just think of the millions of cars in the United States! Most of them are powered by gasoline or diesel fuel.

There are two major ways to conserve energy in transportation. First, the weight of transportation vehicles can be reduced. For example, aluminum and plastic can be used instead of steel. Second, engines can be made more efficient. See Fig. 4-15. The result of these efforts has been a steady increase in fuel economy since the energy shortages of the 1970s.

THE FUTURE

The future of energy use depends on the decisions we make. Should we conserve energy? Should we look into using alternative energy sources? Are we willing to pay the cost of pollution control? Our answers to these questions will decide the quality of life for us and many others.

STUDY QUESTIONS

1. Which energy source(s) cause the most damage to our environment?
2. Define *pollution*.
3. Of the three fossil fuels, which one pollutes the least?
4. Name the four types of air pollution caused by the use of fossil fuels.
5. What are the main "ingredients" in the chemical reaction that forms smog?
6. Identify three ways in which smog is harmful to people.
7. How does carbon monoxide harm people?
8. What are particulates?
9. How do lead particles get into the air?
10. Briefly describe the greenhouse effect.
11. How does carbon dioxide cause earth warming?
12. How does acid rain form?
13. Explain how excess heat released by power plants can pollute rivers, lakes, and oceans.
14. What two kinds of pollution are formed by burning wood?
15. Which is more polluting — wood or coal?
16. What kind of pollution can result from the use of geothermal energy?
17. True or false: Hydroelectric energy is one energy source that has no harmful effects on the environment.
18. Why are wastes from nuclear power plants considered a type of pollution?
19. True or false: As it is used today, solar energy is pollution-free. We can expect future uses of solar energy to be pollution-free, also.
20. What are the two ways in which we can control the pollution caused by energy use?
21. What is the main reason that pollution is not controlled better?
22. Name three areas in which we must plan conservation.
23. Give four conservation steps that home owners can take to reduce energy loss.
24. Who are the largest users of energy?
25. Which energy source is used the most in transportation?

ACTIVITIES

1. Develop an energy conservation plan for your home. Your plan could include the following items:
   - A list of low-cost repairs that would increase energy efficiency (for example, caulking windows and weatherstripping doors)
   - A list of energy-conserving steps that cost nothing to perform (for example, lowering the hot water temperature and lowering the room temperature)
   - A list of major steps that would conserve energy (for example, adding storm windows and adding insulation)
2. Prepare a report on energy conservation materials. Your report could include information on different types of insulation, different window coverings (for example, double- and triple-pane windows and plastic films), and weatherstripping materials.
3. Do a study of the energy-efficient homes that have been developed by utility companies. Identify how each special design feature saves energy. You could also compare the cost of the energy-saving features with the actual savings that they bring about. This comparison will tell you if the energy-saving features are really practical.

Control and Transmission of Power

Now that you know something about energy sources, it's time to learn how we use energy sources to get power. Any kind of system that allows us to do this is called an **energy control system** — or simply a **power system**. We're going to look at three basic types of power systems that are widely used in our society:

- mechanical systems
- fluid systems
- electrical systems.

In each power system description, you'll learn first about the principles underlying system operation. **Principles** are the general ways in which things work. For example, you can balance two people on a see-saw if you move them certain distances apart. Or you can pass a certain amount of electrical current through a wire if your power source is strong enough. You'll see how general principles apply to working systems as you read through Section II.

The last chapter in this section is important because it shows you all three major types of power coming together. The subject of this chapter is automated control, which includes robots, computers, and other interesting developments.

Mechanical Power

5

In your reading of Chapter 2, you learned that mechanical energy is the energy of motion. The wind, flowing water, and a turning crankshaft all have mechanical energy. Many times, we can harness this energy. We can then put it to work and measure it. When we use mechanical energy to do work, it becomes **mechanical power**.

Usually, we have to modify (change) mechanical power in some way before we can actually put it to use. Many changes are possible. We can start the power, stop it, change its direction, or make it stronger or weaker. We can also

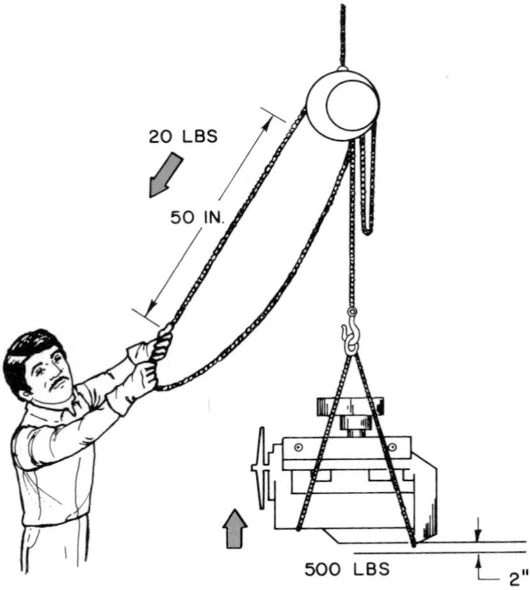

Fig. 5-1 **The chain hoist is a machine that gives the man a mechanical advantage. With it, he can do work that would otherwise be impossible.**

slow the power down or speed it up. The kinds and number of changes depend on the work we want to do.

There are many different devices used to modify mechanical power. These devices are called **machines**. Usually, people think of a machine as being a fairly large and complicated piece of equipment. However, a machine can be as simple as a loading ramp. Later in the chapter, you will learn about the operating principles, or laws of nature, behind six **simple machines**. All "complicated" machines use the same principles in one way or another.

Another important idea in the whole theory (set of ideas) behind mechanical power is the idea of **mechanical advantage**. This idea will help you understand just what is involved in using machines. If you know about mechanical advantage, you can also calculate exactly what you gain and what you lose when you use a machine.

MECHANICAL ADVANTAGE

Would you ever think about lifting a 500-pound engine from a car by yourself? It sounds impossible. However, you *could* lift the engine very easily with the help of a machine. The chain hoist in Fig. 5-1 is just such a machine. It increases, or **multiplies**, the amount of force that you apply to the task. The increase in force that you gain from using a machine is called a *mechanical advantage*.

Notice that the man in Fig. 5-1 can lift the 500-pound engine with only 20 pounds of force. The hoist multiplies his strength by 25 times (20 lbs. × 25 = 500 lbs.). We can say that the man has a mechanical advantage of 25. For every pound of force he applies, he can lift 25 pounds.

There are simple machines all around us. You have probably used many machines to gain a mechanical advantage. For example, door knobs are simple machines. They make it easier for you to turn the latch. Or maybe you have rolled a heavy object up a ramp instead of lifting it. A ramp is another machine that provides a mechanical advantage. In fact, screwdrivers, pliers, nut crackers, wheelbarrows, and many other tools are all machines. They are all devices that provide a mechanical advantage.

How Machines Provide Mechanical Advantage

Chapter 3 presented information on measuring energy and power. This information is directly related to machines and mechanical advantage. If you recall, work is useful motion. We can calculate the amount of work performed by multiplying *force times distance*. This gives us an answer in *foot-pounds*. Power, on the other hand, is work per unit of time.

We can only do a certain amount of work with a certain amount of power. Machines don't actually increase the amount of work performed.

Fig. 5-2 **Rolling a barrel up two different ramps points out how force and distance relate to the amount of work accomplished. In both cases, the work done is the same, but the force and distance are different.**

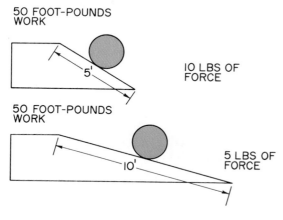

50 FOOT-POUNDS WORK

5'

10 LBS OF FORCE

50 FOOT-POUNDS WORK

10'

5 LBS OF FORCE

They just make it easier to do the *same* amount of work. (Figure 3-9 in Chapter 3 is a good example of this.)

Machine output (the amount of work performed) is always the same as **machine input** (the amount of power applied). In equation form:

Input Work = **Output Work**

or

Force × Distance = Force × Distance

Basically, whenever we use a machine, we change the amounts of force and distance in the input-output equation. For example, look at the two ramps in Fig. 5-2. With one of the ramps, a person must roll the barrel 5 feet to get to the top. The other ramp involves a distance of 10 feet. Rolling the barrel also involves a different amount of force for each ramp. However, in both cases the amount of work performed is the same: 50 foot-pounds.

Input Work   **Output Work**
10 lbs. × 5 ft. = 50 ft.-lbs.
5 lbs. × 10 ft. = 50 ft.-lbs.

You can see that for a set amount of output work, a particular machine can change either the amount of force required or the amount of distance required. If you increase the amount of force in the input work, the distance required decreases. If you increase the *distance* in the input work, the amount of *force* required decreases.

Usually when we talk about mechanical advantage, we are referring to the gain in force and loss of distance that a machine provides. However, as you can see from Fig. 5-2, a machine can also provide a mechanical advantage in distance instead of force.

Calculating Mechanical Advantage

Every machine provides an exact amount of mechanical advantage. Engineers and technicians can calculate these amounts with several different formulas. All of the formulas produce ratios that compare either forces or distances. (A **ratio** is a numerical comparison of two or more amounts.)

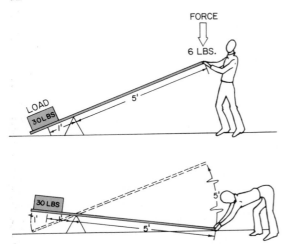

Fig. 5-3 **If 6 pounds of force moves a 30-pound load, the mechanical advantage is 30:6, or 5:1, or 5. However, the 6 pounds of force must move five times as far as the 30-pound load.**

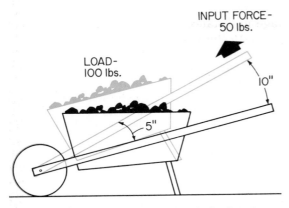

Fig. 5-4 **We can find the mechanical advantage by comparing the input distance to the output distance. Since the input force moves twice as far as the load, the mechanical advantage is 2.**

One way to calculate mechanical advantage is to compare output force to input force. For example, in Fig. 5-3 the man is lifting a load (output force) of 30 pounds with an input force of 6 pounds. The see-saw provides a mechanical advantage of 30:6 ("30 to 6"). This ratio is equal to 5:1, or a mechanical advantage of 5.

We should also look at the *distance* involved in the example in Fig. 5-3. The mechanical advantage multiplies the input force by 5. However, the load (output force) moves only 1/5 as far as the input force.

Another way to calculate mechanical advantage is to compare input distance to output distance. For example, in Fig. 5-4 the input distance is 10 inches. The output distance is 5 inches. Therefore, the ratio is 10:5. This is the same as 2:1, a mechanical advantage of 2.

With any particular machine, the distance ratio and the force ratio will equal each other. For example, let's compare the two ratios involved in Fig. 5-4:

$$\frac{\text{Input Distance}}{\text{Output Distance}} = \frac{\text{Output Force}}{\text{Input Force}}$$

$$\frac{10 \text{ in.}}{5 \text{ in.}} = \frac{100 \text{ lbs.}}{50 \text{ lbs.}}$$

$$\frac{2}{1} = \frac{2}{1}$$

As you can see, the only things needed to calculate a mechanical advantage are the distances involved or the forces involved.

SIMPLE MACHINES

There are basically six simple machines used to control and modify mechanical power. All complex machines use the principles of one or more simple machines. The simple machines are the lever, the wheel and axle, the pulley, the inclined plane, the wedge, and the screw.

Actually, the six simple machines can be grouped according to just two basic principles. These are the principle of the lever and the principle of the inclined plane. As you read on, you will see how these two principles apply to the different simple machines.

The Lever

Generally speaking, a lever is a bar that rests on a **pivot point** (a balancing or turning point). The pivot point is also called a **fulcrum**. At some point along the bar, a force is applied to move a load.

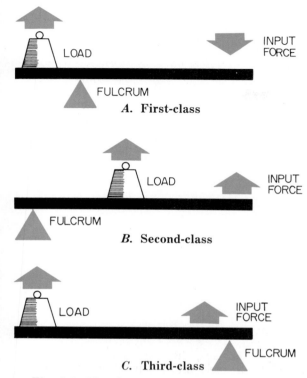

Fig. 5-5 **The three classes of levers.**

levers also change the direction of the force. That is, a downward input force exerts an upward push on the load (just like in a see-saw).

A crowbar is one example of a first-class lever. Another example is a pair of scissors. Scissors are a combination of two first-class levers that use the same fulcrum.

In **second-class levers**, the load is between the input force and the fulcrum. See Fig. 5-5B. Second-class levers provide a mechanical advantage in force. However, they do not change the direction of the force. As with first-class levers, the input force is farther away from the fulcrum than the load is. This arrangement provides the mechanical advantage in force. A wheelbarrow is one example of a second-class lever (see Fig. 5-4 again).

In **third-class levers**, the input force is applied between the load and the fulcrum. See Fig. 5-5C. Third-class levers provide a mechanical advantage in distance instead of force. Like second-class levers, they do not change the direction of the force. The mechanical advantage in distance is caused by the force being closer to the fulcrum than the load is.

In many cases, a third-class lever is used where speed is more important than force. (Speed can be defined as *distance per unit of time.*) Brooms, hammers, and baseball bats are examples of third-class levers.

In operation, a lever has two lever distances. The **force lever arm** is the distance from the fulcrum to the input force. The **load lever arm** is the distance from the fulcrum to the load. We can calculate the mechanical advantage by comparing these distances. In Fig. 5-6, the ratio of the distances is 5:1. This gives a mechanical advantage of 5. Two pounds of force will lift five times that amount (10 pounds). Note that the

There are three types of levers: first-class, second-class, and third-class. See Fig. 5-5. Notice how the placement of the fulcrum, the input force, and the load are different for each lever. Each type of lever provides a mechanical advantage in a different way.

In **first-class levers**, the fulcrum is between the input force and the load. See Fig. 5-5A. Look at the distance from the input force to the fulcrum. It is longer than the distance from the load to the fulcrum. This arrangement provides a mechanical advantage in force. First-class

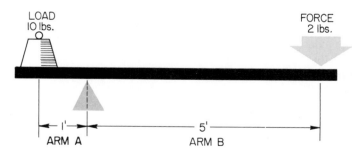

Fig. 5-6 **In this example, the force lever arm is five times as long as the load lever arm. The mechanical advantage is 5:1, or 5.**

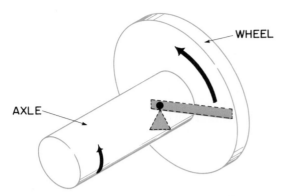

Fig. 5-7 **The wheel and axle works on the same principle as the lever. The wheel acts as a lever, and the center of the axle acts as the fulcrum.**

product of force times distance is the same for both sides of the lever.

The Wheel and Axle

The wheel and axle is another simple machine. It is an axle (a rod or shaft) that is attached to a wheel. This machine works on the same principle as the lever. The radius of the wheel acts as the lever. The center of the axle is the fulcrum. See Fig. 5-7. The wheel and axle can provide a mechanical advantage in either force or distance. Whether the gain is in force or in distance depends on the placement of the input force and the load.

When the input force is applied to the wheel, the wheel and axle acts as a second-class lever. It provides a gain in force with a loss in distance. The winch in Fig. 5-8A is an example of this arrangement. As you can see, the distance covered when the handle is turned in a circle is greater than the circumference of the axle. (The **circumference** is the distance around a circle, or around the outside of a circular object.) Therefore, when the handle is turned, the axle and the load attached to it move a shorter distance. The leverage of the handle multiplies the operator's force. But the operator must turn the handle more to make up for the loss in distance.

When the input force is applied to the axle, the wheel and axle acts as a third-class lever.

It provides a gain in distance with a loss in force. The Ferris wheel in Fig. 5-8B is an example of this arrangement. Here the input force is applied to the axle. The wheel makes one complete rotation, or **revolution**, for each rotation of the axle. As in the winch, the circumference of the wheel is greater than the circumference of the axle. Therefore, the outer edge of the wheel moves much farther than the outer edge of the axle. Here the leverage multiplies the distance,

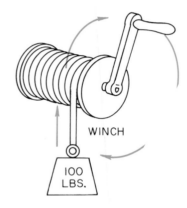

A. **Gain in force**

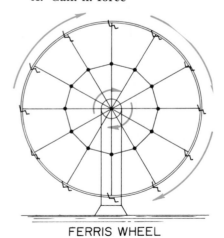

FERRIS WHEEL

B. **Gain in distance, or speed**

Fig. 5-8 **Depending on where the force and load are applied, a wheel and axle can provide a mechanical advantage in either force or distance.**

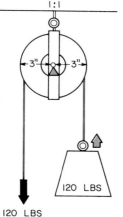

Fig. 5-9 A simple fixed pulley does not provide a mechanical advantage. It only changes the direction of the applied force.

tion of the applied force. We can also use them to multiply either force or distance. (This second use is described later, under *CONTROL DEVICES.*)

Like the wheel and axle, pulleys work on the principle of the lever. The lever in the pulley is either its diameter or its radius. The fulcrum is either the axis or the edge of the pulley. Figure 5-9 shows a simple **fixed pulley**. Here the fulcrum is at the center of the pulley. The lever distances are the same — 3 inches. This means that there is a lever arm ratio of 1:1, which does not provide a mechanical advantage. To raise the load, you must apply a force equal to the load. This type of pulley is used only to change the direction of the applied force.

The pulley in Fig. 5-10A is a **moveable pulley**. It does not change the direction of the applied force. However, it does provide a mechanical advantage in force. Notice that the fulcrum is at the edge of the pulley, opposite the input force. The force lever arm is equal to the diameter of the pulley. The load lever arm is equal to the radius of the pulley. Since the diameter is twice as long as the radius, the mechanical advantage is 2:1.

We can use many different pulley combinations to change the direction of the force and provide mechanical advantage. For example, a

or speed. But more power must be applied at the axle to make up for the loss in force.

We can calculate the mechanical advantage of a wheel and axle in one of three ways. We can compare either the wheel's circumference, diameter, or radius to the corresponding (same) part of the axle. The most common method is to compare the radius of the wheel to the radius of the axle.

The Pulley

A pulley is a wheel that turns on an axis (a center of rotation). Most pulleys are grooved to carry ropes or belts. We can use pulleys to do two things. With them, we can change the direc-

Fig. 5-10 Using additional pulleys can increase the mechanical advantage and provide directional changes. However, every increase in force is balanced by a loss in the distance moved.

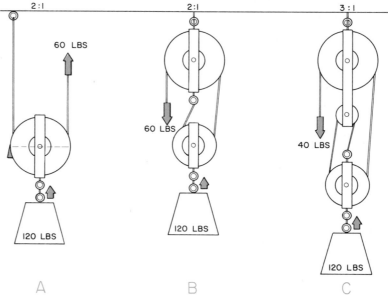

Technique Studios, Inc.

A. **Pike's Peak Road**

B. **Loading ramp**

Fig. 5-11 A "switchback" road and a loading ramp are both examples of a simple machine: the inclined plane.

fixed pulley has been added in Fig. 5-10B. This pulley changes the direction of the force. However, the mechanical advantage stays the same — 2:1.

In Fig. 5-10C a third pulley has been added. The effect is the same as having three separate ropes supporting the 120-pound load. Each rope supports 40 pounds. Therefore, the mechanical advantage is 3:1.

The Inclined Plane

The inclined plane is a machine that makes use of a sloping surface. We usually do not think of a winding mountain road or a loading ramp

as a machine. But they are both examples of an inclined plane. See Fig. 5-11.

Figure 5-12 shows how an inclined plane provides a mechanical advantage. It is difficult for the person to lift the 150-pound barrel and put it on the platform. It is much easier to roll the barrel up the inclined plane. We can calculate the mechanical advantage by comparing the length of the inclined plane with the height of the platform. In this case, the ratio is 6:2. This is equal to a mechanical advantage of 3:1. A force of 50 pounds is enough to push the barrel up the inclined plane. However, notice that this force must be applied for 6 feet. This is three times the distance involved in the lifting method.

Fig. 5-12 Using an inclined plane gives the person a 3:1 mechanical advantage in moving the barrel.

FORCE = 150 lbs.

2 FT.

FORCE = 50 lbs.

150 lbs.

6 FT.

2 FT.

The Wedge

A wedge is made up of two inclined planes. They are placed so that the sloping sides come together at a point. This arrangement makes a simple tool. It can be used to cut or pierce solid surfaces. Nails, axes, and wood-splitting wedges are simple examples of this principle. See Fig. 5-13.

The splitting wedge shown in Fig. 5-14 multiplies the force of a hammer blow. It concentrates the force at the pointed end of the wedge. The mechanical advantage of any wedge is the ratio of its length to its greatest thickness. In this case, the mechanical advantage is 10:2, or 5:1. Of course, the hammer-swinger has to "pay" for the gain in force by providing more input distance. To split the wood apart one inch, the wedge must be driven into the block five inches.

The Screw

A screw is an inclined plane cut in a spiral around a cone or shaft. See Fig. 5-15. This arrangement provides a long and gradual slope around the shaft. The slope gives a very high mechanical advantage in force. For example, a 1/2-inch bolt may have 20 screw threads per inch. This provides a mechanical advantage of more than 30:1.

When you fasten something with a nut and bolt, you apply a great deal of force. However, you have to provide the distance (more turns of the screw) to pay for the gain in force. Figure 5-16 shows different uses of the screw.

Fig. 5-14 **A wedge is made of two inclined planes placed together. The wedge shown here provides a mechanical advantage of 5. This mechanical advantage makes it easy to split the wood.**

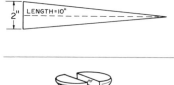

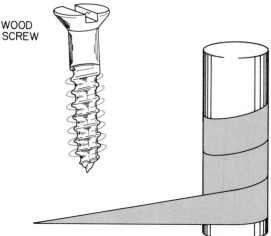

Fig. 5-13 **These wedges concentrate force at the sharp or pointed end to pierce a board or split a log.**

SPLITTING WEDGE

AX

NAIL

Fig. 5-15 **A screw is basically an inclined plane wrapped around a shaft.**

WOOD SCREW

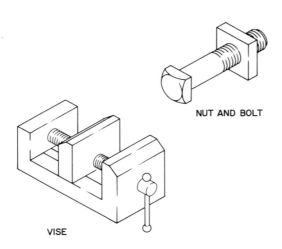

NUT AND BOLT

VISE

Fig. 5-16   **Different applications of the screw principle.**

MECHANICAL POWER SYSTEMS

Sometimes we put mechanical power to work directly, without changing it in any way. For example, the power from a steam turbine can be used directly to run a generator. Another example is a lawn mower. The cutting blade is connected to the power input, the engine's spinning crankshaft.

Usually, though, we have to change mechanical power in some way before putting it to use. We use **control devices**, such as gears, pulleys, and clutches, to make these changes. These devices control and transmit the mechanical power. When we connect them to a power source and use the power to do work, we have a **mechanical power system**. See Fig. 5-17.

If you recall, Chapter 2 first introduced the idea of an energy control system. An energy control system has three basic parts: the energy source, the conversion and transmission of energy, and the eventual use of the energy.

A mechanical power system basically has the same parts, except that we call the parts **input**, **control**, and **output**. The input power comes from some type of power source. Special control devices can change the input power in three basic ways:

- They can switch the power on and off.
- They can change the power's direction.
- They can change the power's force and speed.

The control devices can make one or all of these changes. They can also make one kind of change at several different times. The kinds and number of changes depend on the eventual use, or power output, of the system. The output may be the movement of a car, the spinning of a lawn mower blade, or the work of a complex piece of industrial machinery.

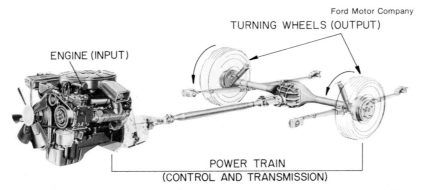

Ford Motor Company
TURNING WHEELS (OUTPUT)

ENGINE (INPUT)

POWER TRAIN
(CONTROL AND TRANSMISSION)

Fig. 5-17   **The automobile engine converts fuel into mechanical power — the motion of the crankshaft. Various control devices then transmit and control this motion so that the car can start and stop, move forward, reverse direction, or accelerate. The movement of the car is the output or use of the entire power system.**

On-Off Switching

Sometimes we need to stop the power output in a system for just a short time. Then we turn it back on. We may need to do this quickly and frequently. We may not want to or be able to completely stop the input power. In these cases we need a **switching control device**. This device connects and disconnects the power output without stopping the input power.

A manual-shift car is an example of a mechanical power system that requires a switching control device. The engine provides the input power for driving the car. At stop signs we need to stop the car. This means we need to switch off the power output. It would be a lot of trouble to turn off the engine at every stop sign. This would also be hard on the car. Instead, we use a control device: a clutch. This device disconnects the input power from the power output. (Clutches are described later in this chapter.)

On-off switching is needed in many types of mechanical power systems. Most transportation devices have controls for on-off switching. Many industrial machines need to be turned on and off again and again. This must be done without stopping the input power.

Changes in Direction

In all power systems, there are basically three types of motion possible: reciprocating, rotary, and linear. **Reciprocating motion** is up-and-down motion. For example, car engines have pistons that move up and down in cylinders. See Fig. 5-18. **Rotary motion** is turning motion, or motion in a circle. The turning of a turbine is an example of rotary motion. **Linear motion** is motion in a straight line. Jets and rockets use linear motion.

In a mechanical power system, the input power is usually in the form of rotary motion. A shaft turning in a certain direction is one example. To do the required work, we may have to change the direction of the motion. There are three common types of directional changes:

- **Reversing** — We often need to make the input motion turn in the opposite direction. For example, cars usually go forward. But sometimes they must go backward. The automobile transmission provides the proper gear arrangement for reverse motion.

- **Turning** — Sometimes we need to *redirect* the power. For example, the rear-end assembly of a car redirects the power 90 degrees from the drive shaft to the rear wheels. See Fig. 5-19.

- **Changing one type of motion to another** — We must often change linear motion into rotary motion, or change rotary motion into linear motion. Other conversions are common, too. For example, the reciprocating motion

Fig. 5-18 **There are three basic types of motion used in all power systems.**

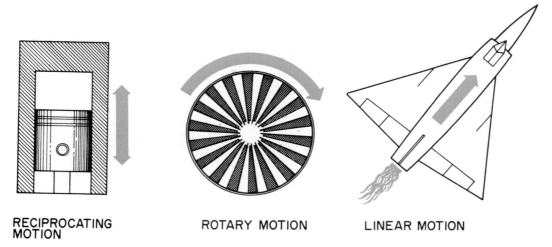

RECIPROCATING MOTION   ROTARY MOTION   LINEAR MOTION

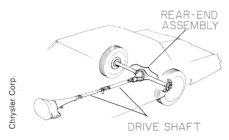

Chrysler Corp.

Fig. 5-19 **A rear-end assembly is a control device that changes power direction. It redirects motion from the engine 90° to the wheels.**

of an automobile piston is changed into rotary motion by the connecting rod and the crankshaft. See Fig. 5-20. When the piston moves down, the crankshaft turns a half-circle. When the piston moves up, the crankshaft keeps moving to complete a rotation.

Changes in Force and Speed

Up to this point, we have mainly talked about mechanical power in terms of force and distance. However, remember that we can also refer to distance as *speed* (distance per unit of

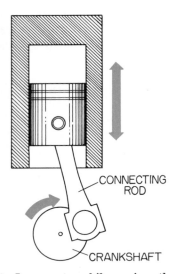

Fig. 5-20 **In an automobile engine, the connecting rod and crankshaft change the piston's reciprocating motion into rotary motion.**

time). Force and speed relate to each other in the same way as force and distance. If your control devices decrease the amount of one, they increase the amount of the other.

Many mechanical power systems require changes in the balance between force and speed. For example, to get moving, a stopped car needs a great deal of torque (turning force) at the drive wheels. But it does not need a lot of speed. After the car starts moving, it needs more speed and less torque. A control device in the car — the **transmission** — produces the proper balance of force and speed at the right time.

CONTROL DEVICES

Mechanical power systems use several major types of control devices. In this section, you will learn about five basic types of control devices. These are gears, pulleys and belts, sprockets and chains, clutches, and couplings.

Gears

Gears are basically wheels that have teeth cut around their outside surface, or circumference. There is also a hole in the center so that a shaft can be attached to the gear.

Gears work on the principle of the lever. See Fig. 5-21. The teeth of one gear mesh (fit in between) the teeth of another gear. When a power source turns one gear, the other gear

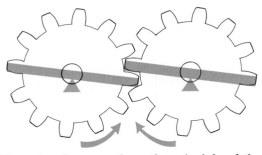

Fig. 5-21 **Gears work on the principle of the lever. A gear turning in one direction causes the meshing gear to turn in the opposite direction. The gears shown here work somewhat like the fixed pulley in Fig. 5-9. They change the input direction but do not provide a mechanical advantage.**

also turns. The gear connected to the power source is called the **drive gear**. The other gear is called the **driven gear**.

We use gears to make all three basic power changes: on-off-switching, directional changes, and changes in force and speed. The kind of change depends on the size, type, and position of the gears.

There are many types of gears. Read on to learn about four of the most common: spur gears, bevel gears, miter gears, and worm gears.

Spur Gears. The spur gear is the simplest and most common type of gear. It has its teeth cut parallel to the shaft opening. See Fig. 5-22. Spur gears can provide directional changes, as in Fig. 5-21. They can also change force and speed.

Figure 5-23 shows how two spur gears can multiply torque. The gray bars show the lever action. The center of each gear acts as a fulcrum. The drive gear (on the left) has a radius of 1 foot. Since the driving force is 10 pounds, the drive gear has a torque of 10 pound-feet (Force × Radius = Torque). This force transfers to the larger gear. The larger gear has a radius of 2 feet. Therefore, it has a torque of 20 pound-feet at its shaft.

With gears, we calculate changes in force and speed in terms of **gear ratio**. This is the ratio of the number of teeth on the driven gear to the number of teeth on the drive gear. For

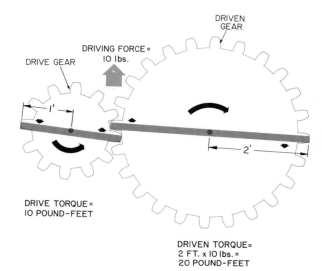

DRIVEN GEAR

DRIVE GEAR

DRIVING FORCE = 10 lbs.

1'

2'

DRIVE TORQUE = 10 POUND-FEET

DRIVEN TORQUE = 2 FT. x 10 lbs. = 20 POUND-FEET

Fig. 5-23 **Different size gears can be used to multiply torque. The grey bars show the lever action of the gears.**

example, the driven gear in Fig. 5-23 has 24 teeth. The drive gear has 12 teeth. Therefore, the ratio for this gear set is 24:12, or 2:1. This means that the driven gear has twice as much torque (turning force) as the drive gear.

A gain in force always means an equal loss in speed. In Fig. 5-23, the large gear makes only one revolution for every two revolutions of the small gear. With gears, we measure speed in **revolutions per minute (rpm)**. By comparing their rpm, we can see that the large gear turns at half the speed of the small gear.

But what happens if we make the large gear the drive gear? In this case, there would be an increase in speed. For example, suppose the large gear turned at 10 rpm. The small gear would turn at 20 rpm. Of course, the small gear would only have one-half the torque of the large gear.

Bevel Gears and Miter Gears. Both bevel gears and miter gears have their teeth cut at a 45-degree angle to the shaft opening. The shafts in a bevel or miter gear set always form a 90-degree angle. Both bevel gears and miter gears provide a directional change from one gear to the other. See Fig. 5-24.

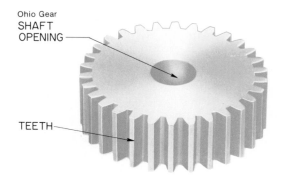

Ohio Gear

SHAFT OPENING

TEETH

Fig. 5-22 **In a spur gear, the teeth are parallel to the shaft.**

A. Bevel gear set   B. Miter gear set

Fig. 5-24 Bevel gears and miter gears provide directional change. Bevel gears also provide mechanical advantage.

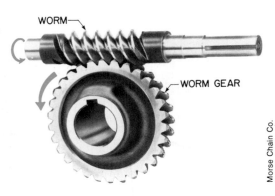

Fig. 5-25 In a worm gear set, the turning worm rotates the worm gear.

In a miter gear set, the gears have the same number of teeth. Therefore, the gear ratio is 1:1, and there is no mechanical advantage. In bevel gear sets, the gears have different numbers of teeth. Therefore, there is always a mechanical advantage in either force or speed.

Worm Gear Sets. A worm gear set consists of a **worm** and a **worm gear**. See Fig. 5-25. Worm gear sets make major changes in torque and speed. They also make a 90-degree change in the direction of the input power.

The input power is usually applied to the worm. The worms acts like a screw to rotate the worm gear. Each time the worm makes one full revolution, the worm gear rotates only one tooth.

Worm gear sets can provide very high mechanical advantages. For example, the worm gear in Fig. 5-25 has 30 teeth. The worm must rotate 30 times to rotate the worm gear once. This is a mechanical advantage of 30:1. The output torque is 30 times greater than the input torque. However, the output speed is 30 times slower than the input speed. This is an example of the main use of worm gear sets: to provide increased torque with reduced speed.

Gear Uses

There are many home, industrial, and transportational uses for gears. The automobile transmission in Fig. 5-26 is one example. The gears in this transmission provide torque, speed, and directional changes.

Gears can also provide on-off switching. We can move them out of mesh. This cuts off the input power from the output gear. For example, automobile transmissions have a "neutral" position. Shifting into this position unmeshes two gears. When this happens, power cannot transfer from the engine to the drive wheels.

Pulleys and Belts

A pulley is basically a metal wheel that is grooved around its circumference. It either drives or is driven by a flexible belt. See Fig. 5-27. In this way, power transfers over a distance from one pulley to the other.

Fig. 5-26 This complex automotive transmission is a mechanical control device. It makes changes in the force and speed of a car's power output.

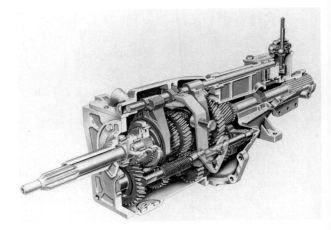

Fig. 5-27 Pulleys and belt.

Fig. 5-28 On this automobile engine, the drive pulleys are attached to the crankshaft. Through V-belts, they drive the alternator, water pump, and other equipment.

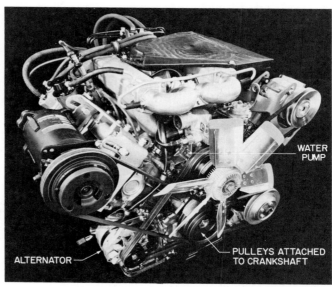

WATER PUMP

ALTERNATOR

PULLEYS ATTACHED TO CRANKSHAFT

Volvo

Fig. 5-29 Sprockets look something like gears. However, they transmit power through a chain, instead of meshing with each other.

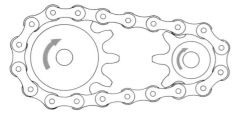

The most common types of pulleys are called **V-pulleys**. These pulleys have a V-shaped groove that accepts a V-shaped belt. Figure 5-28 shows the V-pulleys and belts on the front of an automobile engine. The **V-belts** on the crankshaft pulleys drive the water pump, alternator, and other equipment.

Pulleys and belts can provide all three basic power modifications: on-off switching, directional changes, and force and speed changes. However, they are mainly used to transmit power and to change force and speed. For example, connecting a power source to the large pulley in Fig. 5-27 results in a speed increase at the small pulley. Applying the power to the small pulley would result in a torque increase at the large pulley.

Fig. 5-30 Some automotive engines use sprockets and a chain to transfer crankshaft movement to the camshaft. The positive drive makes sure that the valve moves up and down at the right time.

Sprockets and Chains

Sprockets are much like gears. The difference is that they drive or are driven by chains instead of by other sprockets. See Fig. 5-29. Sprocket-and-chain assemblies provide force and speed changes in the same way as gears. However, they do not change the direction of rotation.

In a sprocket-and-chain assembly, the sprocket teeth mesh with holes in the chain. Because of this, there is no slippage between the drive sprocket and the driven sprocket. We call this type of arrangement a **positive drive**. The two sprockets maintain a set timing relationship to each other. See Fig. 5-30. This feature makes sprockets and chains very useful in situations where timing is important.

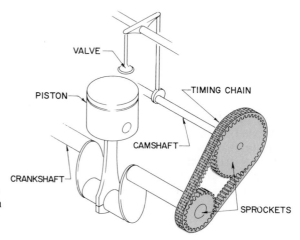

VALVE

PISTON

TIMING CHAIN

CAMSHAFT

CRANKSHAFT

SPROCKETS

Clutches

Clutches are control devices used only for on-off switching. There are many different kinds of clutches. In this section, we will look at two of the most common: the friction clutch and the centrifugal clutch.

Friction Clutch. The friction clutch is the type of clutch used on cars with manual transmissions. The drive unit of the clutch consists of the **flywheel** (a heavy disc attached to the crankshaft) and a **pressure plate**. The driven unit consists of a **friction disc** that is attached to the transmission's input shaft. See Fig. 5-31, parts A and B.

Normally, the two parts of the clutch are engaged, or connected to each other by friction. The pressure plate presses the friction disc against the flywheel. See Fig. 5-31C. The friction disc grips the flywheel and turns with it. In this way, engine power transfers from the flywheel to the friction disc. From the friction disc, the power transfers to the transmission.

To disconnect the power flow, the driver pushes the clutch pedal down. This releases the pressure holding the friction disc to the flywheel. The disc disengages from the flywheel. See Fig. 5-31D. The flywheel then turns freely without moving the disc. As a result, no engine power transfers to the transmission.

Notice that we can apply or release the friction clutch without shutting off the input power. This is one of the clutch's major advantages. We can engage or disengage the input power from the power output at any time. This provides a lot of control over the power flow.

Centrifugal Clutch. Centrifugal clutches work on the principle of centrifugal force. This is the force that causes a spinning or rotating object to move away from the center of rotation. In Fig. 5-32, centrifugal force acts on the stone as the person spins it. To keep the stone going in a circular path, the person must apply force. The person must increase this force to increase the speed of rotation.

Like friction clutches, centrifugal clutches have a drive part and a driven part. In the clutch shown in Fig. 5-33, the drive part consists of two weights attached to a crankshaft.

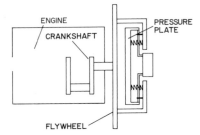

A. Drive unit

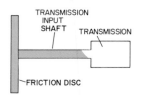

B. Driven unit

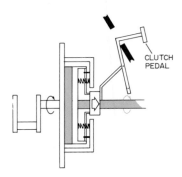

C. Clutch engaged

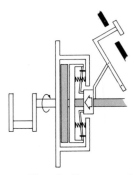

D. Clutch disengaged

Fig. 5-31 **The parts and operation of a friction clutch.**

The driven part consists of a housing attached to an output shaft.

As the crankshaft rotates, the weights move outward. When the rotational speed is high enough, the weights are forced against the housing. The weights "lock on" to the housing. The housing then rotates at the same speed as the weights. The centrifugal force keeps the weights pressed against the housing. Power transfers from the crankshaft, through the weights, to the housing and output shaft.

Some small gas engines use a centrifugal clutch to transfer power from the crankshaft to the power output. Centrifugal clutches are also used in industrial equipment.

Couplings

Couplings are permanent connections used to transmit power. We use them to produce directional changes or to connect lengths of shaft.

There are many types of industrial couplings in common use. **Rigid couplings** are used to connect separate shafts. For example, the coupling in Fig. 5-34 consists of two toothed sprockets and a double roller chain. One sprocket fastens to the end of each shaft being connected. The chain is then wrapped around both sprockets. This locks them together. Power transfers from one shaft to the other through the chain.

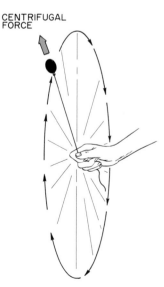

Fig. 5-32 Spinning a stone on a string illustrates centrifugal force. As the stone rotates, centrifugal force tries to move it away from the center of rotation.

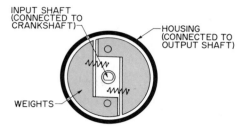

A. When the input shaft rotates slowly or not at all, the springs hold the weights in.

B. As the input shaft spins faster, centrifugal force throws the weights out. They contact the housing and cause it to rotate.

Fig. 5-33 The operation of a centrifugal clutch.

Morse Chain Co.

A. Roller chain coupling

B. Disassembled coupling with protective housing

Fig. 5-34 A roller chain coupling is a rigid coupling that uses sprockets and a chain to transfer power.

Universal joints are couplings that provide for alignment (positioning) changes. A universal joint permits the shafts connected to it to be out of line with each other and still move. This arrangement is not possible with the rigid couplings.

Figure 5-35 shows one type of universal joint. It consists of two sets of bearings on a crosspiece. (Bearings are described in detail later in this chapter.) Each shaft connects to a set of bearings. The power transfers from one shaft to the other through the crosspiece. The bearings permit the shafts to move up and down during operation. This type of "U-joint" is used commonly on automobile drive shafts.

A. Assembled "U-joint"

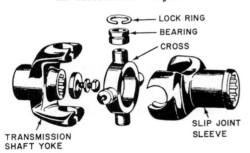

LOCK RING
BEARING
CROSS

TRANSMISSION
SHAFT YOKE

SLIP JOINT
SLEEVE

B. Disassembled joint

Fig. 5-35 **A universal joint allows the connected shafts to be out of line with each other and still move.**

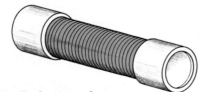

Fig. 5-36 **Springs can be used as flexible couplings.**

Figure 5-36 shows a simpler flexible coupling — a simple spring connection. The coupling is strong and does not need lubrication. It can withstand some shock and flexing (bending).

POWER TRANSMISSION AND FRICTION

Moving power from place to place is called **power transmission**. In a mechanical power system, the main power transmission devices are the control devices already described. These devices move the power to where it is needed. They also modify the power to fit the needs of the system.

Look back at Fig. 5-17 for a moment. Notice that the part of the system that controls and transmits the power is called the **power train**. The automobile transmission contains control devices that modify the engine's power. The drive shaft then transmits the power to the rear-end assembly. The rear-end assembly redirects the power to drive the rear wheels.

The automobile power train seems like a fairly simple way to transmit power. However, there is also a lot of power *lost* during this power transmission. In fact, there are power losses in every type of power system. The more complex the system, the greater the loss. In mechanical systems, the power loss is mostly due to friction.

Friction

Friction is an energy loss in the form of heat. In power systems, friction develops in control devices. It also occurs along the line of transmission. This energy loss can seriously reduce the power at the system's output.

There is friction whenever a moving surface contacts a stationary (fixed) surface. The stationary surface slows down the moving surface (or tries to slow it down). The amount of friction developed depends on the make-up of the two surfaces. For example, surfaces made of ice and rubber produce different amounts of friction when they are moved across the same stationary surface.

Bearings

We use bearings to reduce the amount of friction developed in a mechanical power system. Generally speaking, bearings are cylindrical metal pieces that fit around moving parts, such as a rotating shaft. Bearings can reduce friction in two ways. **Sleeve bearings** provide a special stationary surface that develops less friction than other kinds of surfaces. **Anti-friction bearings** have an **inner race** that actually rotates with the moving part. See Fig. 5-37.

Lubrication

We have another "tool" that we use to reduce friction in mechanical power systems. This is the use of some type of oil, or **lubricant**. The kinds of bearings described above both need lubrication to work at their best and provide a low-friction surface.

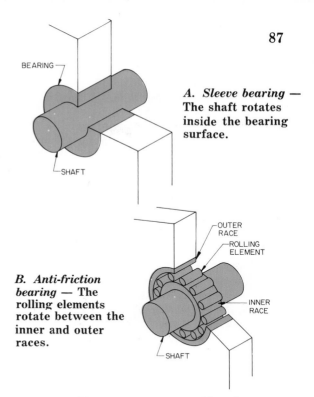

A. Sleeve bearing — **The shaft rotates inside the bearing surface.**

B. Anti-friction bearing — **The rolling elements rotate between the inner and outer races.**

Fig. 5-37   **Two common types of bearings.**

Timken's Tapered Bearings

Henry Timken operated a carriage-building company in the late 1800s. Timken spent much of his time looking for ways to improve the quality of his product. In doing so, he suffered several business setbacks. But he never let the setbacks get him down. He just kept rolling along.

Timken was very interested in the problem of friction in carriage wheels and axles. He looked at the bearings that were widely used, and he thought he could make a better bearing. "First of all," he thought, "I'll try tapering the inner part of the bearing. And I think I'll use small cylinders in the bearing instead of balls."

Timken thought that his new design would reduce friction. It would also spread the wear out over a larger area. This would make the bearing last longer. Henry drew up the plans and gave them to his sons. The men set to work making the first tapered roller bearing by hand. After three years, they perfected the design.

The Timkens obtained a patent for their bearing in 1898, and the Timken Bearing Company was born! By 1903, most automobiles had Timken bearings. Today, the tapered roller bearing is very important. It is used in many products in which shafts and axles turn. Yes, the industrial world rolls on a lot more smoothly because of Henry Timken.

STUDY QUESTIONS

1. What do we call devices that modify mechanical power?
2. What is mechanical advantage?
3. State two ways of calculating the mechanical advantage of a simple machine.
4. Name the six simple machines.
5. Which three machines operate on the same principle as the lever?
6. Which three machines operate on the principle of the inclined plane?
7. Where is the fulcrum on a first-class lever?
8. Where is the force applied on a third-class lever?
9. How are force and distance affected by a wheel and axle when the input force is applied to the wheel?
10. What kind of mechanical advantage does a single moveable pulley provide?
11. List the three basic ways that power can be changed in a mechanical power system.
12. Name the three types of motion possible in a power system.
13. What happens to speed when we increase force by using control devices?
14. Name five basic mechanical control devices.
15. How are the teeth cut on a spur gear?
16. What kind of power change can both bevel gears and miter gears provide?
17. What is the main use of a worm gear set?
18. Briefly describe how an automotive friction clutch operates when it is engaged.
19. Briefly describe the operation of a centrifugal clutch.
20. What is friction?
21. What is the purpose of a bearing?

ACTIVITIES

1. Prepare a written report that traces the transmission of power in a car from the pistons to the rear wheels. Describe the purpose of each part. Emphasize what happens during the following:

   - power stopping and starting
   - directional changes
   - changes in mechanical advantage

2. Find devices in your home that use the principle of a simple machine. List each device and explain how it provides a mechanical advantage. There are many of these devices. They include power tools and appliances. Examples to get you started are screwdrivers, pliers, wrenches, and chisels. Appliances include can openers, knives, clothes dryers, and washing machines. Your listing should look like this:

   - Screwdriver: The handle is larger in diameter than the blade. This provides the mechanical advantage of a second-class lever, or a wheel and axle.
   - Knife: The knife blade provides the mechanical advantage of a wedge (inclined plane).
   - Nut and bolt: The threads provide the mechanical advantage of an inclined plane.

3. Solve these problems:
   A. A simple machine has an input force of 20 pounds, an input distance of 5 feet, and an output distance of 10 feet. What is the output force?
   B. When using a series of pulleys, you discover that you have to pull the rope down 6 feet to move a weight up 1 foot. What is the mechanical advantage of the pulley system?
   C. What is the force necessary to lift a 180-pound man when the mechanical advantage is 9:1?
   D. An automotive engine produces a torque of 120 pound-feet. This torque is transferred to the transmission, which is shifted into low gear. The gear provides a mechanical advantage in force of 3:1. What is the torque of the transmission output shaft?
   E. If the input shaft of Problem D is rotating at 1800 rpm, what is the speed of the transmission output shaft?

Fluid Power

6

Fluid power is the use of pressurized **fluids** (liquids and gases) to control and transmit power. The liquid or gas receives power from an outside source. The fluid then moves through a system of transmission lines and control devices. It arrives at a location where the power is put to work. In this way, fluids work like machines in mechanical power systems. There are many, many uses of fluid power in our industrial world. See Fig. 6-1.

We can use fluids in power systems because we know how they will react under certain conditions. Fluids react to changes in force, temperature, and volume. We can calculate the exact effect of any change.

Much of our knowledge about fluids has developed from fluid power experiments conducted during the past 50 years. Therefore, fluid power would seem to be one of our newest ways of transmitting and controlling power. However, our recent knowledge is built on fluid power experiments dating back more than two thousand years. Early experimenters developed an understanding of the basic principles of **hydraulics** (moving liquids) and **pneumatics** (moving gases). In this chapter, you will learn the most basic and important principles of fluid power.

FLUID PRESSURE

One of the most important measurements of fluid power is pressure. Chapter 3 introduced the idea of pressure as *force per unit of area*. In this chapter, you will learn about atmospheric pressure, vacuum, and other factors that affect the pressure on a fluid.

Fig. 6-1 **Stationary and mobile uses of fluid power.**

A. **Mobile use: The shovel is controlled by hydraulic cylinders.**

B. **Stationary use: Two large hydraulic cylinders provide the power to drive the press down.**

Atmospheric Pressure

Atmospheric pressure is the reference point, or standard, we use to measure fluid pressure.

The air surrounding the earth applies a force of 14.7 pounds per square inch (psi) on the surface of the earth. See Fig. 6-2. We often round off this figure to 15 psi.

Most pressure gages are set so that atmospheric pressure registers as 0 psi. The gage often shows pressures both above and below this point. See Fig. 6-3. Pressures above 0 are recorded in psi. Pressures below 0 are recorded in inches of mercury.

When we use gages in which 0 psi equals atmospheric pressure, we add the word *gage* to the reading. Therefore, atmospheric pressure is *0 pounds per square inch gage (psig)*. The *g* is often dropped from psig for convenience.

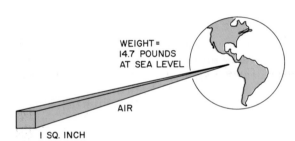

Fig. 6-2 **The air surrounding the earth produces an atmospheric pressure of 14.7 psi. This pressure presses on the entire surface of the earth.**

Fig. 6-3 **This gage registers atmospheric pressure as 0 psi. It measures pressures both above and below zero.**

Whenever you see a pressure gage reading in psi, it means psig.

We usually measure pressure when we are using it to produce motion. In most fluid power systems, there is no motion until the pressure of the fluid overcomes atmospheric pressure. Therefore, we are usually concerned with pressures above 15 psi.

Vacuums

We refer to a pressure less than atmospheric pressure as a *vacuum*. Our measurements of vacuums are based on a method that shows that atmospheric pressure is about 15 psi. Mercury (symbolized as *Hg*) is a liquid metal. It weighs about 1/2 pound per cubic inch. A 1-square-inch column of mercury 30 inches high weighs about 15 pounds. This column of mercury equals atmospheric pressure. Therefore, we can use the column to measure fluid pressures.

Figure 6-4 shows several tubes standing in a dish of mercury. Unlike the other tubes, tube A is open at the top. Atmospheric pressure can enter the tube. Therefore, the pressure inside the tube and the pressure outside the tube are equal.

Tube B is sealed at the top. One-third of the air in the tube has been pumped out. As a result, the pressure drops to 10 psi (1/3 of 15 equals 5). The atmospheric pressure outside the tube then pushes mercury up the tube. It continues to do this until the pressure inside the tube equals the pressure outside the tube. To produce this balance of pressures, enough mercury to produce 5 psi must enter the tube. We know that mercury weighs 1/2 pound per cubic inch. At 1/2 psi for each inch, 10 inches produces 5 psi. Therefore, the mercury column must be 10 inches high. Added to the 10 psi of air pressure, the mercury balances the atmospheric pressure outside the tube.

Tube C of Fig. 6-4 has 10 psi of air removed. The mercury now stands at 20 inches. This height is equal to 10 psi. Together, the pressure from the mercury and the air balance the atmospheric pressure.

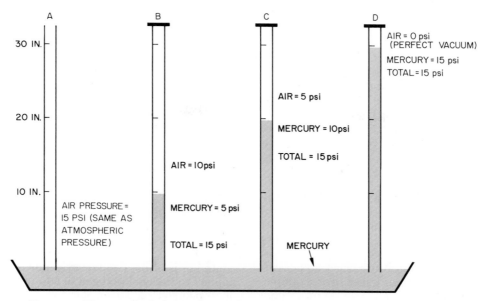

Fig. 6-4 We use the height of a column of mercury to measure a vacuum. A vacuum is any pressure less than atmospheric pressure.

Tube D has all of its air removed. This is an example of a **perfect vacuum**. It now takes a 30-inch column of mercury to balance the atmospheric pressure outside the tube. Thirty inches is the highest level to which atmospheric pressure can lift mercury.

You should now see how we can measure pressure by measuring the height of a column of mercury. Each inch of mercury equals one-half pound of pressure below atmospheric pressure. Using this method, we can measure any vacuum.

Pressure Gages

Figure 6-5 shows how atmospheric pressure registers on three different types of pressure gages. Gage A measures pressures above atmospheric pressure in psi. It measures pressures below atmospheric pressure in inches of mercury. The zero point is atmospheric pressure. This type of gage is called a **compound gage**.

Gage B is the most common type of pressure gage. This gage is similar to gage A. It registers atmospheric pressure as O. However, gage B

Fig. 6-5 These commonly used gages all measure atmospheric pressure. A regular gage can be converted into an absolute gage by adjusting the dial to read 15 psi at atmospheric pressure.

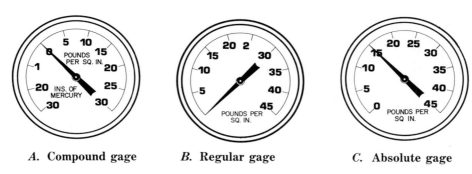

A. Compound gage   *B.* Regular gage   *C.* Absolute gage

91

does not show pressures below atmospheric pressure.

Gage C of Fig. 6-5 is an **absolute gage**. It registers atmospheric pressure as 15 psi. When we read this type of gage, we add the word *absolute* to the reading. Therefore, we read atmospheric pressure as *15 pounds per square inch absolute (psia)*. A perfect vacuum is 0 psia.

CONTROLLING GASES

Gases and liquids have many of the same properties. However, they are different in regard to volume. A fluid's **volume** is the amount of space it takes up. Liquids have a definite volume; gases do not. This means that changes in pressure and temperature affect liquids differently than they affect gases. For example, as the pressure on a gas increases, its volume decreases. However, the same pressure increase will not change the volume of a liquid. See Fig. 6-6. In this section, you will learn what happens to gases when pressure and temperature changes occur. In the examples given, volume is expressed in **cubic feet**.

Compression of Gases

When a gas is put into a container, it expands to fill the container. In fluid power systems, the

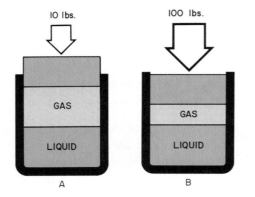

Fig. 6-6 **Both pressure and temperature affect the volume of gases, but not the volume of liquids. For example, when we apply pressure to a container of gas and liquid, only the gas is compressed.**

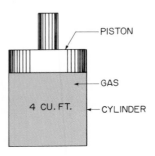

Fig. 6-7 **A gas expands to fill its container. We can add pressure to the gas by applying a force to it with a piston.**

container is often a hollow **cylinder**. A force is applied to the enclosed gas to add pressure to, or **pressurize**, it. Often this force is provided by a **piston**. See Fig. 6-7.

The force of the piston on the gas pushes the gas molecules together. We call this forcing together **compression**.

Gas molecules resist compression. As the piston pushes the molecules closer together, they push back. They try to move further apart. The more the molecules are pushed together, the more they try to move apart.

The effort of the gas molecules to resist compression and move apart produces pressure. Therefore, when we apply force to an enclosed gas, we are actually making it produce its own pressure.

Scientists have found that the volume of a gas is *inversely proportional* to its pressure. This means that any change in pressure produces an equal but opposite change in volume.

For example, if we double the pressure on an enclosed gas, its volume will be cut in half. See Fig. 6-8. (Notice that the pressures in this figure are absolute readings. In calculations, we must always use absolute pressure.)

The relationship between the pressure on a gas and the volume of the gas is known as **Boyle's Law**. This law is named for its discoverer, the English scientist Robert Boyles. Boyle's Law states, *The volume of a gas varies inversely with the pressure applied to it, provided the temperature remains constant.*

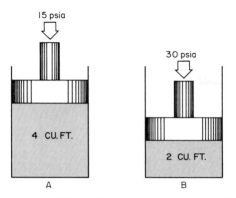

Fig. 6-8 **Doubling the pressure on the gas from 15 psia to 30 psia reduces the volume by one-half.**

This volume-pressure relationship also works in reverse. A change in volume creates an equal but opposite change in pressure. If you increase the volume, the pressure decreases.

Effects of Heat on Gases

You have just read about how pressure affects the volume of a gas. The volume of a gas is also affected by temperature. Whenever we talk about the volume of a gas, we must know the temperature at which the volume was measured.

Two French scientists, Jacques Charles and Joseph Gay-Lussac, are given credit for explaining the relationship between the volume of a gas and its temperature. These men showed that the volume is *directly proportional* to the temperature. That is, any change in temperature produces an equal change in volume. However, this equal change happens only if the pressure stays the same. For example, suppose we double the temperature of a gas. Its volume will double, as long as the pressure on it does not change.

The Absolute Temperature Scale

The relationship between temperature and volume seems fairly simple. However, we have to be careful to use the right temperature scale when calculating the effects of a temperature

change. Like pressure, temperature can be read according to different scales.

We are most familiar with the Fahrenheit and Celsius scales. But to calculate the effects of a temperature change, we have to use an **absolute temperature scale**. On an absolute scale, 0° is the temperature at which a material has no heat. At this **absolute zero**, there is no movement of atoms or molecules. If you recall from Chapter 2, heat is produced by atoms and molecules in motion. Therefore, at absolute zero there is no heat. (Do you see how the absolute temperature scale is like the absolute pressure scale, in which pressure is measured from 0 pounds per square inch?)

Figure 6-9 compares the regular Fahrenheit scale with the absolute Fahrenheit scale. Absolute Fahrenheit temperature is written $°F_A$ **(degrees Fahrenheit absolute)**. To convert a regular Fahrenheit reading into an absolute reading, we simply add 460 degrees.

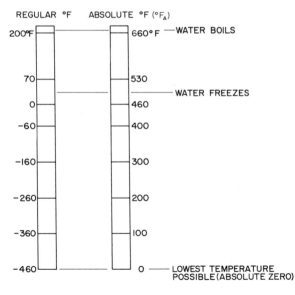

Fig. 6-9 **This chart shows the relationship between the regular Fahrenheit temperature scale and the absolute Fahrenheit temperature scale.**

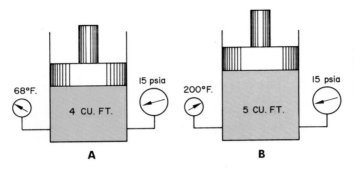

Fig. 6-10 As a gas is heated, its molecules move faster and faster. This increases the volume of the gas if the pressure on it stays the same.

In Fig. 6-10, container A holds 4 cubic feet of gas under a pressure of 15 psia. The temperature of the gas is 68° F (528° F_A). Container B shows what happens if we heat the gas to 200° F (660° F_A) without changing the pressure. The volume of the gas increases to 5 cubic feet. Now, if we compare the regular Fahrenheit temperatures between containers A and B, it seems like the temperature has almost tripled. However, on the absolute scale, the temperature has only increased by one-fourth (from 528° to 660°). Therefore, the volume of the gas changes by only one-fourth (from 4 cubic feet to 5 cubic feet).

Temperature can affect the pressure of a gas in the same way that it affects volume. If the volume stays the same, the pressure increases or decreases in direct proportion to a temperature change. In calculations, both the temperature and the pressure must be measured on an absolute scale.

In most cases, a temperature change affects both volume and pressure. Neither one stays the same. When this happens, the combined change in volume and pressure is directly proportional to the temperature change.

CONTROLLING LIQUIDS

As you learned earlier, liquids have a definite volume. They cannot be compressed. This property of liquids permits a direct and efficient transfer of force. Liquids can also be used to multiply force. Very high mechanical advantages are possible in hydraulic power systems.

Transferring Force through Liquids

Figure 6-11 shows how fluids transfer force. The force applied to piston A produces a pressure on the liquid. The fluid then exerts the same amount of pressure — 50 psi — in all directions. At the piston A end of the cylinder, this pressure is cancelled out by the input pressure. And the wall of the cylinder contains the pressure pushing out on *it*. Therefore, the fluid pressure has only one possible output — piston B. As a result, the force applied to piston A transfers to piston B.

Changes in the size of the fluid's container do not change the transmission of force. For example, in Fig. 6-12 fluid must flow through a small-diameter pipe to reach piston B. However, the diameter of the pipe does not affect the transfer of force. As long as the pipe is large enough to permit fluid flow, the input force and output force will be equal.

The input and output will stay the same even if we change the direction of flow. In Fig. 6-13, the fluid is flowing upward as it pushes on piston B. It still supplies the same input force to piston B (50 pounds).

Fig. 6-11 The physical properties of liquids make them efficient and extremely useful substances for transferring force from one location to another.

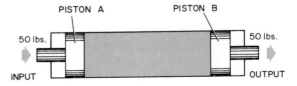

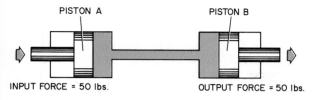

Fig. 6-12 **Compare the transfer of force in this illustration with that in Fig. 6-11. As long as a liquid can flow freely, changes in the size of its container will not change its ability to transfer force.**

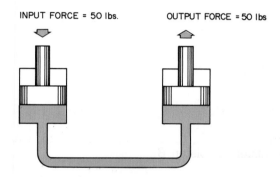

Fig. 6-13 **Changes in direction do not change the transfer of force through a liquid.**

Remember, pressure is *force per unit of area.* In equation form:

$$\text{Pressure} = \frac{\text{Force}}{\text{Area}}$$

Figure 6-14 shows how we can calculate the amount of pressure in a fluid power system. A force of 50 pounds is applied to the 5-square-inch piston. Each square inch of the piston receives 1/5 of the total force. Therefore, the mechanical force is equal to 10 pounds per square inch. (Notice that we have not changed the amount of force — only the way that it is measured.)

Multiplying Force with Liquids

We can use a fluid's ability to transmit force to produce a gain in mechanical advantage. (For an explanation of mechanical advantage, see Chapter 5.) Figure 6-15 shows how this is done through the use of two pistons of different size. A force of 50 pounds is applied to piston A. Piston A has an area of 5 square inches. Therefore, the pressure throughout the fluid is 10 psi.

Piston B has an area of 20 square inches. The force of 10 psi pushes up on this piston. We can calculate the output force with the following formula:

Force = Pressure × Area

The output force is 10 psi × 20 square inches, or 200 pounds. The ratio of output force to input force is 200:50. This is a mechanical advantage in force of 4 to 1.

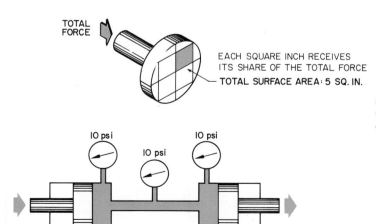

Fig. 6-14 **In this fluid power system, the pressure depends on two factors: the input force and the area of the piston.**

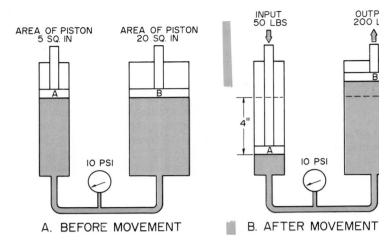

AREA OF PISTON 5 SQ. IN. AREA OF PISTON 20 SQ. IN. INPUT 50 LBS OUTPUT 200 LBS

A B B A

10 PSI 10 PSI

A. BEFORE MOVEMENT B. AFTER MOVEMENT

Fig. 6-15 **We can multiply force by using different size cylinders and pistons. A four-inch downward movement of piston *A* produces only a one-inch upward movement of piston *B*. However, due to the different piston sizes, there is a mechanical advantage in force of 4:1.**

However, you will recall from Chapter 5 that a gain in force always means an equal loss in distance or speed. Figure 6-15B shows how much distance is lost. The 50-pound input force must move 4 inches to move the 200-pound output force 1 inch.

FLUID POWER SYSTEMS

A fluid power system is *an assembly of units that controls the flow of pressurized fluid. It transmits power from an energy source (input) to a power use (output).* Let's see how this definition fits in with the idea of an energy control system.

If you recall, an energy control system has three basic parts: (1) the source of energy, (2)

Fig. 6-16 **This hydraulic hoist can lift up to 9000 pounds. The power source for the fluid system is a 2-horsepower electric motor.**

the conversion and transmission of energy, and (3) the use of the energy. All fluid power systems have the same three general stages, or steps:

- **Source (input)** — Fluid power is always a **secondary** form of power. This means that it is produced from another power source. First an engine or electric motor generates mechanical power. Then the mechanical power is converted into fluid power. See Fig. 6-17.
- **Transmission and control** — Fluid power is transmitted through the fluid system with **pipes** and **hoses**. The pressure and rate of flow are controlled with different kinds of **valves**.
- **Use (output)** — Before it can be used, fluid power must be changed back to mechanical power. In this form, the power can do work.

As you can see, fluid power has more to do with transmission and control than with the power source or the use. For this reason, a fluid power system is always a part of a larger power system. The larger system usually consists of mechanical systems, other fluid power systems, and electrical systems.

Even a very simple fluid power system requires a mechanical input. For example, the hydraulic jack in Fig. 6-18 has a simple mechanical input: a pumping lever. As the operator pumps the lever, the fluid system changes the mechanical power into fluid power.

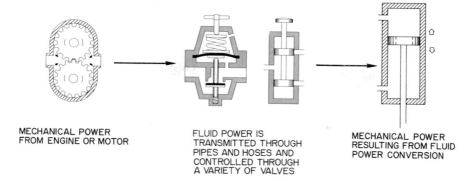

MECHANICAL POWER
FROM ENGINE OR MOTOR

FLUID POWER IS
TRANSMITTED THROUGH
PIPES AND HOSES AND
CONTROLLED THROUGH
A VARIETY OF VALVES

MECHANICAL POWER
RESULTING FROM FLUID
POWER CONVERSION

Fig. 6-17   **A fluid power system is not a source of power. It is a way of transmitting and controlling power.**

MECHANICAL
OUTPUT

FLUID
SYSTEM

(PUMPING LEVER)
MECHANICAL INPUT

Fig. 6-18   **Fluid power jacks can easily provide mechanical advantages of over 100:1.**

The output is in the form of mechanical power: the movement of the metal rod at the top of the jack.

Fluid power systems can provide very high mechanical advantages. For example, with the jack in Fig. 6-18 you could lift the end of a car with an input force of less than 10 pounds. Mechanical advantage and flexibility are two features of fluid systems that make them especially valuable for industrial uses. Fluid systems can produce huge gains in force in relatively small spaces. They can also carry power to wherever a pipe, hose, or piece of tubing can be placed.

Fluid systems provide much of the power for earth-moving equipment used in construction and mining. See Fig. 6-19. Fluid systems provide the power to run production lines and pneumatic tools. They also power everything from dentist's drills to large industrial presses. As you can see, fluid systems are an important way of transmitting and controlling power.

TYPES OF FLUID SYSTEMS

There are two basic types of fluid power systems: hydraulic and pneumatic. **Hydraulic systems** use a liquid, usually oil, to transmit power. **Pneumatic systems** use a gas, usually air, to transmit power.

Most fluid power systems have the following parts:

- **Fluid** — either a gas, a liquid, or both.
- **Reservoir or receiver** — a container that stores fluid.

Fig. 6-19   **Fluid power systems are used in this front-end loader to lift and carry dirt. The dump truck will also use fluid power to dump the full load.**

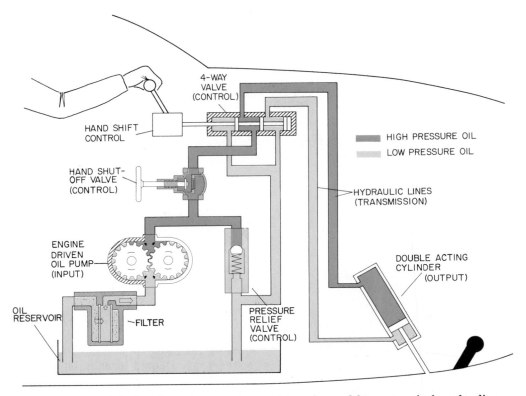

Fig. 6-20 A hydraulic system can be used to raise and lower an airplane landing gear. Here the system is in the *gear down* position.

- **Pump or compressor** — a device that converts mechanical power into fluid power. It supplies fluid under pressure to the system.
- **Filter** — a device that cleans the fluid in the system.
- **Transmission lines** — a system of pipes, hoses, and tubes. These parts contain the fluid so it can be transmitted under pressure.
- **Control valves** — the devices that regulate the fluid pressure, flow rate, and direction.
- **Actuator** — a cylinder, motor, or other converter that changes fluid pressure into the desired mechanical form.

Before learning about the parts of fluid power systems, we will compare the two basic types of systems. We will follow the complete path of the fluid through both systems. Then we will study the individual parts. You should then be able to see how the parts work in the whole system.

Hydraulic Systems

Compared to pneumatics, hydraulics is a newcomer to industry. The first hydraulic presses were developed in the early 1800s. By 1850, two manufacturing cities in England had central hydraulic systems. These systems supplied pressurized water to factories. Inside the factories, the pressurized water was converted into mechanical power. However, these systems were not self-contained. The pressurized fluid always had to be piped in from outside. This means that the uses of fluid power were all stationary.

Self-contained hydraulic systems for industrial use were developed about 1925. These systems are now an important part of almost all transportation vehicles. Many braking devices, automobile transmissions, power-assisting devices, and auxiliary (supplementary or additional) power units are fluid power systems.

Figure 6-20 shows a hydraulic system used to operate the landing gear of an airplane. In this system, the fluid is oil. The same type of hydraulic system is used to operate convertible tops and power automobile windows. Large industrial presses also use this type of system.

Figure 6-20 shows the path of the fluid as the pilot lowers the landing gear. The **oil pump**

pumps the oil from the **reservoir**, through the **filter**, and into the **transmission lines**. The oil travels past the **hand shut-off valve** into a **four-way valve**. The pilot operates the four-way valve by hand. With the valve in the position shown, the oil flows to the upper end of the **double-acting cylinder**. (This cylinder is the **actuator** in the system.)

When the oil reaches the cylinder, it is under pressure from the pump. Therefore, it pushes the piston inside the cylinder downward. This motion activates mechanical devices connected to the piston. The mechanical devices lower the landing gear.

There is also oil in the cylinder *below* the piston. As the piston moves downward, it forces oil out the bottom of the cylinder. The oil travels through a transmission line under low pressure. It passes through the four-way valve. Then it

Fig. 6-21 A hydraulic landing gear system in the *gear up* position.

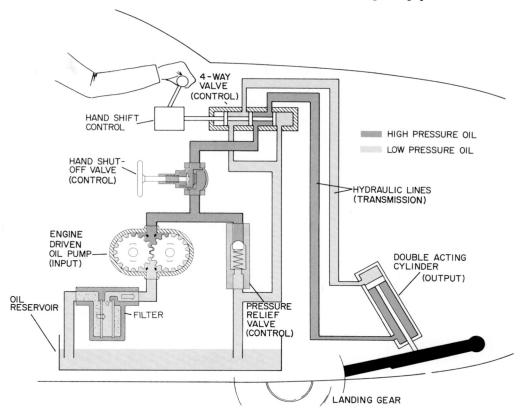

returns to the reservoir. This completes the oil's circuit or path through the system. There is a steady, constant flow of power throughout the system.

Notice the **pressure relief valve** in Fig. 6-20. The oil did not pass through this valve on the path just described. However, the pressure relief valve is a very important part of the system. It protects the system. If the system pressure gets too high, the pressure relief valve opens. This allows high-pressure oil to return to the reservoir, thus reducing the pressure in the system.

To raise the landing gear, the pilot shifts the position of the four-way valve. See Fig. 6-21. The valve directs high-pressure oil to the bottom of the double-acting cylinder. The oil pushes up on the piston. This raises the landing gear. At the same time, the oil in the top of the cylinder flows back to the four-way valve. From there it returns to the reservoir. This completes the circuit.

Pneumatic Systems

Pneumatic systems have served humanity longer than hydraulic systems. Possibly the first use of pneumatics was the blowgun. The **bellows** was a later pneumatic device. It supplied air under pressure for forging and casting. See Fig. 6-22. The modern industrial use of pneumatics has grown over the past century. In this short time, pneumatic systems have replaced great

Fig. 6-22   **The hand bellows was an early pneumatic tool. Here a bellows is used to supply air to a furnace used to melt iron.**

amounts of manual labor. They have also replaced many mechanical and electrical systems. In comparison with fluid systems, mechanical systems are slow and clumsy. Electrical systems are costly and complex.

Figure 6-23 shows the parts of a typical pneumatic system. Automotive garages use this system to lift cars and trucks for servicing and repair. Because the system uses both air and oil, it is called an **air-over-oil** system.

The fluid in this pneumatic system is air. An **air compressor**, powered by an electric motor, pulls in air through the **air inlet** and **primary filter**. The primary filter removes dirt and dust from the air. The compressor then pressurizes the air and pumps it into the **storage tank**, or **receiver**. The receiver stores the pressurized air until the system needs it.

As the system uses air, more air flows from the receiver. It passes through a **shut-off valve**. Then it continues to an **air-processing unit**. This unit acts as a filter to clean the air.

After processing, the high-pressure air travels through a three-way valve. From there it flows into a **pressure cylinder**. There is oil in the bottom of this cylinder. The incoming air forces the oil into the **actuator**, the **hydraulic cylinder**. This raises the ram on which the car is mounted.

Like the hydraulic landing gear system, the pneumatic system has a pressure-relief valve. This valve protects the system from excessive pressure. The valve is at the left of the receiver in Fig. 6-23.

Figure 6-24 shows how the system lowers the ram. The operator shifts the three-way valve. The air in the pressure cylinder then flows back through the valve and into the air outlet. From there the air escapes into the atmosphere.

As the air escapes, the pressure pushing oil into the hydraulic cylinder drops. The combined weight of the car and the ram becomes greater than the air pressure. When this happens, the ram moves downward. It pushes oil back into the pressure cylinder.

On its way back, the oil must pass through a small opening. This restriction controls how fast the lift lowers. The restriction is a safety feature that keeps the hoist from coming down too fast.

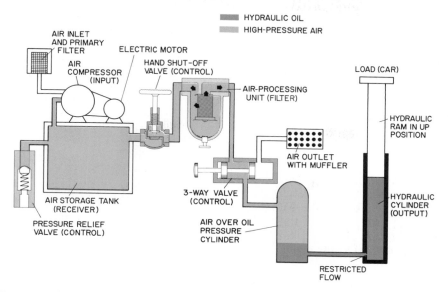

Fig. 6-23 This pneumatic system is used to raise and lower cars and trucks. Compressed air forces oil into the hydraulic cylinder to raise the load.

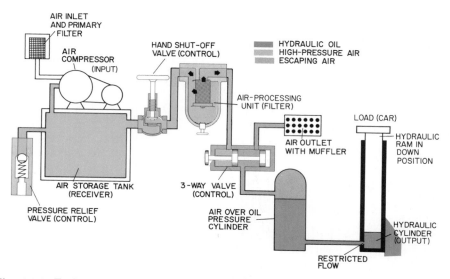

Fig. 6-24 To lower the hoist, the operator shifts the 3-way valve. This allows air to escape from the pressure cylinder. The weight of the ram and load forces the oil back into the pressure cylinder. This in turn lowers the ram.

STORING AND FILTERING FLUIDS

Both hydraulic and pneumatic systems must have a plentiful supply of fluid. Whatever the type of fluid, it must be clean. Otherwise, dirt particles will decrease the efficiency and life of the system. The two types of systems use similar devices to store and filter fluids.

Reservoirs and Receivers

In hydraulic systems, the liquid (usually oil) is stored at atmospheric pressure. A tank called a *reservoir* holds the liquid. See Fig. 6-25. Pneumatic systems use a similar tank. It is often called a *receiver*. See Fig. 6-26. The receiver stores pressurized gas from a compressor. This is the major difference between pneumatic tanks and hydraulic tanks. Pneumatic tanks store fluid under pressure. Hydraulic tanks do not.

The typical hydraulic reservoir has several important functions. Its main purpose, of course, is to store hydraulic liquid. Another purpose is cooling. The air around the reservoir absorbs some of the heat from the fluid. In this way, the reservoir indirectly helps to keep the fluid cool. The reservoir also helps clean the fluid. While the fluid sits in the tank, heavy contaminants separate from the oil. **Contaminants** are dirt and other matter that could clog the system. They drop to the bottom of the reservoir.

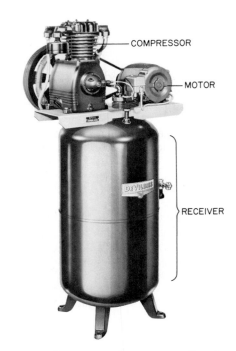

Fig. 6-26 **In pneumatic systems, the storage tank is often called a *receiver*. It receives pressurized air from the compressor.**

Pneumatic receivers work much the same way as hydraulic reservoirs. However, receivers are often larger than reservoirs. This is because gases need more space than liquids. Also, the tank itself needs more space in the system. Therefore, receivers are often farther from the output location than reservoirs are.

DeVilbiss Co.

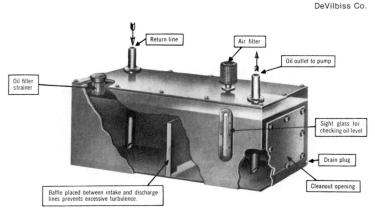

Fig. 6-25 **A cutaway view of a typical oil reservoir.**

Filters

As fluids move through fluid systems, they pick up dirt and dust particles. The smallest bit of dirt or metal can damage a part. It can even cause the system to break down. Contamination will also cause parts to wear out faster. To work at peak efficiency, fluid systems must be kept clean. Filters help keep fluid systems clean by removing contaminants.

Hydraulic Filters. Hydraulic systems have at least one filter. Large systems, or ones that work under extremely dirty conditions, may have many filters.

Hydraulic filters come in a wide variety of materials and designs. However, the basic design and operation are the same for all filters. As liquid passes through the filter, it moves through a screen that has many small holes in

it. See Fig. 6-27. This screen removes contaminants from the liquid. Fluid system mechanics must routinely remove contaminants from the filter. They do this by either cleaning or replacing the screen.

Pneumatic Filtering. The air used in pneumatic systems contains both moisture and dirt. These materials must be filtered out. However, the air also requires other treatment. After it is compressed, the air usually goes through a combination filtering, pressure-regulating, and lubricating device. This device is called an **FRL unit**. See Fig. 6-28.

The first part of the FRL unit is a filter that works by centrifugal force. The internal shape of the filter causes the incoming air to swirl. Centrifugal force throws water and large dirt particles out of the air. It deposits them on the inner surface of the filter bowl. The water and dirt eventually fall to the bottom.

After throwing off the contaminants, the air passes through a **filter element**. This element is a fine screen. It keeps any remaining dirt and moisture from entering the fluid lines. Mechanics can clean out the contaminants in the filter by removing a plug at the bottom.

The next part of the FRL unit is the **pressure control valve**. The operation of this valve is explained further in this chapter under *Pressure Control Valves*.

The third part of the FRL unit is the **lubricator**. See Fig. 6-28 again. Many of the parts in a pneumatic system must be lubricated. Adding a small amount of oil to the air provides this lubrication.

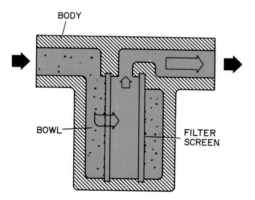

Fig. 6-27 **The operation of a hydraulic oil filter.**

Fig. 6-28 **An FRL unit is a combination of three devices: a filter, a pressure control valve, and a lubricator.**

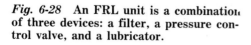
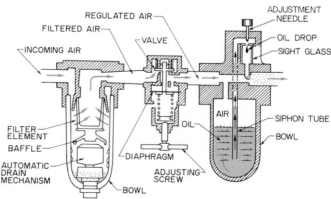

PUMPS AND COMPRESSORS

As mentioned earlier, fluid systems usually receive their power from either an engine or an electric motor. The systems must have some way to convert the mechanical input into fluid movement. In hydraulic systems, this is the job of the pump. In pneumatic systems, a compressor does this work.

Pumps

The most important part of a hydraulic system is its pump. The pump converts incoming mechanical power into fluid power. It also moves the fluid under pressure through the system. We can call the pump the "heart" of the hydraulic system.

There are many different pump designs. In this section, we will look at three basic types. The **piston pump** will show you the basic principles of pump operation. Then you will learn about **gear pumps** and **vane pumps**.

Most hydraulic pumps are **positive-displacement pumps**. This means that they take in a certain amount of fluid. Then they push the fluid through the system under pressure. We measure the output of these pumps in terms of **total fluid displacement**. We say that there are so many cubic inches of displacement per revolution of the mechanical power input shaft. (You will understand more about displacement as you read on.)

Manually Operated Piston Pump. Figure 6-29 shows a manually operated reciprocating piston pump. By pulling and pushing on the lever, the operator raises the piston. This type of pump is used in hydraulic jacks and in some presses.

An upward piston movement creates a low-pressure area in the piston chamber. See Fig. 6-29A. This allows atmospheric pressure to push on the fluid in the reservoir. The fluid forces the **intake check valve** open. It then flows through the charge line and into the piston chamber.

When the operator pushes down on the lever, the piston chamber is filled with liquid. As the piston moves down, it forces the liquid out. See Fig. 6-29B. First the liquid tries to go back into

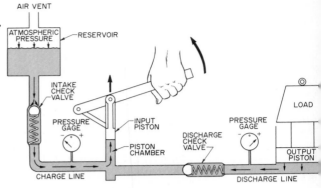

A. Intake stroke: input piston moves up.

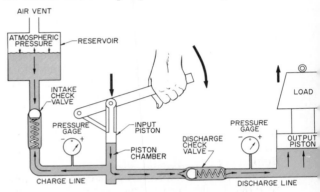

B. Discharge stroke: input piston moves down.

Fig. 6-29 **The operation of a simple manually operated reciprocating piston pump.**

the reservoir. However, the backward flow pushes the intake check valve shut. The liquid then tries to push open the **discharge check valve**. At the same time, the load is pushing on the liquid on the other side of the valve. The input side liquid must push harder than the load does. If this happens, the discharge check valve will open. The liquid then flows into the discharge line. The pressure of the liquid on the output piston then raises the load. When the operator pulls up on the lever, the whole series of actions repeats. The operator keeps pumping until the load is at the desired height.

Gear Pumps. Gear pumps are fairly simple pumps. They consist of two gears inside a housing. See Fig. 6-30. The two gears rotate in opposite directions. An input shaft drives one of the gears. This gear drives the other gear.

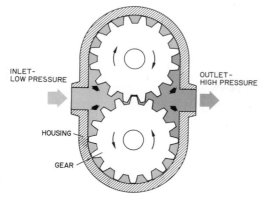

INLET-
LOW PRESSURE

OUTLET-
HIGH PRESSURE

HOUSING

GEAR

Fig. 6-30 **The gears of a gear pump pick up fluid and carry it to the outlet side of the pump.**

As the gear teeth turn past the inlet, they pick up liquid. Each tooth pushes a certain amount of liquid along the inside of the pump housing. The teeth then force the liquid through the outlet. The picking up of fluid from one side and carrying it to the other side is called **displacement**.

Vane Pumps. Vane pumps use sliding vanes to move liquid through the pump. (**Vanes** are flat rectangular pieces of metal.) Vane pumps have four main parts. These are the rotor, the sliding vanes, the drive shaft, and the housing. See Fig. 6-31. The rotor is not centered in the pump housing. It is set to one side. This creates a large open space on the other side of the rotor.

Figure 6-31 also shows how a vane pump operates. The drive shaft turns the rotor clockwise. As the rotor turns, the vanes rotate. At the same time, centrifugal force pushes the vanes out against the housing. Some pumps have small springs that push out on the vanes.

The spaces between the vanes are called **chambers**. Each two vanes form an **inlet chamber** between them. Notice that the inlet chambers increase in size as the rotor turns and they approach the inlet. This action creates a low-pressure area in the chambers. The low pressure allows oil to flow into them. The chambers then carry the fluid to the discharge side of the pump. As the chambers approach the outlet, they decrease in size. The smaller size creates higher pressure in the chambers. This forces the fluid into the discharge line.

Compressors

The air compressor is the pump of the pneumatic system. It operates on the same basic principles as the hydraulic pump. Both types of pumps convert mechanical power into fluid power. However, the air compressor pumps gas instead of liquid. Figure 6-32 shows a typical air compressor. Construction workers use this type of compressor to power pneumatic tools.

Fig. 6-32 **This air compressor is powered by a gasoline engine. Because the compressor doesn't require electricity, it can be used almost anywhere.**

Kellogg-American

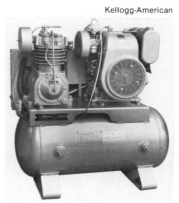

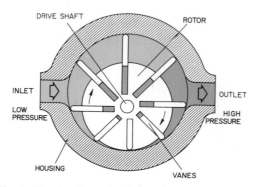

DRIVE SHAFT

ROTOR

INLET

LOW
PRESSURE

OUTLET

HIGH
PRESSURE

HOUSING

VANES

Fig. 6-31 **As the rotor turns in a vane pump, two things happen: (1) the vanes rotate, and (2) they slide out to the inner wall of the housing.**

There are many different designs of compressors used in pneumatic power systems. We will look at a compressor used for many home and industrial uses. It is the **single-piston reciprocating air compressor**.

Figure 6-33 shows how a reciprocating air compressor works. The downward movement of the piston creates a low-pressure area above the piston. When the pressure is low enough, atmospheric pressure pushes the intake valve open. See Fig. 6-33A. Air then flows into the cylinder until the cylinder pressure equals atmospheric pressure. When this happens, the intake valve closes.

As the piston moves upward, it compresses the air into a smaller space. As you know, a decrease in volume causes an equal increase in pressure. This pressure forces the discharge valve open. The air then flows out the discharge valve and into the receiver. See Fig. 6-33B. This series of actions repeats until the receiver is filled with pressurized air.

The operation of the compressor is controlled by the pressure in the receiver. If the power source is an electric motor, the receiver usually has a pressure-operated electrical switch. The switch is pre-set to a certain pressure for the system. When the receiver pressure falls below the set amount, the switch turns the compressor on.

CONTROL VALVES

Many types of valves are used in fluid power systems to control the flow of fluid. The operator uses the valves to start and stop the flow. The valves may also change the fluid's force, speed, and direction. Some valves are pre-set to maintain safe pressures throughout the system.

On-Off Valves

Operators use many different kinds of valves to turn the fluid flow on and off. These valves are often called **two-way valves**. Two-way valves are mainly used as on-off valves. However, operators also use some two-way valves to regulate fluid flow. This section describes three common types of two-way valves.

Globe Valves. Globe valves are simple and very reliable. They are used a great deal in homes and industry. Their main job is to provide on-off control of fluid flow.

Figure 6-34 shows how a globe valve works. Fluid passing through the valve must pass through the **valve seat**. When the operator turns the valve stem down, the valve rests tightly against the seat. This stops the flow of fluid. Turning the stem up releases the flow.

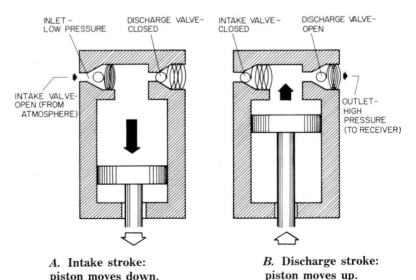

Fig. 6-33 **Operation of a single-piston reciprocating air compressor.**

A. **Intake stroke: piston moves down.**

B. **Discharge stroke: piston moves up.**

Braking with Air, the Westinghouse Way

Mechanically inclined George Westinghouse was born in 1846. Before he died, he had obtained over 400 patents and had founded the Westinghouse Electric Company. George enjoyed solving problems. One problem he saw was the need for a better system for stopping trains.

In the late 1800s, steam trains used a fairly awkward system for stopping. There was a **brakeman** stationed between each pair of cars. At a signal from the engineer, the brakemen would each apply a hand-operated brake to the car wheels. This was a slow way to stop the train. Many serious accidents happened because trains couldn't be stopped quickly.

After checking out the currently used braking systems, Westinghouse worked to develop a better way. He had problems at first. Then he hit on the idea of using a compressed-air system. In this system, air hoses would run the length of the train. It would be connected to brakes on each car. By pull-ing a lever, the engineer would send compressed air to all the brakes at the same time. Fluid power is much more powerful than muscle power. Therefore, the train would stop much more quickly.

Westinghouse was right! In 1868 he produced the Westinghouse airbrake. It proved to be very effective. In fact, today's trains still use the same basic system.

Globe valves force the fluid to change direction twice as it passes through them. This creates some resistance to flow. When the rate of flow is high, this resistance can cause a loss of pressure. Usually the loss is small. In general, the globe valve is an inexpensive and reliable valve.

Gate Valves. The gate valve is as reliable as the globe valve. In a gate valve, the fluid does not change direction as it passes through. This makes the gate valve more efficient than the globe valve. However, gate valves are more expensive. Gate valves are used on systems with a high rate of flow.

Figure 6-35 shows how a gate valve works. Turning the stem up moves the tapered wedge up. This allows the fluid to flow freely. Turning the stem down lowers the wedge and blocks the flow.

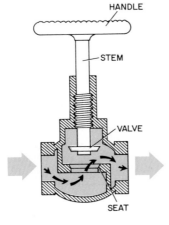

Fig. 6-34 **A globe valve in the open position.**

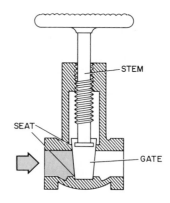

Fig. 6-35 **A gate valve in the closed position.**

Spool Valves. The spool valve has many industrial uses. This valve consists of a plunger that moves up and down inside a housing. See Fig. 6-36. The plunger has spools attached to it. These spools fit tightly against the wall of the housing. This is similar to the way a piston fits inside a cylinder.

Spool valves may be two-way, three-way, or four-way valves. The three- and four-way types are described below under *Directional Control Valves*. The two-way type is most commonly used as an on-off valve.

Figure 6-36 shows how a typical two-way spool valve works. In *A*, the spring holds the plunger in an open position. This allows fluid to pass through the valve freely. In *B*, the plunger has been pushed down. The top spool is now between the inlet and the outlet. This blocks the fluid flow.

Directional Control Valves

Many fluid power systems have several power sources, several outputs, or both. These systems often require a way of changing power sources and redirecting fluid flow to different parts of the system. Directional control valves provide a way to do this. There are three general types of directional control valves. They are three-way valves, four-way valves, and check valves.

Most directional control valves are spool valves. Their main job is to control the direction of fluid flow. However, they can sometimes be used as on-off valves.

Three-Way Valves. A three-way valve has three openings, or **ports**. One type of three-way valve has two input ports and one output port. With this type of valve, the operator can use two different sources of pressure. Another type of three-way valve has one input port and two output ports. The operator can direct pressurized fluid to either output port. The automobile hoist in Fig. 6-23 uses a three-way valve with two output ports.

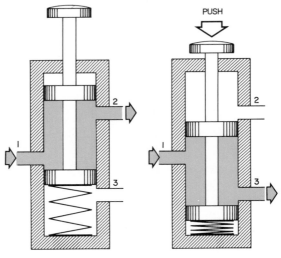

Fig. 6-37 **This is a two-output type of three-way valve. Depending on the plunger position, fluid flows out of either port 2 or port 3.**

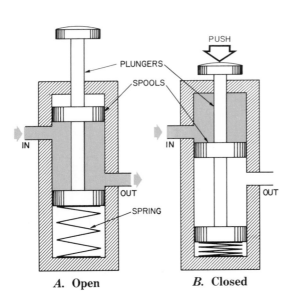

A. **Open**   *B.* **Closed**

Fig. 6-36 **A spool valve in its two common positions.**

A. **Non-actuated position**   *B.* **Actuated position**

Figure 6-37 shows how a three-way valve works. The valve shown is a two-output type. In *A*, the spring holds the plunger up. This is the valve's normal, or **non-actuated**, position. The non-actuated position allows fluid to flow in port 1 and out port 2. The bottom spool blocks the flow from port 1 to port 3.

When the operator pushes the plunger down, the valve moves to the position shown in *B*. The fluid now flows out of port 3. It cannot flow through port 2.

Four-Way Valves. Four-way valves can start the fluid flow, stop the flow, or reverse its direction. They are usually used to control the back-and-forth motion of a piston inside a cylinder. Another common use is to reverse the rotation of a fluid motor. The landing gear system in Figs. 6-20 and 6-21 uses a four-way valve to control the motion of the piston in the double-acting cylinder.

A four-way valve has four *working* ports. The valve in Fig. 6-38 has five ports, as you can see. However, only one discharge port can be used at a time.

A four-way valve allows pressurized fluid to push on a piston in both directions. For example, in Fig. 6-38A, pressurized fluid passes through port 1, then port 3. The fluid enters the top end of the cylinder and forces the piston down. The downward piston movement forces the fluid in the bottom of the cylinder through port 2. It then passes out the lower discharge port.

In Fig. 6-38B, the valve position is reversed. Pressurized fluid passes through port 1, then port 2. The fluid enters the rod end of cylinder and forces the piston up. The upward piston movement forces the fluid in the top of the cylinder through port 3. It then passes out the upper discharge port.

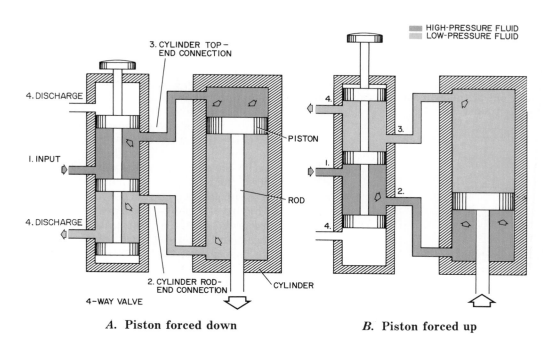

A. **Piston forced down**　　　　　　*B.* **Piston forced up**

Fig. 6-38 **The fluid flow through this four-way valve controls the movement of the piston.**

The four-way valve also has a shut-off position. See Fig. 6-39. In this *closed-centered* position, the spools block the discharge ports and the input port. This stops the movement of the piston. Fluid cannot escape from either side of the cylinder.

Check Valves. Check valves are usually used to allow fluid flow in only one direction. They do not allow the fluid to flow in the opposite direction.

Compressor valves (see Fig. 6-33) are examples of simple check valves. They allow fluid to flow through the compressor in one direction. They also keep the fluid from reversing direction. All directional check valves have the same function. They are often called **one-way valves**.

Figure 6-40 shows how a typical check valve works. Normally, the pressure of a lightweight spring forces a ball against a seat. The valve opens when fluid pressure overcomes the spring pressure, as in *A*. The fluid pressure lifts the ball off the seat. This allows the fluid to flow freely. If the fluid tries to reverse its flow, the valve closes, as in *B*. The reverse pressure exerts force on the back of the ball. This helps to keep the valve closed and sealed.

Fig. 6-39 **This four-way valve is in the closed-centered, or shut-off, position. Fluid cannot move in or out of the cylinder. Therefore, the piston cannot move.**

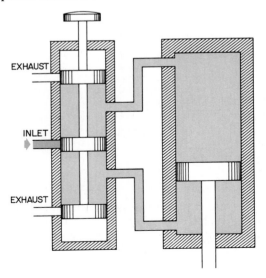

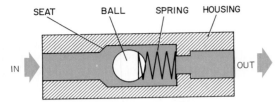

A. **Fluid flows through open valve.**

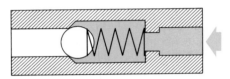

B. **Valve closes when fluid tries to reverse direction.**

Fig. 6-40 **The operation of a check valve.**

Pressure Relief Valves

All parts of a fluid power system are designed to operate most efficiently under certain pressures. The parts must not be forced to work at pressures above their specified maximum pressure. Excessive pressures will make the parts wear out faster. The parts may even break down completely.

Extreme pressure can also harm the engine or motor that supplies power for the system. To avoid these problems, fluid power systems usually have pressure relief valves. These valves direct fluid from the discharge side of pumps to the inlet side. This reduces the fluid pressure.

Figure 6-41 shows how a pressure relief valve works. When the system pressure in the line goes over the pre-set pressure, it pushes up on the ball against the resistance of the spring. This permits fluid to escape through the valve and thus reduce the system pressure. The operator can change the operating pressure of the valve with the adjusting screw. The adjusting screw increases the force of the spring by compressing it. Figures 6-20 and 6-23 show how a pressure relief valve fits into different fluid systems.

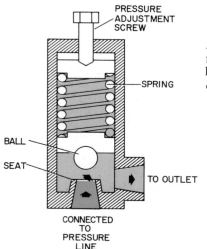

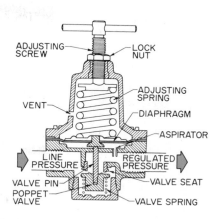

Fig. 6-41 In this pressure relief valve, fluid pressure has overcome the pressure of the spring.

Fig. 6-42 A simple spring-type pressure control valve. The diaphragm is shown in color.

Pressure Control Valves

Pressure control valves are an important part of most fluid power systems. Many systems deliver pressurized fluid to several output units. The different units may require different pressures. However, the system must generate an *overall* pressure that will operate the unit needing the highest pressure.

Pressure control valves are used to provide reduced pressures for lower-pressure units. These valves are placed in the system close to the lower-pressure output units. The valves maintain pressures below system pressure for these units.

There are many types of pressure control valves. One of the most widely used is the simple spring-type valve used in pneumatic systems. This valve has two main parts: a poppet valve and a diaphragm. See Fig. 6-42.

This control valve regulates the line pressure by balancing the following two forces:

- **The force trying to open the valve** — This force results from atmospheric pressure (through the vent) plus the pressure of the adjusting spring. Both forces push *down* on diaphragm.
- **The force trying to close the valve** — This force results from the regulated system pressure plus the valve spring pressure. Both of these forces push *up* on diaphragm.

When the adjusting spring is tightened (compressed), it pushes down on the valve pin. This opens the poppet valve. The regulated pressure increases until it is high enough to balance the added spring force. The result is a higher regulated pressure. If the adjusting spring tension is reduced, the regulated pressure decreases by the same amount. In normal operation, the poppet valve stays open just enough to maintain the regulated pressure at a constant level.

TRANSMISSION LINES AND CONNECTORS

Whether they are liquids or gases, fluids must have a container through which they can flow. In fluid systems, the containers are pipes, tubes, and hoses. These are called *transmission lines*. The entire line of transmission in a system is usually made up of several transmission lines connected together.

Transmission Lines

Transmission lines may be either rigid or flexible. **Flexible lines** are used when an output unit must move or vibrate. These lines are made from a variety of materials. The material used

Fig. 6-43 **Reinforced flexible hose for medium- and high-pressure uses.**

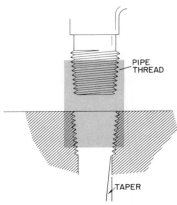

A. **Pipe threads are tapered (cut at an angle). When tightened, the threads come together to form a tight, leak-proof joint.**

PIPE THREAD

TAPER

depends on the pressure in the system. Plastic lines are often used in systems that use low to medium pressures. Higher-pressure systems need stronger lines. Figure 6-43 shows flexible lines reinforced with braided steel wire.

Rigid lines are used whenever flexible lines are not required. Rigid lines cost less. They also provide long, trouble-free service. They are usually made of steel.

Connectors

Transmission lines are connected to stationary parts or other lines with **fittings.** These fittings permit fluids to flow between the part and the line without leaking. Several different types of connectors are used, including **pipe-threaded fittings, flared fittings,** and **quick-disconnect couplings.** Figure 6-44 shows these fittings and describes how they work.

ACTUATORS

Before we can put fluid power to work, we must change it into mechanical power. We use *actuators* to make this change. An actuator is a device that receives fluid power and changes it into mechanical motion. Actuators can produce either linear, reciprocating, or rotary motion. (See Chapter 5 for an explanation of the types of motion.) We use **cylinders** to produce linear motion and reciprocating motion. We use **fluid motors** to produce rotary motion.

Pneumatic and hydraulic actuators have similar designs. They also operate in much the same way. This section describes both types together.

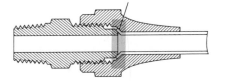

FLARED TUBE SQUEEZED BETWEEN NUT AND FITTING SEALING CONNECTION

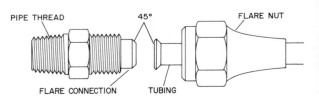

PIPE THREAD   45°   FLARE NUT

FLARE CONNECTION   TUBING

B. **In a flared tubing connection, the seal between the line and the fitting takes place in the colored area.**

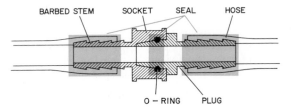

BARBED STEM   SOCKET   SEAL   HOSE

O – RING   PLUG

C. **A quick-disconnect coupling is used to connect two pieces of flexible hose. The colored areas show where sealing takes place.**

Fig. 6-44 **Three common types of connectors used with fluid transmission lines.**

Cylinders

Figure 6-45 shows a typical cylinder and piston used as a linear actuator. When pressurized fluid enters the right port, the piston moves to the left. When pressurized fluid enters the left port, the piston moves to the right. In a pneumatic system, the gas can be released into the atmosphere after use. In hydraulic systems, the fluid is piped back to the reservoir and reused.

Many kinds of jobs can be done with the piston's reciprocating motion. Forming metals and plastics, moving objects, and raising convertible tops are just a few.

There are two basic types of cylinder actuators: single-acting and double-acting.

Single-Acting Cylinders. In single-acting cylinders, pressurized fluid exerts force at only one end of the cylinder. Figure 6-46 shows a single-acting cylinder used on hydraulic jacks and automobile lifts. A pump supplies fluid to the cylinder under pressure. The fluid raises the ram and lifts the load (A). When the operator releases the pressure, the combined weight of the ram and the load pushes the ram down (B).

Double-Acting Cylinders. Figure 6-47 shows a double-acting cylinder. In this cylinder, pressure produces piston motion in both directions. A rod extends from the cylinder to do work. This rod can perform both a push (A) and a pull (B).

A three- or four-way valve is used to control the flow of fluid to the cylinder. In Fig. 6-46A, the valve provides pressurized fluid to the right

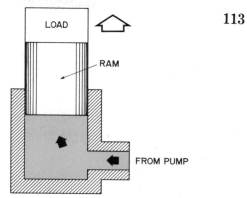

A. Ram extends.

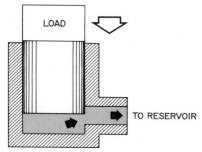

B. Ram retracts.

Fig. 6-46 A single-acting cylinder used as a hydraulic jack.

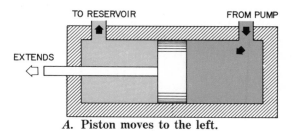

A. Piston moves to the left.

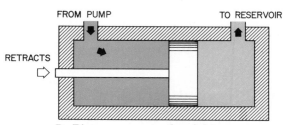

B. Piston moves to the right.

Fig. 6-47 A double-acting cylinder.

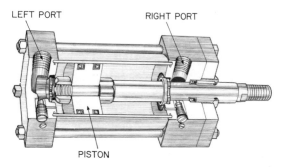

Fig. 6-45 This cutaway view shows the inside of a double-acting cylinder.

side of the cylinder. This forces the piston to the left. The rod extends out of the cylinder. After the motion to the left is complete, pressure is applied to the left side of the cylinder (*B*). The piston returns to its original position. This cylinder is the type used for lifting the landing gear in Figs. 6-20 and 6-21.

Fluid Motors

We use fluid motors to convert fluid power into rotary motion. Fluid motors are actually fluid pumps in reverse. Instead of using mechanical power to produce fluid motion, a fluid motor uses fluid power to produce mechanical motion. Some fluid pumps can be used either as pumps or motors.

A common type of fluid motor is the **vane-type motor**. This motor is similar to the vane pump shown in Figure 6-31. As you recall, vane pumps hold the vanes against the housing with centrifugal force. Vane motors hold the vanes out with springs.

Figure 6-48 shows how a fluid motor works. Pressurized fluid is directed against the vanes. This turns the rotor to which the vanes are attached. The pressure exerts force against one or more vanes at all times. Therefore, the rotor turns at a steady rate. A shaft attached to the rotor can be used to do work.

Other types of hydraulic and pneumatic motors are available. In general, they have the same designs as the pumps with similar names. (For example, a gear-type motor has the same basic design as a gear-type pump.)

Fluid pressure determines the torque of a fluid motor. As long as the pressure is constant, the torque is constant. Fluid motors can also produce more horsepower than electric motors of the same size. Motor speed is determined by the flow of fluid. The greater the rate of flow (gallons or cubic feet per minute), the faster the motor turns.

Fig. 6-48   **The operation of a vane motor.**

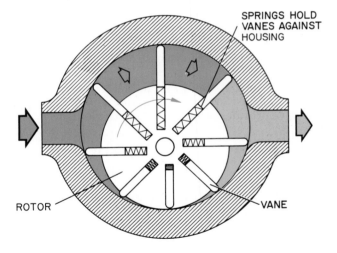

STUDY QUESTIONS

1. What is fluid power?
2. How much pressure does the atmosphere exert on the earth?
3. What is the pressure reading of atmospheric pressure on a gage that registers in pounds per square inch gage?
4. What is a vacuum?
5. How are vacuums measured on a compound gage?
6. What is the pressure reading of atmospheric pressure on an absolute gage?
7. What is the pressure reading of a perfect vacuum on an absolute gage?
8. State Boyle's Law.
9. If the pressure on a gas stays constant and the temperature increases, what happens to the volume of the gas?
10. What is absolute zero?
11. On the regular Fahrenheit scale, what is the temperature at which all molecules stop moving?
12. Describe the relationship between the pressure and the temperature of a gas when the volume remains constant.
13. Describe in your own words how a fluid transfers force.
14. True or false: In fluid power systems, a gain in force means an equal loss in distance or speed.
15. What are the three general stages or steps in all fluid power systems?
16. List the seven parts that most fluid power systems have.
17. What is the major difference between a fluid receiver and a fluid reservoir?
18. Why are filters needed in fluid systems?
19. What is an FRL unit?
20. Name three basic types of hydraulic pumps.
21. Name three types of on-off valves.
22. Name three types of directional control valves.
23. Why are pressure relief valves needed in fluid systems?
24. Why are pressure control valves needed in fluid systems?
25. Name the two basic types of fluid transmission lines.
26. Name three types of fluid system fittings.
27. What is a fluid power actuator?
28. Name two common types of fluid power actuators.

ACTIVITIES

1. Name the different fluid power systems on an automobile. Most automobiles have more than one. These systems include hydraulic brakes and power steering, to name two. What other systems exist?

   Choose one of the systems and prepare a brief report describing how it operates. In the report, tell how the system provides mechanical advantage. Also, try to figure out why fluid power was used instead of a mechanical or electrical system.
2. Build a simple fluid power system. With the help of your teacher, obtain a double-acting cylinder, a four-way valve, a pressure regulator, and lines. Build a simple operating system using an air compressor or compressed air from the school shop as the supply of fluid power.

   SAFETY NOTE: Set the regulator to a low pressure.

3. Solve the following problems:
   A. A gage reads *30 psig*. What would the reading be in absolute pressure (psia)?
   B. The pressure at the top of Pike's Peak is 9 psia. What would the gage pressure (psig) be?

Electrical Power

Electricity has made our modern way of life possible. We need power on a continuous basis. Electricity is the main form in which power is transported. Electricity is now a part of almost every American home. It provides us with light, heat, and useful motion. For example, electric motors convert electricity into motion. We use electric motors to operate refrigerators, mixers, washing machines, dishwashers, disposals, and home workshop tools. We use heat produced by electricity to cook our food, dry our clothes, and heat our homes. Electricity also produces light whenever we need it. See Fig. 7-1.

Electricity also serves industry. Electric motors operate many tools and machines. Electric lighting allows workers to do their jobs at any time. Industry also uses heat from electricity to produce many materials, such as steel, glass, plastic, and rubber.

Electricity powers radios, televisions, calculators and computers, digital watches, and telephones. These and many other communication devices make our lives easier and more enjoyable.

ELECTRON THEORY

To understand electricity, you must first know something about atoms. Atoms are the tiny particles that make up matter. There are 106 known kinds of atoms called **elements**. These elements combine in different ways to form all of the materials on earth.

In simple terms, the **electron theory** states that atoms consist of three kinds of particles.

Fig. 7-1 **Electrical power serves us by providing motion, heat, and light.**

A. Motion

B. Heat

C. Light

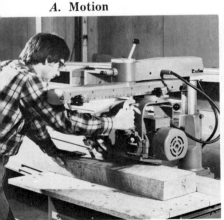

Rockwell International

Whirlpool

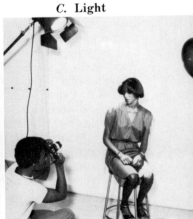

These particles are protons, neutrons, and electrons. Together, the **protons** and **neutrons** form the center of the atom. This center area is called the **nucleus**. The **electrons** revolve around the nucleus.

Different atoms have different numbers of protons, neutrons, and electrons. See Fig. 7-2. All atoms of a particular element have the same number of protons. The number of electrons circling the nucleus is normally the same as the number of protons. When we study electricity, we are concerned only with protons and electrons. Neutrons do not affect electricity.

Charged Particles

Electrons have a **negative** charge. Protons have a **positive** charge. A charged particle either attracts or repels other charged particles. See Fig. 7-3. Particles with the same charge have **like charges**. Like charges repel each other. Particles with different charges have **unlike charges**. Unlike charges attract each other. When an electron and a proton attract each other, they are in balance. Together, they neither attract nor repel other particles.

An atom is normally in a state of balance, or **neutrality**. The protons give the nucleus a positive charge. This charge is balanced by an equal number of electrons rotating around the nucleus. The attraction between the protons and electrons keeps the orbiting electrons from being thrown off by centrifugal force.

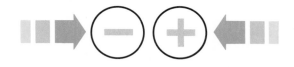

Fig. 7-3 A positive charge and a negative charge will attract each other. Two positive charges or two negative charges will repel each other.

Electron Shells and Electron Flow

The strength of the attraction between the protons and electrons in an atom depends on the distance between them. This distance varies depending on the number of electrons. Electrons arrange themselves in orderly **shells** around the nucleus. Each shell can hold only a certain number of electrons. For example, the first shell holds two electrons, the second shell holds eight electrons, and the third shell holds 18 electrons. In an atom with more than 28 electrons (2 + 8 + 18), the additional electrons will go into larger shells.

Let's look at how the electrons are arranged in a copper atom. Copper has 29 electrons. Therefore, the electrons are arranged in four shells. See Fig. 7-4. Notice that the outermost

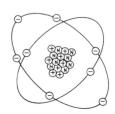

HELIUM
2 PROTONS, NEUTRONS, AND ELECTRONS

OXYGEN
8 PROTONS, NEUTRONS, AND ELECTRONS

Fig. 7-2 Atoms are the building blocks of the universe. The difference between an oxygen atom and a helium atom is in the number of protons, neutrons, and electrons.

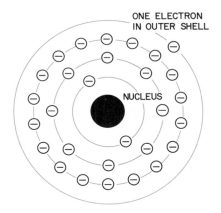

ONE ELECTRON IN OUTER SHELL

NUCLEUS

Fig. 7-4 Here you see the arrangement of electrons in a copper atom. The outermost electron can easily be pulled free of the atom.

shell contains only one electron. Also, it is the farthest away from the nucleus. Therefore, the electron in the fourth shell has the weakest attraction for the protons in the nucleus. As a result, it can easily be pulled from its orbit to become a **free electron**.

When a copper atom loses an electron, it becomes positively charged. This is because there are not enough electrons to balance the protons' positive charge. Any positively charged atom will try to attract a free electron from another atom. When an electron moves from one atom to another, there is **electricity**. We can define electricity as *the movement of electrons from one atom to another*.

With its negative charge, an electron always moves toward a positively charged atom. Electron flow is always from negative to positive.

Conductors

There are other elements besides copper that hold their outermost electrons with a weak force. These elements are called **conductors**. We also use the word **conductor** to refer to a wire made of a conducting element.

For atoms to conduct electricity, many of them must be placed next to each other. This is basically how copper atoms are arranged in a copper wire. See Fig. 7-5. Notice that the outer orbits of the atoms overlap. This allows the electrons to transfer easily from atom to atom.

To produce a flow of electrons, a conductor must be attached at one end to a source of free electrons. The other end must attach to a place for the electrons to go. Often the electrons return to the same place they started.

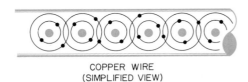

COPPER WIRE
(SIMPLIFIED VIEW)

Fig. 7-5 **In a conductor, the atoms touch each other. This allows free electrons to pass from one atom to the next.**

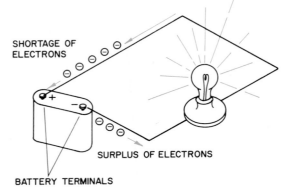

SHORTAGE OF ELECTRONS

SURPLUS OF ELECTRONS

BATTERY TERMINALS

Fig. 7-6 **Through chemical changes, a battery develops a shortage of electrons at one terminal and a surplus at the other. When a wire is connected to the terminals, the electrons flow from the negative terminal to the positive terminal.**

The source of free electrons may be either a generator or a battery. The parts of the electron source to which the conductor attaches are called the **terminals**.

The source works by providing an imbalance of electrons. It develops a surplus of electrons at one terminal and a shortage of electrons at the other terminal. See Fig. 7-6. Electrons flow through the conductor from the negative terminal to the positive terminal. We call this flow of electrons an **electrical current**.

CURRENT FLOW

Let's look a little more closely at electrical current. The battery in Fig. 7-6 moves electrons through a wire to light a bulb. We can view the battery terminals as two containers. One container (the negative terminal) holds many electrons. The other container (the positive terminal) holds only a few. See Fig. 7-7. This creates an **electrical pressure difference** between the negative and positive terminals.

Current begins to flow when the negative terminal pushes an electron into one end of the conductor. The conductor can hold only a set number of electrons. Therefore, one other electron gets pushed off the other end. This keeps the number of electrons in the wire constant. However, the

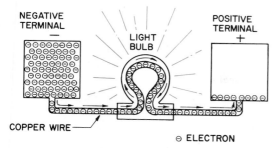

Fig. 7-7 We can think of a battery as having two containers. One container is full of electrons. The other is empty. Pressure from the full container (the negative terminal) forces electrons through a wire toward the empty container (the positive terminal). This causes an electric current.

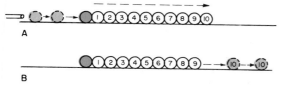

Fig. 7-8 A row of billiard balls demonstrates electrical current. When the cue ball strikes the row of balls, it stops. The impact is transferred to the last ball, which rolls away. In the same way, electrons are added and released from a conductor.

addition and removal of a single electron produces a current through the entire wire.

Figure 7-8 shows another way of looking at electron flow. There is a line of 10 billiard balls. A single ball added to one end produces an almost immediate movement of the ball at the other end. The moment the first ball strikes, its force is transmitted through all the balls. This force drives the last ball from the other end.

This condition is also true with electrons producing an electric current. Actually, electrons move through a wire quite slowly. However, because the movement is transmitted through the entire wire, the resulting current flow is very rapid. Current travels at nearly the speed of light. The speed of light is 186,000 miles (*300,000 km*) per second.

Fig. 7-9 **Magnets attract and hold objects that contain iron.**

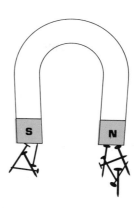

MAGNETISM

Magnetism is another very important idea to understand when learning about electrical power. We can best understand magnetism by observing its effects.

Magnets can attract iron and steel. See Fig. 7-9. The two ends of a magnet are different. One end is called the **north pole** (abbreviated **N**). The other end is called the **south pole** (abbreviated **S**). The poles are named for the direction they will point if they are allowed to move freely. A north pole will always point toward the earth's north pole. It is often called a *north-seeking pole*. A south pole will always point toward the earth's south pole. It is called a **south-seeking pole**.

The poles are similar to electrical charges. See Fig. 7-10. Like poles repel, and unlike poles attract.

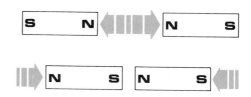

Fig. 7-10 **Like poles of magnets repel, and unlike poles attract.**

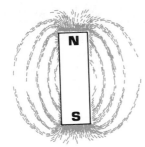

Fig. 7-11 When iron filings are sprinkled on a sheet of paper above a magnet, they form the pattern shown here.

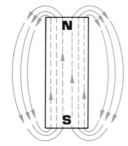

Fig. 7-12 The magnetic field consists of magnetic lines of force. These lines form a continuous circular path as they travel inside and outside the magnet.

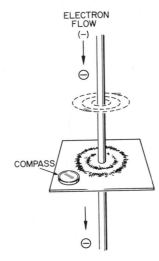

ELECTRON FLOW (−)

COMPASS

Fig. 7-13 Electrons flowing through a conductor produce a magnetic field. This is shown by passing a current-carrying wire through a sheet of paper and sprinkling iron filings on the sheet. The compass shows the direction of the magnetic lines of force.

A magnet develops a **magnetic field** around it. This field is the area in which magnetic attraction or repulsion takes place. We can "see" the magnetic field by placing a piece of paper over a magnet and sprinkling iron filings on the paper. See Fig. 7-11. More iron filings gather at the ends of the magnet than in the center. This shows that the attraction for iron is strongest at the ends of the magnet. Repulsion between like poles will also be strongest at the ends.

The magnetic field is made up of **magnetic lines of force**. These lines have both force and direction. See Fig. 7-12. Outside the magnet, the lines of force travel from north to south. Inside the magnet, they travel from south to north.

Magnetism and Current Flow

When an electrical current flows through a wire, a circular magnetic field develops around the wire. See Fig. 7-13. The strength of the magnetic field depends on the amount of current flowing through the wire.

Notice the compass in Fig. 7-13. If the current flow is reversed, the compass needle will also reverse. This shows that the magnetic lines of force have **polarity**, or direction.

Strengthening Magnetic Fields

Figure 7-14 shows two wires placed close together. The wires are conducting electrons in opposite directions. In this situation, one conductor has clockwise lines of force. The other conductor has counterclockwise lines of force. However, the lines of force *between* the conductors are in the same direction. This creates a strong magnetic field between the wires. The strong field will tend to push the wires apart to weaken the field. Later in this chapter, you will learn how this action is used to produce motion.

A different condition exists when two neighboring wires carry equal currents in the *same* direction. See Fig. 7-15. The lines of force around each conductor will be in the same direction. However, the lines of force between the conductors will be in opposite directions. These opposing lines of force will cancel each other out. No magnetic field will be left between the wires. One large magnetic field will surround both wires. This strong field will push the wires toward each other.

The field surrounding both conductors will be equal to that of one conductor carrying twice the electricity. See Fig. 7-16. Many wires can

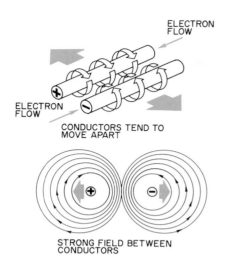

ELECTRON FLOW

ELECTRON FLOW

CONDUCTORS TEND TO MOVE APART

STRONG FIELD BETWEEN CONDUCTORS

Fig. 7-14   Conductors carrying electrons in opposite directions will tend to move apart.

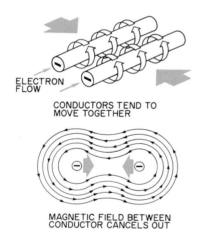

ELECTRON FLOW

CONDUCTORS TEND TO MOVE TOGETHER

MAGNETIC FIELD BETWEEN CONDUCTOR CANCELS OUT

Fig. 7-15   Conductors carrying electrons in the same direction will tend to move together.

be placed side by side. This increases the magnetic effect even more. One large field will surround all the wires.

The magnetic field can be increased even more if the wire is coiled. A coiled wire has the same effect as many wires carrying current in

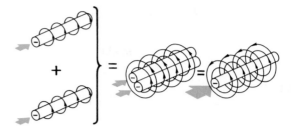

Fig. 7-16   **The magnetic fields of conductors carrying current in the same direction can be conbined to produce a stronger field. Combining two conductors produces the same magnetic field as passing twice as much current through a single conductor.**

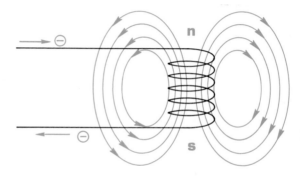

n

s

Fig. 7-17   **A single coiled wire has the same magnetic effect as several conductors placed close together.**

the same direction. See Fig. 7-17. Notice that the coil's magnetic field is just like the field of the bar magnet in Fig. 7-12.

Placing a piece of soft iron inside a coil strengthens the magnetic field even more. See Fig. 7-18. The lines of force can pass more easily through the iron core than through the air. This concentrates the lines of force within the coil. The iron core increases the strength of the magnetic field over 300 times. It does this without any additional current.

Electromagnets

A piece of soft iron with a current-carrying wire coiled around it forms an *electromagnet*.

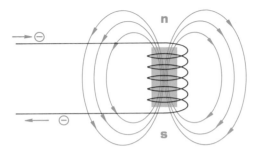

Fig. 7-18 An iron core placed inside a coil concentrates the magnetic lines of force within the coil. This strengthens the magnetic field a great deal.

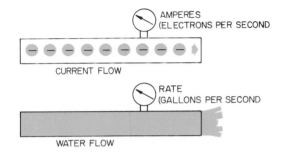

Fig. 7-19 We can compare amperage to the rate at which water flows through a water line.

The current turns the iron core into a strong magnet. When the current is stopped, the core loses almost all of its magnetism.

Magnetism is the link between mechanical energy and electrical energy. We can produce motion with electric current by applying the principles of magnetism. The electric motor is just one of many devices that use magnetism to produce motion. You will learn about these devices later in this chapter.

MEASURING ELECTRICITY

Electricity is the flow of electrons. Therefore, we can measure electricity by finding the number of electrons moved from one place to another. Electrons are very small. An enormous number of electrons must move to generate a small amount of electricity.

The basic unit of measurement for electricity is the **coulomb**. One coulomb is equal to 6,280,000,000,000,000,000 (or 6.28×10^{18}) electrons. This is obviously a large number of electrons! However, it is still only a small amount of electricity. One coulomb is enough to light an average light bulb for one second.

Measuring Current

We cannot describe an electrical current with coulombs alone. We need three other basic units. These are the ampere, the volt, and the ohm. We use these units to measure an electrical current flowing through a conductor.

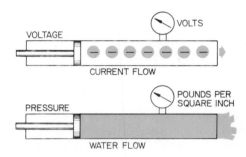

Fig. 7-20 We can compare voltage to the pressure on the water flowing in a pipe. Voltage provides the force to move the electrons.

Amperage is the rate at which current flows through a conductor. One ampere (or "amp") is equal to one coulomb of electricity passing a given point in one second. We can compare it to the rate at which water flows through a pipe. See Fig. 7-19. In calculations, the sign for amperage is **I**. We measure amperage with an **ammeter**.

Voltage is the pressure pushing the current through the conductor. One volt is the pressure exerted by one coulomb of electricity. We can compare voltage to the pressure in a water pipe. See Fig. 7-20. Voltage is also known as **electromotive force**. In calculations, the sign for voltage is **E**. We measure voltage with a **voltmeter**.

Resistance is the opposition to current flow through a conductor. Copper is an excellent con-

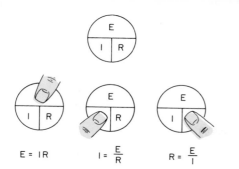

$E = IR$ $I = \dfrac{E}{R}$ $R = \dfrac{E}{I}$

Fig. 7-21 You can use the Ohm's Law circle to find voltage, amperage, or resistance. Cover the value you want to find. This will identify the equation to use.

ductor of electricity. However, energy is lost as electrons move from atom to atom. The attraction of the nuclei for the free electrons produces resistance to current flow. This resistance produces heat. It also produces a drop in the voltage pushing the current through the conductor. We measure resistance in **ohms** with an **ohmmeter**. In calculations, the sign for resistance is **R**.

Amps, volts, and ohms are all mathematically related. A German physicist, Georg Ohm, discovered this relationship. Ohm's Law states, *It takes one volt to force one ampere of current through a resistance of one ohm.* Written as a mathematical formula:

Voltage = Amperage × Resistance
or
$$E = I \times R$$
The formula can also be written with amperage or resistance as the first value:

$I = \dfrac{E}{R}$ or $R = \dfrac{E}{I}$

An easy way to remember Ohm's Law is to use the Ohm's Law Circle. See Fig. 7-21. If you know any two values, you can find the third. For example, if a circuit has a voltage of 120 and an amperage of 10, what is the resistance?

$R = \dfrac{E}{I}$

$R = \dfrac{120}{10}$

$R = 12$ ohms

If a circuit has a resistance of 20 ohms and a voltage of 120, what is the rate of current flow (amperage) through the circuit?

$I = \dfrac{E}{R}$

$I = \dfrac{120}{20}$

$I = 6$ amperes

Measuring Electrical Power

Power is energy (work) per unit of time. We measure mechanical power in foot-pounds per second. We measure electrical power in **wattage**. Wattage is a measurement of the power produced by a certain flow of current under a certain amount of pressure. To measure the wattage, we must know both the voltage (pressure) and the amperage (rate of flow). *One watt of power is produced by a flow of one ampere at a pressure of one volt.* In equation form:

Wattage = Amperage × Voltage
or
$$W = I \times E$$

$W = I \times E$
$W = 0.5 \times 120$
$W = 60$ watts

The wattage formula is the same type of relationship as in Ohm's Law. If you know two of the values, you can find the third. For example, the current passing through the light bulb in Fig. 7-22 has a pressure of 120 volts and a rate of flow of 0.5 ampere. What is the wattage of the bulb?

Fig. 7-22 A circuit is the path of an electric current.

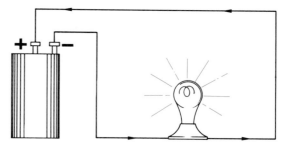

One watt is a small amount of electrical power. It takes 746 watts to equal one horsepower. The **kilowatt-hour (kWh)** is the basic measuring unit used by power companies. One kilowatt-hour is 1000 watts of electricity used in a period of one hour.

ELECTRICAL CIRCUITS

A **circuit** is a system of conductors and electrical devices through which electrical current moves. See Fig. 7-22. The word *circuit* means *a circle*. Electricity flows in a circle. It both moves from a power source and returns to the source. The power source may be a battery, as in Fig. 7-22. The source may also be an electrical generator.

An electrical circuit depends on the electron imbalance set up by the power source. In Fig. 7-22 the source is a battery. The battery's terminals are connected by a conductor and a lamp. Electrons move inside the battery by chemical action from the positive terminal to the negative terminal. This results in a surplus of electrons at the negative terminal and a shortage of electrons at the positive terminal. This is an unbalanced condition.

The conductor and lamp allow electrons to flow through the circuit from the negative terminal to the positive terminal. This current flow balances the electron distribution. The battery keeps the current flowing. It continues to move electrons internally from the positive terminal to the negative terminal. Therefore, the electrons move in a complete circle. The current will continue to flow as long as the battery maintains an internal unbalanced condition.

Types of Circuits

There are three basic types of electrical circuits: series, parallel, and series-parallel.

Series Circuit. The simplest circuit is the series circuit. The series circuit provides only one path for current flow. See Fig. 7-23. Electrons flow through a single conductor. They flow from the negative side of the source to the positive side. If the circuit is interrupted or broken, no electricity will flow. For example,

if one of the light bulbs in Fig. 7-23 burns out, current will stop flowing to the other light bulb.

An interruption of current flow is called an **open circuit**. An open circuit anywhere in a series circuit will cause the whole system to go dead. For example, the switch connected **in series** in Fig. 7-23 allows you to control the whole circuit. When you turn the switch on, current flows through the circuit. When you turn the switch off, current flow stops.

Parallel Circuit. Parallel circuits have more than one path for current flow. In Fig. 7-24 the bulbs are arranged **in parallel**. In this type of circuit, current can flow equally along different paths. A parallel circuit can be open at one point yet continue to conduct electricity. In Fig. 7-24, if one bulb burns out, the other bulb will keep

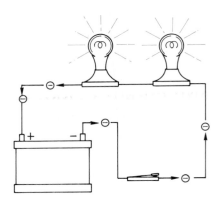

Fig. 7-23 **A series circuit provides a single path for current flow.**

Fig. 7-24 **A parallel circuit provides at least two different paths for current flow.**

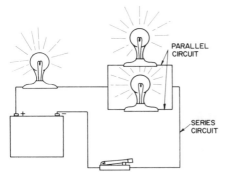

Fig. 7-25 A series-parallel circuit is a combination of series and parallel circuits.

burning. Current will flow as long as it has a complete path to follow.

Series-Parallel Circuit. In a series-parallel circuit, both series and parallel circuits are used. See Fig. 7-25. In this circuit, the switch is in series with the light bulbs. If one bulb burns out, current flow will continue. However, if the switch is opened, all current flow will stop.

ALTERNATING CURRENT AND DIRECT CURRENT

The current flow through a circuit may be one of two types. One type is **direct current**, or simply **DC**. This type has been used in all the circuits shown in this chapter so far. In direct current, the electrons flow in only one direction. The power source must continue to supply electrons at the negative terminal by taking them

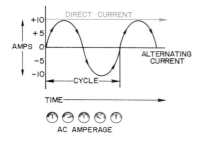

Fig. 7-26 In a 10-amp circuit, AC current and DC current will conduct the same amount of current. However, the AC current is a back-and-forth movement of electrons. The DC current is a steady flow in one direction.

away from the positive terminal. DC current is used to power many portable electrical and electronic devices, such as flashlights, radios, cassette players, and calculators. It is also the type of current used in cars for starting, ignition, and accessories.

Alternating current (AC) is the most commonly used type of current. It is used in both household and industrial circuits. In alternating current, the polarity (direction) of the current changes rapidly. See Fig. 7-26. Electrons flow first in one direction. They stop, then flow in the opposite direction. Again they stop, then flow in the first direction. This sequence makes up one **AC cycle**.

Figure 7-26 will help you understand the difference between AC current and DC current. In a certain circuit, say 10 amps, a DC current will maintain a steady 10-amp current in one direction. In the same circuit using AC current, on the other hand, the current will move back and forth in the conductor very rapidly. The ammeters below the graph in Fig. 7-26 show how the current moves from 0 to 10 amps to 0 to −10 amps, and so on. This back-and-forth movement of electrons is so fast that it produces the same effect as a steady DC current. For example, the lights in our homes seem to shine steadily. Actually, they are flickering on and off very fast.

An AC cycle always takes place in a definite period of time. Usually, there are a certain number of cycles each second. In the United States, alternating current completes 60 **cycles per second**. This is fast enough to produce a steady movement of electrons. The term *hertz* is often used instead of cycles per second. One **hertz** is one cycle per second. We can identify alternating current in the United States as *60 hertz*.

ELECTRICAL POWER SYSTEMS

Electrical power systems are all around us. We have all used both simple and very complex electrical power systems. For example, a flashlight is a simple power system. See Fig. 7-27. One or more batteries act as the *source* of power. An on-off switch *controls* the electrical current. And the *output* is the light.

The electrical system in your home is much more complex. This system gets its power input from a power company. The electrical power is then directed into circuits throughout the home. You control it with switches and other devices to operate lights and provide heat. Electricity also powers our washing machines, refrigerators, and stoves. And, it is the power source for electronic devices such as radios and televisions.

Electrical power has many other uses. It powers electrical tools such as power saws and drills. It is used in industry to operate assembly lines and machinery. Cars and trucks have electrical power systems to operate the starting and ignition systems, the lights, and the accessories. See Fig. 7-28.

Electrical power has one big advantage over mechanical power and fluid power. It can be transported over long distances with little power loss. You can find electrical transmission lines in all parts of the country. Sometimes they are hundreds of miles from the nearest power plant. The fact that electricity can be easily transported makes it our major form of power.

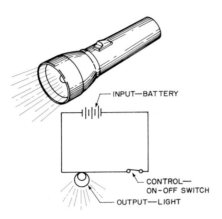

Fig. 7-27 A flashlight is a simple electrical power system. The drawing below the flashlight is a *schematic drawing* of the electrical circuit.

ELECTRICAL POWER INPUT

There are two basic types of electrical input sources. A source may be *self-contained*, or it may be a device that converts mechanical energy into electrical energy. Cells and batteries are self-contained sources. Generators are conversion devices that require a mechanical input.

Fig. 7-28 An automobile has a complex electrical system. The battery and alternator supply the electrical input. This power is controlled to produce outputs such as starting, ignition, and lighting.

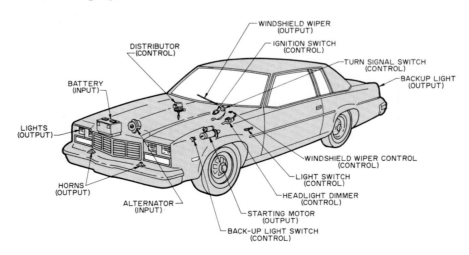

Fig. 7-29 **Common primary cells.**

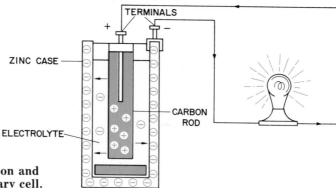

Fig. 7-30 **The construction and operation of a large primary cell.**

Cells and Batteries

Cells and batteries can be as small and thin as a dime. They can also be huge and bulky. Most are sizes somewhere in between. Whatever their size, cells and batteries play a major role in our lives. They can store energy and deliver it on demand. Cells and batteries are also easy to move. These features have made many industrial and consumer products possible. However, cells and batteries have the following disadvantages:

1. The amount of electricity available from them is quite small.
2. Electricity produced by cells and batteries is expensive.

Many people confuse the terms *cell* and *battery*. A **cell** is a device that converts chemical energy into electrical energy. Two or more cells grouped together form a **battery**. There are two basic types of cells: primary and secondary.

Primary Cells. A primary cell acts as a basic source of electricity. It produces power by direct chemical conversion. The power lasts as long as the chemicals in the cell react. The cell is then thrown away. This type of cell is referred to as *primary* because it does not require another input to produce electricity. However, once the chemicals in the cell are used up, the cell cannot be recharged.

Primary cells come in many different sizes and shapes. Figure 7-29 shows several different types of primary cells.

A typical primary cell has three parts. These are a carbon rod, a zinc case, and a paste-like substance called the **electrolyte**. See Fig. 7-30. The electrolyte reacts chemically with the zinc

to produce an electrical current. This reaction takes place whenever an outside connection permits electrons to flow from the negative terminal to the positive terminal.

The chemical action inside the cell makes the zinc case negative. The carbon rod becomes positive. There are more electrons in the case than in the rod. This represents an electrical pressure difference. This pressure difference produces voltage and causes current to flow in the circuit.

As you know, electrons always flow from negative to positive in a circuit. Inside the cell, however, electrons flow from positive to negative. This maintains the electrical pressure difference and the resulting voltage. Current continues to flow until the zinc in the cell is used up.

The flashlight "battery" is one of many primary cells commonly used. Primary cells also power calculators, digital watches, and many other items. See Fig. 7-31.

Secondary Cells. A secondary cell has three functions. First, it converts electricity into chemical energy. Second, it holds the energy in this form until electricity is needed. Third, when current is needed, the chemical energy is changed back to electricity. This type of cell is referred to as *secondary* because it must be recharged by another power source after its chemical energy is used up.

Secondary cells are often grouped together to form **storage batteries**. The most common type of storage battery is the large battery used in cars. Car batteries are usually made up of six cells connected together. Each cell consists of two groups of lead plates in an electrolyte. The

127

Fig. 7-31 **Products powered by primary cells.**

electrolyte is a mixture of water and sulfuric acid. See Fig. 7-32.

The storage battery supplies electricity to start the engine. While the engine is running, an alternator (an AC generator) *recharges* the battery. That is, the alternator supplies electricity to the battery. The battery then converts the electricity into chemical energy. It holds the energy until the engine needs to be started again.

Future Cells and Batteries

Research in developing large-capacity storage cells and batteries has increased in recent years. Power companies are looking for ways of storing electricity. They want to store it during low-use periods at night. The stored electricity would help meet high demands during the day.

Scientists are also trying to develop light-weight capacity batteries. These batteries could be used to power cars. This would help reduce our dependency on fuels such as gasoline and diesel oil.

Generators

A generator is a device that converts rotary motion into electricity. It does this by using the principle of magnetism. Earlier in this chapter, you learned that magnetism is the link between mechanical energy and electrical energy. We can use magnetism to convert motion into electricity. We can also use it to convert electricity into motion.

You also learned that passing a current through a conductor produces a magnetic field. Just the opposite happens in the generation of electricity. This time, a conductor is moved through a magnetic field. This action produces a current in the conductor.

Fig. 7-32 **A phantom view of a modern automobile battery. In this battery, six secondary cells are connected to produce about 12 volts of electricity.**

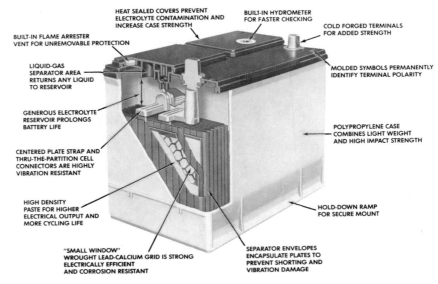

HEAT SEALED COVERS PREVENT ELECTROLYTE CONTAMINATION AND INCREASE CASE STRENGTH

BUILT-IN HYDROMETER FOR FASTER CHECKING

BUILT-IN FLAME ARRESTER VENT FOR UNREMOVABLE PROTECTION

COLD FORGED TERMINALS FOR ADDED STRENGTH

LIQUID-GAS SEPARATOR AREA RETURNS ANY LIQUID TO RESERVOIR

MOLDED SYMBOLS PERMANENTLY IDENTIFY TERMINAL POLARITY

GENEROUS ELECTROLYTE RESERVOIR PROLONGS BATTERY LIFE

POLYPROPYLENE CASE COMBINES LIGHT WEIGHT AND HIGH IMPACT STRENGTH

CENTERED PLATE STRAP AND THRU-THE-PARTITION CELL CONNECTORS ARE HIGHLY VIBRATION RESISTANT

HIGH DENSITY PASTE FOR HIGHER ELECTRICAL OUTPUT AND MORE CYCLING LIFE

HOLD-DOWN RAMP FOR SECURE MOUNT

"SMALL WINDOW" WROUGHT LEAD-CALCIUM GRID IS STRONG ELECTRICALLY EFFICIENT AND CORROSION RESISTANT

SEPARATOR ENVELOPES ENCAPSULATE PLATES TO PREVENT SHORTING AND VIBRATION DAMAGE

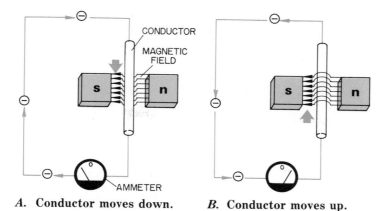

A. Conductor moves down.   B. Conductor moves up.

Fig. 7-33 **An electric current can be produced by passing a conductor through a magnetic field.**

Figure 7-33 shows a simple way of producing current in a conductor. In *A*, the conductor is moved down through a magnetic field. (The N and S poles shown are from a single magnet. The rest of the magnet is not shown in order to keep the drawing simple.) As the conductor cuts through the lines of force, a current is developed, or **induced**, in it. The movement of the ammeter needle shows that current is flowing.

Notice that a current is also generated when the conductor is moved up through the magnetic field (Fig. 7-33B). However, this current moves in a direction opposite the current in *A*. The ammeter needle shows this by moving in the opposite direction. Because the current flows in one direction, then the other, we have an alternating current.

Fig. 7-34 **Rotating a loop through a magnetic field produces current that continually reverses direction.**

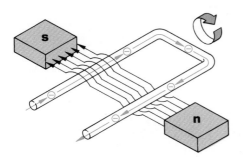

A more efficient way to generate alternating current is to rotate a loop of wire between the poles of the magnet. See Fig. 7-34. Rotating the loop produces an almost continuous current (depending on how fast the loop is rotating). As one side of the loop cuts through the magnetic field, a current is generated. As the other side cuts the field, a current is generated in the opposite direction.

Figure 7-35 shows the working of an AC generator in more detail. It also shows how the alternating current is collected from the generator. Notice the **slip rings** and **brushes**. Each slip ring is attached to one end of the conducting loop. The rings rotate with the loop as it turns. The brushes are special conductors that rest on the slip rings but do not turn with them. The brushes provide a path for the current output. There is always current flowing in or out of the brushes, depending on the position of the rotating loop.

To get a good idea of how the generator works, follow one side of the loop as it rotates. Do you see how the current reverses as the loop rotates? Remember, the slip rings turn with the loop.

What we have, then, with our AC generator, is a power source for an electrical circuit. As the loop rotates, current flows through the circuit at the speed of light — first in one direction, then the other. The reversal of current is so fast that the current seems to be continuous.

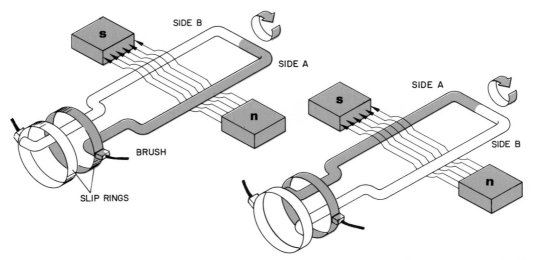

A. Side A moves down through the magnetic field. The current flows in the direction shown.

B. Side A moves up through the magnetic field. The current flows in the opposite direction.

Fig. 7-35 A simple AC generator consists of a magnet, a conductor, slip rings, and brushes. Follow side A and its current flow as the conductor rotates.

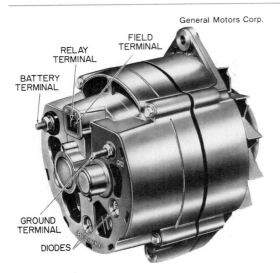

General Motors Corp.

Fig. 7-36 Modern cars produce electrical power with an alternator. The engine's fan belt drives the alternator.

Fig. 7-37 These AC generators are housed in Hoover Dam Power Plant in Nevada. Each generator can produce 82,500 kilowatts of electrical power.

Probably the most common AC generator is the **alternator** used in most modern cars. See Fig. 7-36. Alternators generate AC current. Then control devices called **diodes** inside the alternator change the AC current into DC current. Auto electric systems run only on DC current. (Diodes are explained later in this chapter.)

There are generators that produce direct current. In fact, DC generators were used extensively in cars at one time. However, AC generators are simpler and produce more electricity.

Bureau of Reclamation

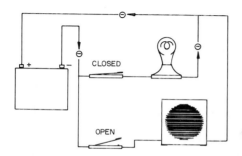

Fig. 7-38 In this circuit, the electrical output units (the lamp and the heater) are placed *in parallel*. The switches in the circuit are placed *in series* with the output units.

Fig. 7-39 A single-pole, single-throw knife switch.

Fig. 7-40 Common on-off switches.

MCI

CONTROL OF ELECTRICAL ENERGY

To be able to do work with electricity, we must have some way to control it. We do this with circuits and special control devices.

As you know, a circuit is the path that electricity takes as it travels to and from the power source. The electricity must pass easily through the circuit. Therefore, circuits are made of electrical conductors.

Copper is an excellent conductor. It is also the one most commonly used. Aluminum is another common conductor. However, to conduct the same amount of current as copper wires, the aluminum wires must be larger. Aluminum wires do have one advantage over copper wires, though. They weigh less. Therefore, aluminum is a better conductor than copper — by weight.

Aluminum wires are especially useful in long-distance power transmission. With aluminum wires, the support towers can be further apart. This makes the cost of an aluminum transmission system lower.

We maintain our control of electrical current with many different control devices. These devices provide four basic types of control:

- On-off control
- Directional control
- Overload control
- Current and voltage control

On-Off Control

Circuits usually have their electrical output units placed in parallel. This way, the units can work independently of each other. For example, the circuit in Fig. 7-38 has two output units. An on-off control device (a switch) is placed in series with each output unit. The switch that controls the lamp is closed. This allows current to flow through the lamp. The switch in line with the heater is open. This cuts off the current to the heater.

We will look at three basic types of on-off switches. These are manual switches, relays, and transistors.

Manual Switches. Manual switches are the most common electrical control devices. Figure 7-39 shows the simplest type of manual switch: the **knife switch**. This switch is a **single-pole, single-throw switch**. The moveable arm attached to one contact is the single pole. To close the switch, you "throw" the arm to the single remaining contact.

On-off switches come in many styles. See Fig. 7-40. In **push-button** types, current flows only when the button is pushed. Other switches operate with spring tension. The light switch used in most homes is a good example. This switch is called a **toggle switch**. Spring tension opens and closes the contact points quickly. This "snap action" keeps the contacts from burning during opening or closing.

Relays. A relay is used to control a circuit from a remote (distant) location. Relays can also serve as automatic control devices.

Figure 7-41 shows the construction of a typical relay and how it fits in a circuit. The relay has three basic parts. These are a soft iron core inside a coil of wire, a moveable piece of metal called an **armature**, and a set of switch contacts.

When you have a relay in a circuit, you basically have two switches. When the operator closes the main switch, current flows through the coil. The iron core becomes an electromagnet. The electromagnet pulls the armature down so that it makes a "bridge" across the contacts. This allows current to flow through the entire circuit and start the motor.

The magnetic attraction stops when the operator opens the main switch. A strong spring then pulls the armature up to open the contacts. This breaks the circuit and stops the motor.

Large relays are used to control circuits that use large amounts of current. In these cases, the hand switch that operates the relay can be located some distance from the motor or other output unit. The relay itself needs very little current to operate. Figure 7-42 shows an automotive voltage regulator. It uses two relays to control the current and voltage of the car's electrical system.

Relays can also be operated by switches controlled by clocks. Thermostats, light meters, and other sensing devices are also used to activate relays. These devices work by connecting two contacts to complete a circuit. This activates the relay. Many automatic circuits operate in this way. Relays operate burglar alarms, furnaces, refrigerators, and many other electrical devices.

Transistors. Transistors are electrical devices that can be used as on-off switches. They control circuits just like relays do. However, transistors have no moving parts. This is a great advantage. Moving parts can wear out or require servicing. Transistors operate entirely on electricity. They have an almost unlimited useful life.

Transistors are one of several types of **solid-state** devices. This means that they are made entirely of solid material. There are two kinds of solid material used in transistors: insulators and semiconductors.

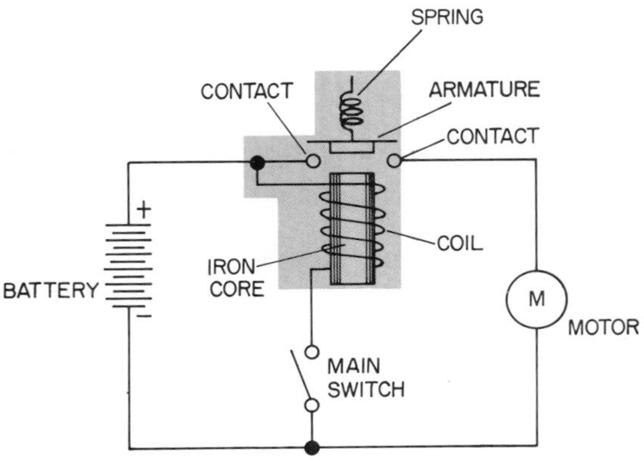

Fig. 7-41 **This schematic drawing shows how a relay fits in an electrical circuit. (The relay is made up of all the parts in the shaded area.)**

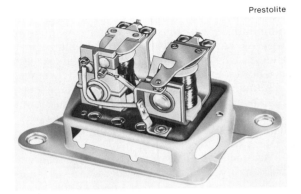

Prestolite

Fig. 7-42 **An automotive voltage regulator consists of relays that control the current entering and leaving the alternator.**

Fig. 7-43 A modern transistor.   **133**

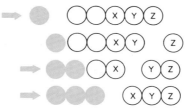

Fig. 7-45 **The direction of hole flow is opposite that of electron flow. This is shown by the row of billiard balls. When a ball (electron) is added on the left, a ball leaves on the right. This forms a hole. As more balls are added, the hole moves to the left.**

SILICON
WITH
ALUMINUM
IMPURITY

P   N   P

SILICON
WITH
ARSENIC
IMPURITY

EMITTER   COLLECTOR
(E)   (C)

BASE (B)

E   C

B

A. Construction   B. Symbol

Fig. 7-44 **This figure shows the make-up of a silicon transistor. The transistor symbol is used in electrical notation.**

Up to this point, we have looked only at materials that are good electrical conductors. However, there are also many materials that are just the opposite of conductors. These **insulators** resist the flow of electrons. Insulators have no free electrons. Therefore, they cannot transmit electric current.

There is also a group of materials between conductors and insulators. This group has some of the properties of conductors and some of the properties of insulators. We call these materials **semiconductors**. Scientists did not find a practical use for semiconductors until 1948. Up to that time, they thought that semiconductors were both poor insulators and poor conductors. There was no need for an "in-between" material.

Today, semiconductors are used in transistors and many other devices. The two semiconductors most commonly used are silicon and germanium. Both of these elements form crystals. Figure 7-43 shows a modern transistor. Figure 7-44 shows the construction of a silicon transistor.

In its pure form, silicon is an insulator. However, the silicon used in transistors is not pure. Small amounts of an impurity are added. Arsenic and aluminum are two common impurities used in transistors. Arsenic causes silicon to free some electrons. This makes the silicon a conductor. The free electrons are able to flow. They also give the silicon a negative charge. Therefore, the part of the transistor that contains arsenic is labeled **N.** See Fig. 7-44 again.

Adding *aluminum* to silicon makes it react in a different way. This form of silicon can *accept* free electrons. It is labeled **P** (positive) in Fig. 7-44. Notice that there are two pieces of P material and one piece of N material. These materials produce a **PNP** transistor. Transistors may also be made with two pieces of N material and one piece of P material (NPN). The three materials make up the **emitter (E), base (B),** and **collector (C).** There are three wires, one connected to each of the three parts.

Transistors and Electron Flow. Before you learn how transistors and other solid-state devices work, you must review the movement of electrons through a conductor. Remember, electrons flow from negative to positive. As one electron enters the negative side of a conductor, one electron leaves the positive side.

Solid-state devices operate on the principle of **hole flow** instead of electron flow. Hole flow is in the opposite direction of electron flow. The principle of hole flow is illustrated in Fig. 7-45.

Each billiard ball represents an electron. As one ball joins the line of balls, one ball moves from the right. This leaves a hole that is filled by the next ball. As more balls (electrons) are added, the balls (electrons) combine to move to the right. As you can see, electrons move to the right while the hole moves to the left.

Transistor Operation. Figure 7-46 shows how a transistor works. The **emitter-collector (E-C) circuit** is the main path of the current flow. For current to flow through this path, a small amount of current must flow through the **emitter-base (E-B) circuit**. If no current flows through the E-B circuit, no current flows in the E-C circuit. Therefore, the E-B circuit can be used as a switch. It turns the E-C circuit on or off.

Fig. 7-46 **A transistor operates when a small amount of current flows in the E-B circuit. The E-B circuit acts as a switch to activate the E-C circuit. The arrows show the direction of *hole flow*, not electron flow.**

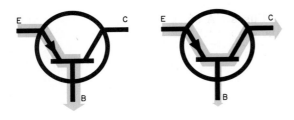

Figure 7-47 shows a transistor in a circuit. When the main switch is closed, electrons flow from the battery to the base of the transistor. From there, the electrons flow to the emitter. This completion of the E-B circuit activates the E-C circuit, which in turn allows current to start the motor. Notice that the transistor serves the same purpose as a relay.

Compare the use of the transistor in Fig. 7-47 with the use of the relay in Fig. 7-41. Do you see how much simpler the transistor is? Transistors take up less space, are lighter, and have no moving parts. Many jobs formerly done with relays are now done with transistors.

Transistors have many uses besides acting as on-off switches. They have replaced the vacuum tubes that were once used in radios, televisions, and early computers. Hundreds of transistors can fit on tiny chips. These chips are used in digital watches, calculators, and computers. Many products can be made almost completely with transistors and other solid-state parts.

Directional Control

Another type of solid-state device is the diode. **Diodes** permit current to flow through them in only one direction. They act much like fluid power check valves. These valves allow liquid or gas to flow in only one direction.

Figure 7-48 shows several typical diodes. Figure 7-49 shows diode construction and the diode symbol. The arrow in the symbol points in the direction of *hole movement*, not electron flow.

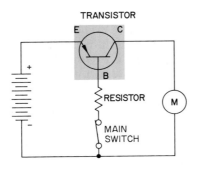

Fig. 7-47 **Using a transistor as a switch in an electrical circuit. The resistor reduces the current flow to the base of the transistor.**

Schauer Manufacturing Corp.

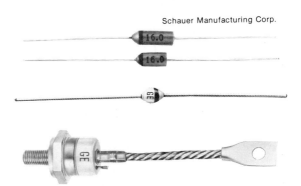

Fig. 7-48 **Modern diodes are available in many shapes and sizes.**

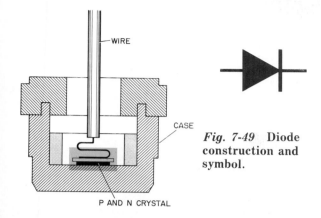

Fig. 7-49 **Diode construction and symbol.**

Diodes are made by joining a small N-type semiconductor to a P-type semiconductor. In Fig. 7-49, the P and N crystal is below the wire near the bottom of the diode. The case and surrounding parts protect the crystal from damage.

A diode has only two connections. The wire leading up from the diode acts as one connection. The case acts as the other. In many diodes, the crystal is insulated from the case. This type of diode has a second wire leading from the case.

Diodes can be used to change alternating current (AC) into direct (DC). This process is called **rectification**. As you know, AC current moves back and forth. Diodes allow only the electrons moving in one direction to flow through the circuit. The resulting DC current can be used to charge batteries. See Fig. 7-50. This use for diodes has made rechargeable batteries practical for many devices. Computers, electric shavers, carving knives, and electric toothbrushes are examples. The most common use of rechargeable batteries is in cars.

Overload Control

Fuses and **circuit breakers** are switches that open quickly when too much current passes through a circuit. Excessive current can result from a short circuit or an overload.

A **short circuit** is produced whenever the electrons are allowed to return to the source without passing through the complete circuit. That is, the electrons *bypass normal circuit resistance.* See Fig. 7-51. Generally speaking, any electrical device, such as a motor, lamp, heater, or radio, is referred to as a **resistor**. In Fig. 7-51 the light bulbs are the resistors.

When resistors are bypassed, the electron flow through the circuit is very heavy. Excessive current flow can also be caused by "loading" the circuit with too many output units. In either case, the result is an **overload** of current.

An overload produces excessive heat. In turn, the heat can burn the insulation from wires. In a home or industrial plant, the heat developed by an overload can ignite the insulation and cause a serious fire.

Fuses. A fuse protects a circuit from dangerous overloads. Figure 7-52 shows how a fuse works. In *B* the amperage of the shorted circuit is greater than the capacity of the fuse. The heat produced by the current melts a metal link inside the fuse. This opens, or "breaks," the

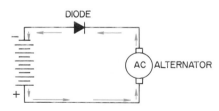

Fig. 7-50 **Batteries made up of secondary cells can be recharged by forcing direct current through them from negative to positive. The diodes in an automobile alternator change AC current to DC current to charge the battery.**

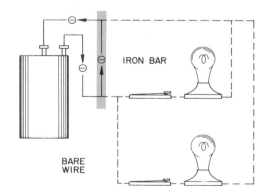

Fig. 7-51 **A short circuit permits current to bypass the normal circuit resistance. In this circuit, the resistors are the two light bulbs.**

circuit. All current flow stops. You must replace the fuse before the current will flow again.

There are several different types of fuses. Figure 7-53 shows two types used in household circuits.

Circuit Breakers. A circuit breaker has the same job as a fuse. However, you can reset it by hand after the circuit is broken. See Fig. 7-54.

There are several circuit breaker designs commonly used. One kind works like a relay.

Fig. 7-52 The connection inside a fuse breaks whenever the circuit is overloaded.

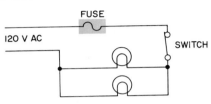

A. Current flow normal — fuse completes circuit.

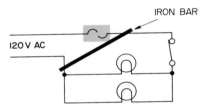

B. Current flow excessive — fuse melts and breaks circuit.

Fig. 7-53 Two common types of fuses.

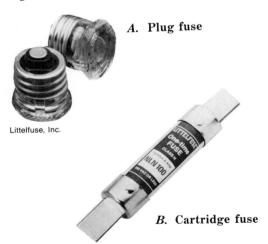

Littelfuse, Inc.

A. Plug fuse

B. Cartridge fuse

Fig. 7-54 This type of circuit breaker is used in many home electrical circuits. It operates much like a relay.

General Electric Co.

Fig. 7-55 The operation of a circuit breaker.

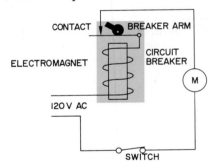

A. Circuit breaker closed — motor operates.

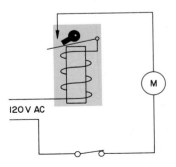

B. Circuit breaker open — motor stops.

Figure 7-55 shows a circuit that uses this type of circuit breaker. In *A* the current is flowing and the motor is running. The current flow may become too great. In this case the magnetic field of the circuit breaker coil will become strong enough to pull down the breaker arm. See Fig. 7-55B. When this happens, the circuit breaks and the motor stops. To start the motor, some-one must reset the circuit breaker. This is done by moving a switch or pushing a button on the circuit breaker.

All circuits are protected by either fuses or circuit breakers. If you replace either kind of device, you must use a fuse or circuit breaker of the proper capacity. An over-capacity fuse provides little protection against fire or other damage.

Current and Voltage Control

Besides starting and stopping current flow, we must often vary the amount of voltage and current. Several types of devices have been developed to provide this control. They include transformers and resistors.

Transformers. A transformer changes the ratio of the voltage and amperage in a circuit. For example, electrical power plants generate electricity at about 2200 volts. This voltage is too low for long-distance power transmission. Therefore, the power company uses transformers to boost the voltage to about 220,000 volts. At the same time, the transformers lower the cur-

Fig. 7-57 **This small transformer reduces voltage from 120 volts to 10 volts. The lower voltage is used to operate a doorbell.**

Trine Manufacturing Co.

rent's amperage. High-voltage current loses less power over long distances than high-amperage current does. When the electricity reaches its area of use, other transformers reduce its voltage and raise its amperage. For home use, the voltage is reduced to 240 volts and 120 volts. Whenever there is a change in voltage, there is a change in amperage. If voltage increases, amperage decreases. If voltage decreases, amperage increases.

Transformers are available in many sizes. Figure 7-56 shows a very large transformer. A power company uses it to change the voltage of millions of watts of electrical power. Figure 7-57 shows a small transformer used to obtain the voltage and amperage needed to operate a doorbell.

Transformer Operation. Transformers operate on the principle of *induction*. They transmit energy from one circuit to another. In the process, they modify the energy.

Transformers have two sets of windings. These windings are called the **primary coil** and the **secondary coil**. Both coils are usually wrapped around a soft iron core.

Figure 7-58 shows how a transformer is part of two different circuits. The circuit that includes the primary coil is called the **primary circuit**. The circuit that includes the secondary coil is called the **secondary circuit**.

Note that the primary coil is not connected to the secondary coil. However, the two coils are very close to each other. When the switch is closed, a magnetic field builds up in the primary coil. This field is called the **primary field**. As it builds up, the primary field cuts through the secondary coil. This action induces a current in the secondary circuit. The current lights the lamp. See Fig. 7-59.

Fig. 7-56 **A large transformer in an electrical substation. Electric fans cool the transformer during operation.**

Illinois Power Co.

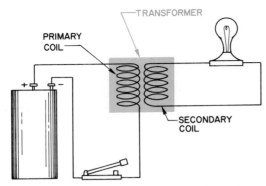

Fig. 7-58 A transformer connects two circuits. Notice that each coil is part of a different circuit.

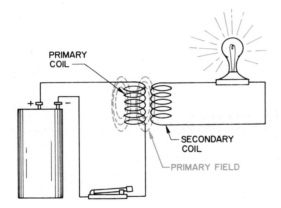

Fig. 7-59 Induction takes place when the primary circuit is connected or disconnected by a switch. Each build-up or collapse of the primary field lights the bulb for an instant.

Induction takes place only when the electromagnetic field moves. Therefore, the lamp lights up only for a moment. Opening the switch will cause the primary field to collapse. As the field collapses, it will again cut through the secondary coil. This will cause the light to burn for another instant.

To induce a current, the magnetic field must either build up or collapse. That is, the current to the primary coil must be switched on and off. Therefore, to get a constant induction of electricity, a transformer circuit cannot use uninter-

rupted direct current. This type of current would not induce electricity in the secondary circuit.

Using alternating current, on the other hand, will result in a continuous induction of current. Alternating current constantly changes direction. Therefore, it produces an electromagnetic field that is either building up or collapsing at all times. This happens so fast that the transformed electrical output seems to be continuous.

Step-up Transformers. In any transformer, the primary coil and secondary coil have different numbers of windings. The ratio of the number of windings is one thing that affects the output voltage. The other factor is the amount of input voltage.

In Fig. 7-60 the ratio of the primary and secondary windings is 10:100, or 1:10. The voltage induced in the secondary coil will be 10 times the voltage in the primary coil. On the other hand, the amperage of the secondary coil will be only 1/10 of the input amperage. A transformer that changes current in this way is called a **step-up transformer**. It increases (steps up) voltage and decreases amperage. However, as you can see from Fig. 7-60, it does not change the amount of *power* in the system. (You will remember that wattage equals volts times amps.)

Try to keep this point in mind: Control devices can change the *characteristics* of power, such as voltage and amperage. However, con-

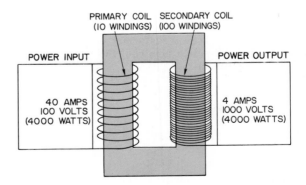

Fig. 7-60 A step-up transformer increases voltage and decreases amperage. However, it does not change the amount of *power* in the system.

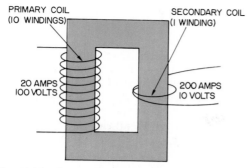

PRIMARY COIL
(10 WINDINGS)

SECONDARY COIL
(1 WINDING)

20 AMPS
100 VOLTS

200 AMPS
10 VOLTS

Fig. 7-61 A step-down transformer decreases voltage and increases amperage.

Ohmite Manufacturing Co.

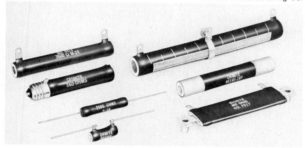

Fig. 7-62 Common resistors.

trol devices cannot change the total *amount* of available power. The only way to increase power is to provide additional energy from another source.

In transformers, some power is lost during induction in the form of heat. However, this loss is quite low. Transformers can change voltage and amperage with very little loss of energy.

Step-down Transformers. Figure 7-61 shows a step-down transformer. This type of transformer has fewer loops in the secondary coil. This arrangement decreases (steps down) the voltage from the primary coil to the secondary coil. At the same time, the transformer increases amperage. The wattage remains the same.

Resistors. In a circuit, resistance converts electrical energy into heat energy. All electrical output devices offer resistance to current flow. However, resistance can also be deliberately added to a circuit to provide control. Special resistors are used for this purpose. These devices have one main function: to reduce current flow.

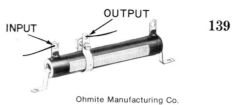

OUTPUT

INPUT

139

Ohmite Manufacturing Co.

Fig. 7-63 On this type of variable resistor, the resistance is varied by moving the center terminal back and forth.

Figure 7-62 shows different types of commonly used resistors. Most resistors consist of an insulated and sealed coil of wire. The wire resists current flow.

Resistors are rated in watts. The watt rating indicates the amount of heat that the resistor can safely transfer to the surrounding air. The watt rating also determines the capacity of the circuit in which the resistor can be used. For example, a 10-ohm, 10-watt resistor provides 10 ohms of resistance. However, it can be safely used only in circuits carrying up to 10 watts.

Some resistors provide an adjustable resistance. This type of resistor is called a **variable resistor**. Other names for a variable resistor are **potentiometer** and **rheostat**.

Figure 7-63 shows the basic construction of a variable resistor. The farther the current must travel through the coiled wire, the greater the resistance. Moving the sliding contact toward the input reduces the length of wire through which the current must pass. This in turn reduces the resistance.

TRANSMITTING CURRENT

Electricity is transmitted more efficiently than any other common form of power. As a result, electricity is used to carry power to all parts of our nation.

Most current-carrying wires consist of a conductor surrounded by an insulating material. See Fig. 7-64. The nonconducting insulation prevents current loss through short circuits. Most plastics make good insulators. They are used to insulate the wiring in electric motors. Often, "bare" copper wire is really wire insulated with a clear plastic.

Although most wires are insulated, some are not. Large power lines are not insulated. They

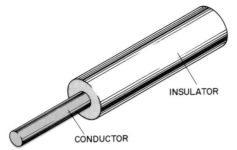

Fig. 7-64 **An insulated electrical conductor.**

Illinois Power Co.

Fig. 7-65 **Electrical power is transmitted over long distances through large-capacity conductors.**

are only insulated from the towers supporting them.

A high-amperage flow produces heat in a conductor. This is due to the friction produced by the moving electrons. Power companies always transmit large amounts of electricity long distances at very high voltages and low amperages. See Fig. 7-65.

ELECTRICAL POWER OUTPUT

We can group the uses of electrical power into four major categories:
- Motion
- Heat
- Light
- Communications

Notice that three of the categories listed above involve converting electrical energy into another form. Electricity can be converted into mechanical energy, heat energy, or light energy. Read on to find out just how these conversions are carried out.

Conversion into Mechanical Energy (Motion)

Electricity is itself a form of motion — the movement of electrons. We can convert this motion into two kinds of mechanical motion. Two different types of electrical devices are used to make the conversion. **Solenoids** change electricity into linear motion. **Motors** change electricity into rotary motion.

Solenoids. A solenoid is a simple variation of the electrical relay. It is basically a relay with a moveable core, or **plunger**. See Fig. 7-66. The plunger is normally held away from the center of the coil with a spring. When current is sent to the coil, a magnetic field is created. The magnetic lines of force flow around and through the coil.

Fig. 7-66 **A solenoid consists of a coil and a metal plunger. When current passes through the coil, an electromagnetic field is created. The magnetic field then pulls the plunger inside the coil.**

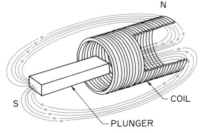

A. **Non-activated**

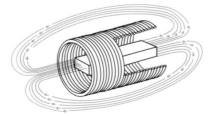

B. **Activated**

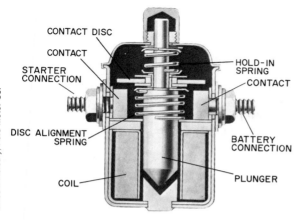

CONTACT DISC
CONTACT
STARTER CONNECTION
HOLD-IN SPRING
CONTACT
DISC ALIGNMENT SPRING
BATTERY CONNECTION
COIL
PLUNGER

Lincoln-Mercury, Ford Motor Co.

Fig. 7-67 This automobile solenoid is operated by the ignition switch. When the driver turns the key, a small current flows to the coil and creates a magnetic field. The field pulls the plunger and contact disc down. A strong battery current then flows from the battery contact to the starter contact.

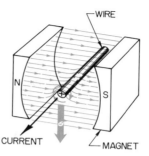

WIRE
N
S
CURRENT
MAGNET

Fig. 7-68 Inside a motor, the lines of force surrounding a current-carrying conductor react with another magnetic field. This reaction produces a weaker field below the conductor. The conductor then moves toward this weaker field.

As described earlier, magnetic lines of force travel more easily through metal than they do through air. They also try to take the shortest path through the field. When the solenoid is switched on, the lines of force first take the longer path through the iron plunger. See Fig. 7-66A. Then the magnetic field develops enough force to pull the plunger into the center of the coil. See Fig. 7-66B. The two actions take place very quickly.

There is a lot of force pulling on the plunger. We can use this force to perform linear work. For example, solenoids are used to control fluid power valves and to engage gears. Another common use is to make contact between two heavy electrical contact points. For example, cars use a solenoid to make the connection between the battery and the starter. See Fig. 7-67. This connection allows 100 to 300 amps to pass from the battery to the starter.

Motors — Principle of Movement. Electric motors are very similar to electrical generators. There is one major difference. Generators convert mechanical energy into electrical energy. Motors do just the opposite.

Figure 7-68 shows a magnetic field with a current-carrying conductor placed inside of it. A current-carrying wire has its own magnetic field. When you place this wire inside another magnetic field, the two fields react to each other.

In Fig. 7-68 the lines of force around the wire are in the same direction as the lines of force above the wire. This causes the lines of force to combine and form a strong magnetic field above the wire.

The opposite thing happens below the wire. Here the wire's lines of force are in the opposite direction of the magnet's lines of force. This causes the lines of force to cancel each other out. With a strong field above it and a weak field below it, the wire moves down.

Suppose we take the wire in Fig. 7-68 and shape it in a loop. The current passing through the wire will generate lines of force in opposite directions on the two sides of the loop. See Fig. 7-69.

If we then place the loop inside the magnet's field again, it will start to rotate. See Fig. 7-70.

Fig. 7-69 Passing a current through a wire loop develops lines of force around the wire. These lines of force travel in directions opposite each other.

DIRECTION OF
LOOP ROTATION

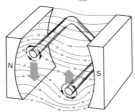

Fig. 7-70 **A current-carrying wire loop placed between the poles of a magnet will rotate in the direction shown here.**

One side of the loop reacts just like the wire in Fig. 7-68. It is forced down by the strong force above it. The other side of the loop reacts in the opposite way. It is forced *up* by the strong magnetic force *below* it. The two sides react differently because of the opposite directions of their lines of force.

However, the loop will not rotate all the way around. It will stop halfway through the rota-

tion. To see why this is, we have to look at the lines of force in Fig. 7-70 again. As soon as the left side of the loop tries to start up on the other side, its lines of force strengthen the field above it. Therefore, it will be pushed down instead of up. The same thing happens with the right side of the loop as it tries to complete a rotation.

To keep the loop rotating, we must have some way of reversing the lines of force when the loop is halfway through a rotation. To reverse the lines of force, we have to reverse the current flow.

Motors — Current Reversal. The method used to reverse the current flow depends on the type of motor. DC motors use a **commutator**. A commutator is a rotating conductor. Figure 7-71 shows a simple commutator. It consists of two halves of a metal ring. The commutator is attached to the ends of the loop and rotates with the loop. Special conductors called **brushes** ride against the commutator. One brush delivers

---

Tesla and the Induction Motor

Every time you use something powered by an AC motor, you should thank Nikola Tesla. This man was an electrical engineer who lived from 1856 to 1943. Even as a child, Tesla experimented with electricity. He dreamed of someday inventing a machine that would make work easier. Later, in the 1880s, Tesla made the dream a reality by inventing the alternating-current induction motor.

Between 1885 and 1887, Tesla worked for Thomas Edison. Edison thought that everyone should use DC current and DC motors. Tesla tried to convince Edison that AC motors would be best. However, Edison disagreed, and the two men went their separate ways.

What was so great about the induction motor? Well, basically it was simpler than a DC motor. It did not require a commutator. Instead, the induction motor used alternating current to energize **field magnets** arranged around a central **armature**. The current produced a rotating magnetic field in the magnets. This field induced another magnetic field in the armature and caused it to turn.

In 1888, George Westinghouse heard about Tesla's new motor. He paid Tesla *one million dollars* for the right to manufacture AC motors! Tesla made several

improvements in the motor, and Westinghouse sold thousands of them.

Nikola Tesla truly had a part in shaping the world as it is today. The induction motor is still one of the major power devices used in industries around the world.

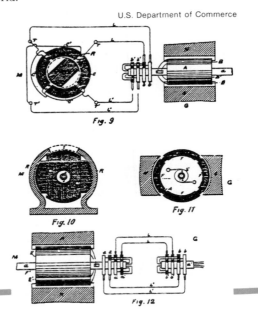

U.S. Department of Commerce

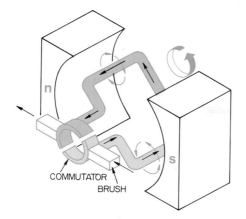

COMMUTATOR
BRUSH

Fig. 7-71 In a DC motor, a commutator reverses the direction of current flow in the wire. This reversal happens when the rotating wire is 90° vertical to the brushes.

current to the loop. The other brush carries current away from the loop.

When the loop is halfway through a rotation, the current flow stops briefly. This is because of the split in the commutator. The loop keeps moving a short distance because it has a strong force behind it. But before it starts up on the other side, the current flow reverses. This in turn reverses the lines of force around the sides of the loop. Therefore, the loop continues to rotate. Once the loop gets going, the current

reversal is very fast. You really can't notice a break in the current flow.

AC motors do not need a special device to reverse the current. This is because AC current constantly reverses itself. Read the box story on page 142 to find out more about AC motors.

Motors — Basic Construction. The magnetic field of the conductor in a motor can be strengthened in two ways. First, the loop can be coiled. If you recall, a coiled wire has the same magnetic effect as several conductors placed close together. Second, an iron core can be added. This will concentrate the magnetic field. Together, the iron core and coiled conductor form an electromagnet.

Figure 7-72 shows a conductor that has been strengthened in both of the ways mentioned. The coil and iron core have also been mounted on a support shaft so that the assembly can rotate. The entire moving piece is called an **armature**. Because the magnet provides a magnetic field, it is simply called the **field**.

The support shaft for the armature provides the mechanical output of the motor. We can attach gears or pulleys to the shaft to put the rotary motion to work.

Figure 7-73 shows an actual motor that can run on either AC or DC current. As you can see, the arrangement of the parts is different from the figures we have been looking at. The **field windings** provide an electromagnetic field in which the armature rotates. The armature itself is mounted lengthwise on the **armature shaft**.

Fig. 7-73 The armature shaft of this motor provides rotary motion. The motion is transferred through gears to an output shaft.

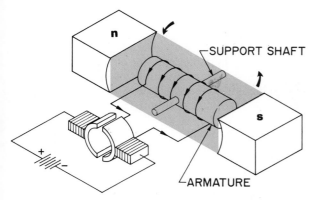

SUPPORT SHAFT

ARMATURE

Fig. 7-72 Wrapping a current-carrying conductor around an iron core provides a strong electromagnetic field. The entire assembly — the *armature* — rotates inside the other magnetic field provided by the magnet.

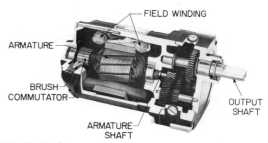

FIELD WINDING
ARMATURE
BRUSH
COMMUTATOR
ARMATURE SHAFT
OUTPUT SHAFT

Bodine Electric Co.

Notice that the armature has many windings. The commutator also has many divisions. This allows it to reverse the current flow through each set of wire loops. Each loop has its own pair of commutator segments.

Conversion to Heat and Light

In the home, we use electrical heat to prepare food, dry clothes, and heat air and water. Heat is used in industry to liquify metals and plastics. Heat can also change the properties of many other materials. Therefore, it is important in many manufacturing processes.

We use **high-resistance wire** to change electricity to heat. Passing current through this type of wire produces heat. For example, heaters and toasters use resistance wire to produce heat. Other items, such as electric stoves, use **resistance coils**. See Fig. 7-74.

Passing current through high-resistance wire can also produce light. A light bulb has a piece of high-resistance wire in it. Passing a current through this **filament** produces enough heat to make it glow white-hot. See Fig. 7-75. As you know, lighted bulbs get very hot. The filament is sealed in a vacuum. The vacuum keeps oxygen from coming into contact with the filament. Oxygen would oxidize and break the filament.

Use in Communication

Electricity plays an important part in helping us communicate. Many of our modern ways

Fig. 7-75 **In a light bulb, current passes through a thin wire, or filament. The filament's resistance to current flow makes it white-hot. This produces light.**

General Electric Co.

of transmitting information directly involve electricity. Telephones, radios, televisions, and computers all use electricity. This book does not have room for descriptions of how these devices operate. However, you can go on to study communication devices in electronics courses.

Doorbells, buzzers, and door chimes are simple communication devices. They all use an electromagnetic coil. (The electromagnetic coil was described earlier under *Relays*.)

In a simple doorbell circuit, current passes through a set of contacts connected to the clapper arm. From the contacts, the current flows through the coil, then back to the source.

When the current flows, the coil pulls the clapper arm against the core. This causes the clapper to hit the bell. This movement also opens the contacts, which in turn disconnects the circuit. With the coil field weakened, a spring returns the arm to its original position. This brings the contacts together. Current then restores the coil's magnetic field. This cycle repeats as long as electricity is available. The result is a continuous ringing of the bell.

A buzzer works in the same way. In a buzzer the contact strip vibrates against the coil's core. This makes a buzzing sound.

A door chime uses a solenoid instead of a relay type of electromagnet. When current flows, the solenoid coil quickly pulls on the plunger. The plunger shoots into the center of the coil. This movement causes a hammer to strike a chime bar or tube.

Doorbells, buzzers, and chimes are all controlled by a switch in another location. The small coils and push-button switches operate on a low-voltage current. A small transformer reduces the 120-volt house current to 10-24 volts.

Fig. 7-74 **An electric range produces heat by passing large amounts of electricity through resistance coils. The user controls the amount of electricity with the knobs.**

The Tappan Co.

STUDY QUESTIONS

1. Name the three tiny particles that combine to make up atoms.
2. What is the electrical charge of an electron?
3. What is electricity?
4. What is electrical current?
5. Describe in your own words how an electrical source produces a flow of electricity in a conductor.
6. How fast does electricity travel through a conductor?
7. How do unlike magnetic poles react to each other? How do like poles react to each other?
8. What is a magnetic field?
9. Tell how the magnetic field of a coil of wire can be strengthened.
10. Define *amperage*.
11. Define *voltage*.
12. How is electrical resistance measured?
13. What is the mathematical relationship between volts, amps, and ohms?
14. What does it take to produce one watt of power?
15. What is the difference between a parallel circuit and a series circuit?
16. What is the difference between direct current and alternating current?
17. What characteristic of electricity makes it our major form of power?
18. What is the difference between a cell and a battery?
19. What is the difference between a primary cell and a secondary cell?
20. What is a generator?
21. In your own words, describe how a generator changes rotary motion into electric current.
22. Why is aluminum used for long-distance power transmission instead of copper?
23. True or false: Circuits usually have their electrical output units placed in series.
24. What is a relay used for?
25. Why is a transistor called a *solid-state* device?
26. What is a semiconductor?
27. What is the difference between electron flow and hole flow in an electrical current?
28. What is a diode?
29. What is the function of fuses and circuit breakers?
30. What is the function of a transformer?
31. What principle do transformers operate on?
32. What is the function of a resistor?
33. What is a solenoid?
34. Briefly describe how a motor produces motion from electricity.
35. What is used to change electricity to heat?

ACTIVITIES

1. Prepare a list of different home devices that operate on electricity. Place each device in a category according to how the device uses electricity. The five categories are:

   - Motion (such as an electric mixer)
   - Heat (such as a toaster)
   - Light (such as a lamp)
   - Communication (such as a radio)
   - Combination uses (such as a dryer, which produces both motion and heat)

2. Find an electric motor in your home. Read the information plate. In a written report, explain what each item on the plate means.
3. Solve these problems:

A. An automobile circuit has a voltage of 14 volts. The ignition system draws (pulls) 7 amps of current through the circuit. What is the resistance of the circuit?
B. If the headlights of the automobile of Problem 4A have a resistance of 1 ohm, what is the amperage? (Remember, the voltage is 14.)
C. The current flow in an automotive recharging circuit is 28 amperes. The resistance is 0.5 ohm. What voltage is the alternator producing?
D. During starting, an automotive starting motor uses a 9-volt, 180-amp current. How many watts of power does the motor use?

Automated Control Systems

Did you ever stop to think what the word *automatic* means? All around us, we have examples of things happening *automatically*. For example, our home refrigerators keep the ice cream frozen and the eggs fresh. Unless there is a power blackout, we can count on the refrigerator keeping things cool all day and night.

Another example of automatic operation is the home heating system. When the living room gets a little chilly, we set the thermostat to a higher temperature. The furnace then "kicks on," and we get cozy. The heating system keeps the temperature at a constant comfortable level.

Let's try one more example — the automatic washing machine. We put a load of clothes in the machine, add detergent, punch a couple of buttons, and close the lid. The machine does the rest: washing, rinsing, and spin-drying. And it does all this without our having to give it a lot of constant directions. It does its job *automatically*.

Have you figured out what *automatic* means? It means **self-acting** or **self-adjusting**. See Fig. 8-1. Keep this in mind as you read on.

AUTOMATED CONTROL

Just how do refrigerators, heating systems, and washing machines run themselves? Where does the automatic control come from?

In refrigerators and heating systems the automatic control is very simple. A refrigerator

works by removing heat from the food compartment. A **thermostat** measures the temperature of the compartment. When the temperature is low enough, the thermostat shuts off the refrigerator.

The food compartment gradually warms up. When it gets to a certain temperature, the thermostat turns the refrigerator back on. The thermostat maintains a constant temperature. It provides *automatic control*.

The thermostat on your home heating system provides automatic control in much the same way. See Fig. 8-2.

Washing machines have more complex controls. When you turn on the washer, a **timer** starts working. The timer operates several switches that control the different actions of the machine.

Fig. 8-2 **The thermostat senses temperature changes and turns the furnace on or off.**

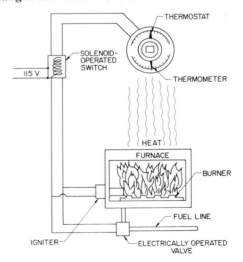

Auto (self) + Matos (mind) = Automatic (self-minding)

Fig. 8-1 **The meaning of *automatic*.**

146

First the timer opens water valves. Water flows through the valves and fills the machine. A sensing device, or **sensor**, tells the timer when the washer is full. The timer then shuts off the water.

Next, the washing action begins. When the clothes have been agitated enough, the timer stops the machine. A new action begins. The dirty water drains, and the washer fills with clean rinse water. The timer and other sensors control the whole washing cycle until the clothes are clean and spun-dry.

So, what is automated control? From the above examples, you should see that it is *continuous automatic operation done to perform a task.*

AUTOMATION

The control devices in home appliances are pretty simple, compared to the controls used in industry. In industry, automated control is called *automation.* Automation is *a set of operations that are performed or controlled in a planned order.* Here's another definition: *the production of goods by self-controlled machines.*

Think about the title of this section of the book — "Control and Transmission of Power." So far you've read about mechanical power, fluid power, and electrical power. Automation uses all three kinds of power. Now, it takes *brains* to handle all three kinds of power at once. This is why complex automated systems use **computers** to direct and control operating procedures. The computer is the "brain" of the automated system.

Automation also requires *senses*, just like a human being needs the senses of sight, hearing, and so on. In automated systems, the senses are things like the thermostat. The thermostat "senses" when a room is warm enough. Then it "tells" the furnace to shut off.

Fig. 8-3 **The computer "runs the show" in automation. But it needs information from sensing devices to direct operations properly.**

Sensors are also called **instruments**. In complex automation, the instruments can be very sensitive, and there can be a lot of them. They tell the computer many things about how an operation is going. The information that instruments give the computer is called **feedback**. See Fig. 8-3.

Well, so far we have a brain and some senses. What else do we need to do anything? That's right — something that actually *moves*, like a hand. We need something like *muscles*.

In automation, the "muscles" are mechanical, fluid, or electrical devices. These devices do the actual work in the system. Sometimes all three types of devices are combined into a **robot**. A robot can do many things much faster and better than a human being.

Do you see how it all fits together? Computers, instruments, and power devices are the "brains," "senses," and "muscles" of automation. See Fig. 8-4.

The rest of this chapter goes into detail on the three major parts of automation. First we'll look at instruments, then computers, then robots and other automated manufacturing devices.

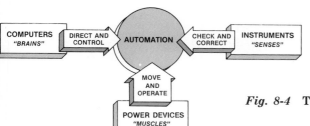

Fig. 8-4 **The three basic parts of automation.**

INSTRUMENTS

Let's start with a definition. An instrument is *a device that senses, measures, and displays information.*

Tonight, take a close look at the thermostat in your home. As described earlier, the thermostat measures the heat in the living area. Often a red needle shows what this temperature is. See Fig. 8-5. A thermostat is a simple instrument. It senses temperature, and it measures and displays the temperature.

Honeywell, Inc.

Fig. 8-5 Practically every home has one of these instruments in or near the living area.

Fig. 8-6 The automobile dashboard displays information needed by the driver. These instruments only provide information; they don't control operation. For example, the speedometer tells the driver the speed of the car. However, the driver controls the speed.

One popular television program, "Knight Rider," featured a completely automated car. Some new cars may have a few of the features of the Knight 2000. However, most cars use several simple instruments to inform the driver about the car's operation. See Fig. 8-6. The driver must take it from there. The most common instruments are:

- A **speedometer**, which measures the speed of the car.
- An **odometer**, which shows how many miles the car has been driven.
- A **fuel gage**, which shows how much fuel is in the tank.

Some cars also have gages to measure oil pressure, coolant temperature, and the battery charging rate. Other cars only flash a warning light when there is trouble in these areas.

Almost all transportation devices use instruments to provide information to the driver. The driver then changes the speed and/or direction of the vehicle. In air transportation, the pilot must also control the **altitude** (height above the ground).

Some airplanes have an **automatic pilot**. The human pilot sets the speed, direction, and altitude of the plane. At the push of a button, the automatic pilot "locks on" to these settings. Instruments check the plane's course as it flies. They report back to the automatic pilot, which makes small changes to keep the plane flying as it should. See Fig. 8-7.

Obviously, an automatic pilot must use some sort of "brain" to "read" information from the instruments and make changes with power devices. You'll soon learn more about computers.

Chevrolet

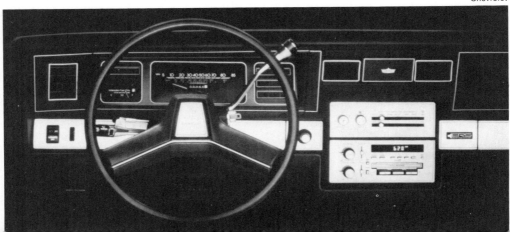

Fig. 8-7 "Look, Ma, no hands!" An automatic pilot can fly and even land a plane.

McDonnell Douglas

must know where they are at any particular time. If a location sensor breaks down, the robot might drop a load of parts on the floor. There are many types of location instruments. Some use laser beams, some use a kind of radar, and some use TV cameras. Of course, a location sensor can also be very simple, like a mechanical switch.

All of the things listed above are only a few of the conditions that instruments can keep track of. Automation could not work at all without instruments.

But first, let's go through a list of conditions that instruments can check and report on:

- **Temperature** is often measured and controlled by thermostats. Temperature control is important in many industrial processes. Steelmaking, metal forming, and welding all require careful temperature control.
- **Pressure** must be measured and controlled in all fluid power systems. Pressure gages and pressure control valves are used for this purpose.
- **Speed** control is important in both transportation and manufacturing. Speed must be controlled for efficiency and safety.
- **Time** is important in many automated control systems. Timers are used both in home appliances — such as ovens and washers — and in manufacturing. Production on assembly lines depends on accurate timing, too. Each part must be delivered to the assembly line at just the right time.
- **Dimensions** are the measurements that describe the size of objects. For example, a metal box has width, length, and height. Special instruments check the dimensions of objects as they are manufactured. Without this checking, automated machines could not accurately form metal, drill and tap holes, or do many other machining operations.
- **Location** sensors are also needed in automated production. Robots that handle parts

COMPUTERS

Computers are the brains of automated control systems. They can store and remember information. They also accept new information from instruments, and they control operations. Figure 8-8 shows how the computer fits into the whole system. Like other systems, an automated control system has inputs, control or processing, and output.

There are two basic types of input. The first type, **programming** consists of operating instructions and **data** (information) placed in the computer. The second type of input is data provided to the computer by instruments.

The processing or control of data is done in the computer itself. In fact, this is where the term **data processing** comes from.

After "considering" the data inputs, the computer gives directions to power devices. It tells them how to perform to bring about the desired

Fig. 8-8 **Automated control is a *system*. Like other systems, it has inputs, control or processing, and an output.**

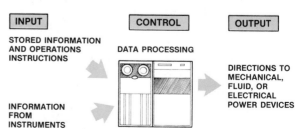

INPUT	CONTROL	OUTPUT
STORED INFORMATION AND OPERATIONS INSTRUCTIONS	DATA PROCESSING	DIRECTIONS TO MECHANICAL, FLUID, OR ELECTRICAL POWER DEVICES
INFORMATION FROM INSTRUMENTS		

result. These directions and results are the output of the automated control system.

Here's an example of computer control: More and more modern cars have small computers to control engine operation. See Fig. 8-9. These computers are programmed to include the following information:

- The correct air-fuel mixture for a variety of driving conditions.
- The procedures or directions for changing the air-fuel mixture according to the driving condition.
- The correct time for the spark plugs to fire for each condition.
- The procedures or directions for changing the time when the spark plugs fire.

During engine operation, instruments tell the computer what is happening. For example, when the driver presses on the accelerator, an instrument reports the accelerator position to the computer. The computer then adjusts both the air-fuel mixture and the spark timing for maximum efficiency.

There are many different instruments in the engine. They constantly report information to the computer. This results in the best performance, fuel efficiency, and pollution control.

Many other transportation and manufacturing systems use similar types of automated control.

Types of Computers

There are many different types of computers for different uses. However, there are four basic categories that computers fall into: microprocessors, micro-computers, main frame computers and mini-computers.

Microprocessors. The word *microprocessor* means *small processing unit*. Sometimes a microprocessor is called a **single-purpose computer**. Microprocessors provide fairly limited control of a machine or other piece of equipment. For example, the small computer used to control auto engine operation is a microprocessor.

There are many uses for microprocessors in industry. They are used to control machine operation and electrical power generation. Microprocessors are also used in some home appliances. Many people expect microprocessor use to increase greatly in the years to come.

Microcomputers. PCs, or **personal computers**, are becoming very popular. These *microcomputers* are used in homes, schools, and industry. See Fig. 8-10.

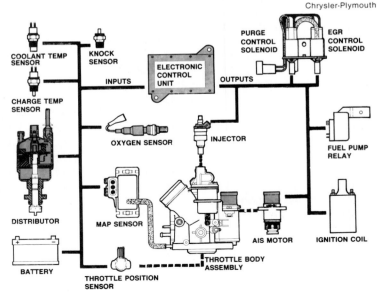

Chrysler-Plymouth

Fig. 8-9 **Many new cars use a computerized fuel-injection and ignition control system. The electronic control unit (computer) is programmed by the manufacturer. It also constantly receives input data from sensors located throughout the engine.**

Apple Computers

Hewlett Packard

Fig. 8-11 **In a few seconds, a small calculator can work out a problem that might take you an hour of "pencil-and-paper" work.**

Fig. 8-10 **This personal computer is used in homes and small businesses.**

Microcomputers have many more uses than microprocessors. They can be used for entertainment, calculations, word processing, and data management. The following are a few examples:

- **Mathematics** — You can use a PC at home to work out the family budget and income tax reports. Small companies use PCs to maintain all of their records and accounts.
- **Word processing** — PCs can help you prepare and edit written materials. As a matter of fact, some parts of this book were written through the use of a PC.
- **Information management** — In your home, you can use a PC to store your appointment schedule, your favorite recipes, and your automobile maintenance record. In business and industry, PCs are used to maintain inventory records and to schedule activities.

Another familiar type of computer is the **personal calculator**. See Fig. 8-11. Many people use personal calculators to work out math problems quickly and easily. In fact, some people use them so often that they forget how to do the calculations with a pencil and paper! This is an example of how we must use technology wisely. Technology is good for us as long as we remember that it is a *tool*. However, we must not become so dependent on it that we are lost without it.

Main Frame Computer. What about the agencies and companies that have to handle

International Business Machines Corp.

Fig. 8-12 **A main frame computer can process an enormous amount of data.**

enormous amounts of information, like the Internal Revenue Service and the major telephone systems? These agencies use a "super computer," or *main frame computer*. See Fig. 8-12. Main frame computers have tremendous memories, and can store vast amounts of data.

Some super computers are used mainly for research and mathematical calculations. For example, the National Aeronautics and Space Administration (NASA) uses a main frame computer to direct the space shuttle flights. The major airlines use computers mainly for data management — for example, recording reservations. Large banks use main frame computers to keep records of all their savings, loan, and checking accounts.

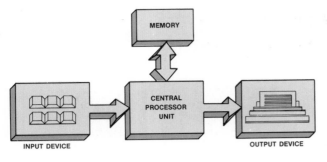

Fig. 8-13 **Every computer has four main units.**

Minicomputers and Specialized Computers. Many companies use *minicomputers* to process data. These computers are mid-way in size and ability between main frame computers and microcomputers.

There are also many types of specialized computers. Have you ever heard the term *CAD/CAM*? This refers to **Computer-Aided Design** and **Computer-Aided Manufacturing**. Engineers use CAD/CAM computers to design and plan the manufacture of industrial products.

Computer Operation

A computer has four main parts: an input device, a central processor unit, a memory, and an output device. See Fig. 8-13. There may be more than one input or output device.

Input Devices. Take a look back at Fig. 8-10. The person is providing input (information) to the computer by tapping on the keyboard. Therefore, the keyboard is one kind of input device. The input shows up on the **video display screen**.

Input devices can also be discs or magnetic tapes that contain data. For example, to work out certain calculations, you insert a **computer program** into the computer. The program is a set of operating instructions.

Input can also come from sensing devices. For example, a computer-controlled car engine uses sensors to check the air-fuel mixture and the spark timing. These instruments provide the input.

How Many Bits in a Byte?

This tiny **computer chip** can store a vast amount of information. Computer memory is organized into units called **bytes.** One byte contains eight bits of information.

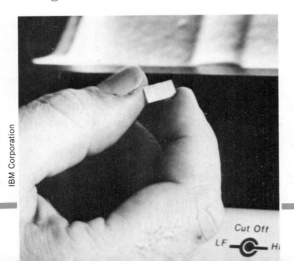

IBM Corporation

A **bit** is the smallest unit of memory. It is actually just a very small transistor. As you recall from Chapter 7, a transistor is an electronic switch. It is either triggered (on) or not triggered (off). The on-off switching of many transistors provides information to the central processor in the form of a code.

The computer chip shown is a type of megabit chip. It has an area of less than 1/10 of a square inch. Yet it can store over one million bits of information. The exact number is 1,048,576. All the information from over 50 pages of this book could be stored on a single chip.

Now, how many bytes of memory are on the chip? Use the exact number of bits given above. Look below for the answer.

(Answer: The chip can store 131,072 bytes of information.)

GENICOM Corp.

Fig. 8-14 **A printer is an automatic typewriter that takes information from the computer and prints it for use.**

Memory. The computer memory stores information. It does this by using solid-state devices such as transistors and diodes. Each solid-state device is so small that you would need a powerful microscope to see it!

The electronic parts are assembled on a **memory chip** *(see boxed story)*. The chip, with its huge storage capacity, is the most important part of the computer memory.

Central Processor Unit. The central processor unit, or **CPU**, is the center of computer operation. It is connected to all three other parts of the computer. See Fig. 8-13 again.

The CPU receives information from the input device(s). It then **processes** (analyzes or arranges) this information. To process the information properly, the CPU needs help from the program stored in the memory.

After processing the data, the CPU sends the results to the output device(s).

Output Devices. Output can be very similar to input. For example, both input and output information can appear on the same video display screen. Output can also be placed on tape or computer discs for storage or use on other computers.

The most common form of output, however, is the **computer printout**. The output information runs from the CPU to a **printer**. The

printer types the information out very rapidly. See Fig. 8-14.

Another kind of output is the actual control of a machine or transportation device. For example, an automatic pilot provides output in the form of altitude and direction control.

Hardware and Software

You have probably heard the word *hardware* and *software* in reference to computers. But just what is hardware, and what is software?

Basically, **hardware** is the hard physical equipment used in a computer. **Software**, on the other hand, is the information put into the computer to control its operation.

Software can either be built directly into the computer or placed in the computer by the operator. For example, you may have a personal calculator with division and multiplying buttons. This means that the calculator has instructions for dividing and multiplying built right into it. You don't have to tell the calculator how to divide one number by the other. You simply enter the numbers and use the button.

The kind of software that you put into a computer yourself is called a **program**. For example, a video game cartridge is a program. You plug it into a personal computer, and the program goes into the computer's memory. You can then **interact** with the computer by pushing buttons or working a "joystick." The video game program responds to your input.

Microprocessors do not use programs. All the instructions they need are built right into them. The same goes for small calculators. However, larger computers can take all kinds of programs, from simple word processing to complex math calculations.

Types of Computer Programs

Most computer programs fall into two categories: information processing and mathematics.

Information processing includes the handling of all kinds of data, such as mailing lists, airline reservations, school registrations and records, and so on. Data processing is one of the major uses of computers in business, industry, and education.

Mathematics programs include instructions for solving all kinds of math problems. This was one of the first uses of computers, and it has only been improved with time.

The potential for computer programming and use has only been developed to a small extent, compared to what it could be. We are now seeing the development of a "second age" in computer use. During this second age, computers will be used to solve complex problems. This is something that only human experts could do in the past. The new computers with their complex software are called **artificial intelligence** or **expert systems**.

Figure 8-15 shows the use of one new expert system. The man standing behind the computer screens is an expert on diesel-electric locomotives. The computer has been programmed with this man's knowledge, or **expertise**. When the computer is hooked up to input devices on the locomotive, it can pinpoint problems. It then displays a procedure for correcting the problem.

As expert systems are developed, we will see their use in problem solving grow. The new age of computers should be exciting and interesting!

Fig. 8-15 **These people have developed an expert system for discovering and correcting locomotive operation problems.**

General Electric Research and Development Center, Schenectady, NY.

ROBOTS AND AUTOMATED MANUFACTURING DEVICES

Let's review the main parts of an automated control system: (1) *Instruments* provide the information, (2) *computers* decide what needs to be done, and (3) *robots or other power devices* do the work.

Robots and automated manufacturing devices are the muscles of industry. These devices use the principles of mechanical, fluid, and electrical power. For example, electric motors power moveable robots. Fluid power gives automatic machines great strength. And mechanical arms use the principles of levers to duplicate human motion.

Robots

When someone says "robot," many people think of a machine that resembles a human being. But this is not usually the case. Robots *do* perform tasks that were formerly done by humans. But the tasks are usually very specialized. For example, a common task for robots used on auto assembly lines is to weld auto bodies. See Fig. 8-16. Another robot task is to sort out parts from a bin. See Fig. 8-17. These jobs don't require a robotic human being. They only require a robotic arm and special sensing devices.

Fig. 8-16 **Robots are used at this auto assembly line to weld parts of auto bodies together.**

Ford Motor Co.

Robot Arms for the Handicapped

Several years ago, Johnny Bowen was a worker at the General Motors auto assembly plant in Fremont, California. Then he was involved in a serious car accident that left him paralyzed. Johnny cannot move from the neck down.

Now Johnny is getting a helping hand, or rather both an arm and a hand. It belongs to a robot — the PUMA-250 — that was designed to do industrial assembly. The difference now is that the robot responds to voice commands from Johnny.

The disabled man and the industrial robot are working together in a research project at the Palo Alto Veteran's Administration Hospital in Palo Alto, California. Using complex computer instructions, researchers program the robot to pick up and move objects on Johnny's command. In the photo, the robotic arm is responding to Johnny's order to pick up a container of salt.

Jim Mendenhall/Mercury News

At this time, the research project is in the early developmental stages. The researchers hope to develop a portable robotic device that can be used in a handicapped person's home. This is very good news to Johnny Bowen. The robot will act as an extension of himself. It will give him the ability to do many things that most people take for granted.

In the future, we will see more and more of this meeting, or **interface,** of humans and machines, an interface that has the potential to be beneficial to us all.

The robots shown in Figs. 8-16 and 8-17 are parts of a complete automated control system. The system has the following parts:

- A **computer** to control the operation of the robot.
- **Instruments** to monitor robot operation. For example, a parts-handling robot needs sensing instruments to locate parts.
- **Mechanical devices** to do work. Robotic arms and legs (or wheels) use fluid, mechanical, and electrical devices to produce motion.

Automated Manufacturing Devices

We have had automated machines in industry for over 20 years. These machines perform boring, repetitious jobs. See Fig. 8-18. They are controlled by computers and use instruments to monitor their operation. They can change tools and adjust themselves without human assistance.

General Electric

Fig. 8-17 The robot can locate and pick up parts for movement and assembly.

Once you add robots to the assembly line, it is possible to have fully automated production. For example, a robot could bring a metal casting to a machine. The machine could drill, cut, and finish the casting. The robot would then carry the casting to another assembly line. Other robots would attach other parts to the casting to make the final product.

THE FUTURE OF AUTOMATED SYSTEMS

We can expect an increased use of robots and other automated devices. They will perform many jobs that are boring or dangerous for humans. The result should be high-quality products that are less expensive.

Fig. 8-18 Here, automatic drilling machines are being used to produce automobile engine blocks.

One of the main social concerns about automation is its displacement of human workers. Thirty or forty years ago, a factory might have needed hundreds of human workers. Now, with automation, the same plant can be run by only a few people. These people supervise and maintain the automated system.

Employment is a very real concern. As automation is used more and more, there will be fewer of the traditional "blue collar" jobs available in the heavy industries. There will be a growing need for people to operate computers and maintain automated systems. However, the number of jobs created by automation will be less than the number displaced.

Automated control systems are a good example of the advantages and disadvantages that technology can bring with it. They certainly will make our lives easier — in some ways. But they can also force us to make some very tough decisions about the direction our careers will take.

Technology is changing our world rapidly. To keep up with it, and to find our places in the world, we must explore our personal resources. We must also research potential careers and set our sights on realistic goals. Keep these points in mind as you continue to learn about technology in this book. The last section of the book, "Career Planning and Development," will help you get started in discovering your own personal future!

STUDY QUESTIONS

1. What does *automatic* mean?
2. What is automated control?
3. Give a definition of automation.
4. What are the "brains," "senses," and "muscles" of automation?
5. Name four conditions that instruments can check or monitor.
6. List the four basic categories of computers.
7. List the four main parts of a computer.
8. What is the difference between computer hardware and computer software?
9. What are the two basic categories of computer programs now being used?
10. What name is given to computers that have the ability to solve complex problems?
11. Name the three parts of a robotic automated control system.
12. What kind of work do automated machines do?

ACTIVITIES

1. Develop a list of the computers used in your school or in your home. Remember, computers can range from small microprocessors to large main frames. The following examples might get you started:

   - Microprocessors used in ovens and washing machines
   - Minicomputer used to store school records

2. Visit a computer store in your area. Look at four different computers, and list the following for each one:

   - **Memory** — The memory capacity will be identified as a number followed by a K, such as 2K or 8K. K usually means *kilo*, or one thousand. In "computer talk," K means 1024. So 2K means 2 × 1024, or 2048 bits.
   - **Hardware** — For example, a keyboard, display screen, printer, and so on.
   - **Software** — Check out some programs that you would be interested in using someday. Which of the computers can the programs be used on?

Transportation

Transportation is part of your everyday life. You may be surprised to find out how dependent you are on transportation. It's also important to know how this technological system contributes to our economic system. You'll find out all about these things as you read Sections III and IV.

Section III will get you started on transportation. You'll discover the five basic ways of moving passengers and cargo from place to place. These ways of moving are called **transportation modes**. You'll learn how the different modes depend on each other, and more. For example, you'll see how transportation acts as a *system*. The system, like many others, has three main parts:

- *inputs*
- *processes*
- *outputs*

You'll learn what happens in the system between departure and destination.

Transportation Basics

Have you ever wondered why there are so many trucks on the highways? What about the freight trains that pass through the cities, or the jet airliners that fly overhead? These **vehicles** (transportation devices) are obviously different from each other. But they all have the same purpose: to move people and things from one place to another. This moving of people and things is called **transportation**.

WHAT TRANSPORTATION IS ALL ABOUT

Transportation can be very complex. It can also be very simple. For example, lifting this book and carrying it across the room is a simple form of transportation. So is walking down a flight of stairs to get to your next class. In the first example, you are transporting your

Fig. 9-1 **Everything you see in this store required transportation to get there.**

Evans Product Company

textbook. In the second example, you are transporting yourself.

In this book, you will learn about forms of transportation that are more complex. For example, the vehicles mentioned above are used in complex systems to move people and things. This is how we will look at transportation. **Transportation** is *the movement of people and things, using vehicles such as buses, trucks, airplanes, and railroad cars.*

We usually take transportation for granted. We don't think about it much or see how it affects our lives. But just stop for a moment and think about where you are. Were you sitting in the same place yesterday? How about last Saturday?

Take a look around and name six different things you can see. Were those things always there, or did they have to be transported to where they are now? See Fig. 9-1. Maybe you're starting to see the picture. Transportation affects many things around us. And it affects *us*! It influences where we go, what we wear, what we do, and even what we eat.

The clothes you are wearing right now were transported from the factory to the store where you bought them. What about the corn flakes and milk you had for breakfast? Well, first the corn and raw milk had to be transported from farms to processing plants. After processing, the products were on the move again. They were hauled from the processing plants to the supermarket by truck or train. The move could have been a short distance. It could also have been clear across the United States! Thousands of products are moved in the same way every day.

What about the question of how you got to school this morning? Did you ride a bus, or did you ride the subway? Think about how you spent your last summer vacation. What vehicles did you use? Maybe you flew in a jumbo jet to your relatives' house. Or maybe you rode a bus across town every day to get to the swimming pool. As you can see, transportation involves the movement of both things *and* people.

Another way to start thinking about transportation is to look at the word *transport* itself. *Trans* is a Latin word that means *across*. *Port* is a shortened form of an old French word that means *to carry*. From the two root words, we get the present meaning of *transport*: to carry something across a distance.

As you learn about transportation, you will see that it involves everything needed to move something somewhere. Transportation can be an entire industry. For example, the major airlines use many planes to transport many passengers and much cargo. However, transportation can also involve just one person and one vehicle — like someone making a quick trip to the store. But it still takes a lot of things to make any kind of transportation happen. It takes people, money, vehicles, highways, and much, much more. As you study about transportation, you will start to see just how big and complicated the system is. You'll also see more and more just how it affects our lives.

Moving Passengers and Cargo

Everything that is transported can be classified as either passengers or cargo. **Passengers** are people who are being transported from one place to another. Most passenger travel is done in private cars. However, passengers also travel by airplane, bus, taxi, train, and ship. See Fig. 9-2.

People are passengers. Everything else is classified as **cargo**. Cargo is also called **freight**. Petroleum, ping-pong balls, and pulpwood are all cargo. Computers, cucumbers, and cattle are also cargo. See Fig. 9-3.

There are two basic types of cargo. Depending on how it must be handled, cargo is either bulk cargo or break bulk cargo.

Bulk cargo is any kind of *loose* cargo. This kind of cargo doesn't have to be separated into separate units. All of the cargo can be handled at one time.

Fig. 9-2 Millions of passengers travel millions of miles each year.

Fig. 9-3 This ship carries a cargo of iron ore pellets.

Bulk cargo can be dry solid material, liquid, or even a gas. Coal, soybeans, and cement are examples of dry bulk cargo. Liquid bulk cargo could be milk, gasoline, or corn syrup. A gaseous bulk cargo might be nitrogen or oxygen.

Many raw materials used in manufacturing are shipped to factories in bulk. For example, iron ore is transported in large ships. Soft drink flavoring is shipped in tank trucks. Oil is pumped through underground pipelines. See Fig. 9-4. Bulk cargo can be moved in very large quantities with very little handling.

Break bulk cargo is any kind of freight that consists of separate units. For example, a load of TV sets is break bulk cargo. Bicycles, books, and baseball bats are other examples. See Fig. 9-5.

You can see that break bulk cargo needs more handling than bulk cargo. Loading and unloading break bulk cargo usually takes more time.

Coca-Cola USA

Fig. 9-4 Tankers such as this carry syrup for making Coca-Cola to nearly 300 bottling operations across the United States.

Certified Grocers of California, Ltd.

Fig. 9-5 These boxes of detergent are an example of break bulk cargo. The worker must handle each box separately.

The Economic Value of Transportation

Have you ever heard the word *utility*? It has several different meanings. One meaning is *usefulness*. Transportation provides a special kind of usefulness that is called **place utility**. Place utility actually adds value or usefulness to something.

Let's see how place utility works. Suppose you live in New York and you are thinking how great it would be to have an orange. Now, oranges don't grow in New York. They grow in states with warmer climates, such as Florida. But the amazing thing is that you can go to the supermarket and buy a bag of oranges! Transportation makes this possible. Of course, you have to pay more for oranges in New York than you would have to in Florida. But you are willing to pay more because the cargo transportation saves you a trip to Florida. The transportation system actually adds value to the oranges by making them available in a different place. This is where the *place* in *place utility* comes from.

Types of Transportation

By now, you are probably starting to see why transportation is so important. You know about passengers and cargo and place utility. Now let's look at three basic types of transportation.

The three basic types of transportation are personal transportation, commercial transportation, and governmental transportation. Different groups of people use the different types of transportation. As you read on, you'll see that for each type, people can both use transportation services and provide them. That is, the people are both **users** and **providers**. See Fig. 9-6.

Fig. 9-6 People involved in the basic types of transportation both *use* transportation services and *provide* transportation services.

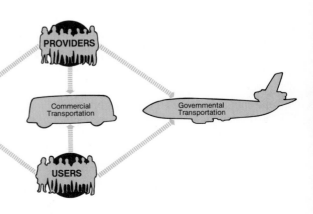

Personal Transportation. Personal transportation is the moving of people and things by individuals (persons) for their own benefit. For example, riding your bike to school is personal transportation. So is a ride in the family car to a football game. Personal transportation is usually for personal business or personal use of spare time.

How are people both users and providers in personal transportation? Well, suppose your father is driving the whole family to another state for a vacation. Since he is driving the car, he is the *provider* of the transportation. However, he is also getting the benefit of being transported himself. Therefore, he is also a *user* of the family transportation service.

Commercial Transportation. We all know that a commercial is something that interrupts a TV show about four times an hour. But the word *commercial* also means *having to do with commerce (business)*.

Some companies are in the transportation business. These companies charge money to move you or your cargo. They make money by charging you more for the transportation than it costs them. The money they have left over after paying for the transportation is called **profit**.

Another name for commercial transportation is **for-hire transportation**. When you buy a bus ticket, you are "hiring" the bus company to transport you. See Fig. 9-7. When a clothing

U.S. Air Force

Fig. 9-8 **This giant U.S. Airforce plane will be used to transport tanks.**

manufacturer wants to ship a load of blue jeans, it hires a trucking company. Railroad companies are commercial transporters. So are package delivery services like United Parcel Service and Federal Express.

Governmental Transportation. Governmental transportation is the moving of passengers and cargo for national defense, public service, or the protection of special persons. Many different government agencies both use and provide transportation. The federal government moves soldiers and military equipment to maintain our country's national defense. See Fig. 9-8. Mail delivery is another common example of transportation by a federal agency.

State and local governments also provide and use transportation services. For example, the city sends garbage trucks every week to haul away our garbage. Another common example is the city-provided ambulance that transports accident victims to the hospital.

The Effects of Transportation

Many things have happened in our country as a direct result of transportation. If you study U.S. history, you will see that the development of the country is very closely tied to developments in transportation. Canals, highways, and railroads all played important parts in opening up new sections of land. See Fig. 9-9. As newer and better transportation services became available, thousands of people travelled to and settled in new places.

Fig. 9-7 **Buying a ticket is like signing a contract with the bus company. You agree to pay for transportation. The company agrees to provide the transportation.**

Trailways, Inc.

Fig. 9-9 In 1869 the Central Pacific railroad met the Union Pacific railroad at Promontory Point, Utah. This established the world's first *trans-continental* rail line. It connected the east and west coasts of North America.

Fig. 9-10 Modern highways like this one make highway travel fast and convenient.

Today there are many, many ways to travel across the country. More people travel to more places each year than ever before. Business people travel thousands of miles each month. They fly or drive to one city, and then on to another. Highway transportation in general is

much easier because of the interstate highway system. Families on vacation can drive long distances on the interstates in less time. Therefore, they have more time to spend relaxing at their vacation spot. Our overall quality of life is enhanced (made better) because of advances in transportation.

As you know, transportation is one of the large systems that make up our whole technological system. All of the technological systems have both advantages and disadvantages. The air pollution caused by automobile engines is one bad side-effect of personal transportation. The noise caused by jet airliners taking off and landing is a side-effect of commercial transportation. And of course, there are the thousands of people that are killed or injured in accidents involving transportation. We must seriously think about these things in deciding what type of transportation is **appropriate** (right for a particular situation).

MODES OF TRANSPORTATION

There are five basic ways of transporting passengers and cargo. These ways are called **modes**, and they include the following:

- Highway transportation
- Rail transportation
- Air transportation
- Water transportation
- Pipeline transportation

In this chapter, you will learn a little about each transportation mode. There are many different kinds of vehicles used in each mode. You will learn more about the modes and their vehicles in Section IV of this book.

Highway Transportation

Cars, buses, and trucks all travel on roads and highways. These vehicles can move both cargo and passengers on the interstate highway system, on other federal and state highways, and on local roads. The many roads and highways make it possible to travel almost anywhere in the continental United States. See Fig. 9-10.

Rail Transportation

Trains travel across the country on sets of steel rails. There are two basic types of rail transportation: short-distance and long-distance. More passengers are transported for short-distances than cargo. However, more cargo is transported long distances than passengers. See Fig. 9-11.

The next time you see a freight train go by, read the names and other information on the sides of the boxcars. See if you can tell what line (company) owns the car. You might even be able to figure out what is being transported.

Air Transportation

Thousands of miles of air "highways," or **routes**, criss-cross the country. Air travel is faster than other modes of transportation. However, it also costs the transporting company much more to operate.

In air transportation, more passengers are transported than freight. Therefore, air trans-portation is basically a business of **personal time-saving**. For most people, air travel is also an exciting adventure. Two hundred years ago, it would take days just to travel from New York to Virginia. Now people can travel from New York to San Francisco in just a few hours. See Fig. 9-12.

Water Transportation

Ships and barges are used to transport passengers and cargo on water. Much of water transportation is the business of shipping freight. Barges are pushed by **towboats**. These boats move cargo from one city to another on rivers and canals. This is a slow way of transporting goods. However, it also costs less than other methods. There is a great deal of barge traffic on the Mississippi River.

Large ships are used to transport different types of cargo across oceans to and from other countries. See Fig. 9-13. Passengers generally travel on ocean liners or cruise ships.

Fig. 9-11 **Rail transportation provides an efficient way to transport heavy loads across land.**

Fig. 9-12 **Air transportation is the fastest way to travel long distances.**

Illinois Central Gulf

Boeing Commercial Airplane Co.

Fig. 9-13 **Ships such as this one are important for transporting cargo to and from other countries in the world.**

Fig. 9-14 Since pipelines are usually underground, we don't usually see them. But they are a very important mode of transportation.

Pacific Gas & Electric

Pipeline Transportation

Freight is the only thing that is transported through pipelines. At one time, only liquid or gaseous cargo was moved through pipelines. But today, some small particle materials, like crushed coal and sugar, can be moved through pipelines. There are over one and one-half million miles of underground and above-ground pipelines in the United States. They move cargo back and forth across the entire country. Pipelines can be as small as three inches in diameter, or as large as six feet in diameter.

Most pipelines are underground. And the freight is moved through them without the use of a vehicle. Therefore, we usually see only part of the whole pipeline system. We see either the **terminals** (beginning and end points) or the pumping stations. See Fig. 9-14.

INTERMODAL TRANSPORTATION

The different modes of transportation usually don't operate independently (separate from each other). Usually, more than one mode is used to transport passengers or cargo. Using more than one mode is called **intermodal transportation**. (*Inter* means *between*, so *intermodal* means *between modes*.)

Intermodal transportation has to be carefully coordinated, or directed. It has to be planned and scheduled (set up) in advance.

Intermodal Passenger Transport

Some passenger transportation is intermodal. For example, suppose you wanted to travel somewhere by plane. First you would have to get to the airport. To get there, you would ride in a car or bus. Then, at the airport, you would board (get on) the plane and travel to another airport. At the other airport, you would get off the plane, get into another car or bus, and go to your final destination. See Fig. 9-15.

Do you see how air travel and highway travel are used together? Actually, any combination of air, highway, rail, or water transportation can be used to move passengers from one place to another.

Intermodal Cargo Transport

Most large-scale intermodal transportation takes place in the movement of cargo. Cargo is very commonly moved through the use of two or more modes. For example, trucks (highway transportation) carry cargo to and from planes (air transportation). Trains (rail transportation) carry cargo to docks to be loaded onto ships (water transportation).

The transportation of oil is another good example. The oil might first be transported from Texas to Ohio through a pipeline. In Ohio the oil could be loaded into a railroad tank car and moved to West Virginia. In West Virginia the oil might be unloaded into a tank truck and hauled to a factory for use as a fuel. Three different modes are used in this example.

Fig. 9-15 Airplane passengers may be taken to their final destination on a shuttle bus.

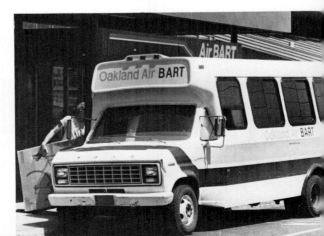

Fig. 9-16 In intermodal transport, two or more transportation modes are used. Here a barge is carrying cargo from Lake Charles, Louisiana, to San Juan, Puerto Rico. How many modes do you think are used on the entire trip?

Almost any cargo that is moved is first carried by truck. From the truck, the cargo might be loaded onto a railroad car, a ship or barge, or even into a pipeline. See Fig. 9-16. When the cargo reaches its destination city, it is usually loaded onto another truck for the final delivery.

Advantages of Intermodal Transport. Intermodal transportation usually allows for better and more flexible service. In transportation, **flexibility** refers to how many different routes there are to a certain destination. For example, highway transportation is much more flexible than rail transportation. There are roads to almost every town in the United States. However, there are much fewer railroad lines. The advantage of rail transportation is that it provides fast, direct transport to and from major cities. This type of transport is much cheaper than hauling cargo just with a truck.

As you read on, you'll see how intermodal transport uses the best features of each mode. Many times, intermodal transport also saves energy. Certain modes are more energy-efficient than other modes under certain conditions. This usually means lower costs for the shipper.

Piggyback Service. Have you ever seen truck trailers being carried on railroad flatcars? This kind of transportation is properly called **Trailer On Flat Car (TOFC).** Most people just call it "piggyback."

In piggyback transportation, the loaded trailer is first lifted or driven onto a special flatcar. Workers fasten the trailer to the flatcar, and it stays there for most of its journey. See Fig. 9-17. At the end of the railroad trip, the trailer is hitched to a **tractor.** (A tractor is the part of the truck that actually pulls the load.) Then a driver hauls the trailer to the final destination.

The main advantage of piggyback service is its lower cost. It is much cheaper to use this kind of transportation than to haul the load with a tractor for a long distance.

Fig. 9-17 One economical form of intermodal cargo transport is piggyback service. Loaded truck trailers are carried on railroad flatcars and then are pulled over the highway to their final destinations.

Union Pacific Railroad

Sea Land Service

Fig. 9-18 Special lightweight containers are designed to be loaded and hauled easily on many different vehicles and modes of transportation.

Union Pacific Railroad

Fig. 9-19 Containers can be unloaded very quickly from flatcars using this special crane.

Containerization. An important development in intermodal transportation is the use of **containers**. A container is basically a large metal box. Most containers are eight feet wide and eight feet high. Containers are usually 40 feet long.

Containers can be used in several different ways. They can be made into trailers by placing them on sets of wheels. They can also be placed on railroad flatcars. This kind of transportation is called **Container On Flat Car (COFC)**.

Another way to move containers is to load them onto barges or ships. In fact, ocean-going **containerships** are made especially to carry containers. See Fig. 9-18. Containers can also be loaded into airplanes.

Advantage of Container Use. The main advantage of containerization is reduced handling of cargo. Let's look at an example. Suppose you own a factory in Fort Wayne, Indiana, that manufactures small gas engines. One of your customers is a company in Memphis, Tennessee, that manufactures lawn mowers. To ship your products to Memphis, you might decide to use regular intermodal transportation. This would mean using trucks, railroad cars, and barges.

First your workers would load the engines into a truck trailer. A driver would then deliver the

truck to a rail yard. There the engines would be unloaded from the trailer and loaded into a boxcar.

A train would deliver the boxcar to the river port at Cincinnati, Ohio. There the engines would have to be unloaded again and loaded onto a barge. The barge would take the load all the way to Memphis. But once there, the engines would have to be unloaded again for final delivery. That's a lot of handling!

Now compare the above method with container transport. At your factory, workers would load the engines into a container. The container would be locked and sealed. It would be set on wheels, hitched to a truck tractor, and hauled to the rail yard. Once there, the container would be lifted off the wheels, lifted onto a flatcar, and secured in place. When the container reached Cincinnati, it would be lifted off the flatcar and placed on a barge. When the container reached Memphis, it would once again be set on wheels and hitched to a tractor. See Fig. 9-19. A truck would then haul the load to the lawn mower factory.

Do you see the reduced handling? The more times cargo is handled, the greater the chance of theft or damage from dropping or bumping. Containerization keeps this from happening.

Users of Intermodal Transportation

There are many companies that use intermodal transportation a great deal. Basically, two types of companies use this service: freight forwarders and small package express companies.

Freight Forwarders. Have you ever heard of a "middle man"? The middle man or middle person is the one who is in-between the person who ships (sends) something and the person who receives something.

Note: To *forward* something means to receive something at one point in its journey and then send it *forward* (further along) to another point.

Freight forwarders are the middle persons in transportation. They collect many small loads from different shippers. Then they **consolidate** the loads (put them together) into a large load. Next, the freight forwarder ships the large load to the receiver.

Most freight forwarders don't actually own their own **line-haul** (long-distance) vehicles. Instead, they hire other transportation companies to haul the consolidated load. The reasons for this arrangement might be a little hard to understand. However, it provides the cheapest way to transport the load.

Surface freight forwarders use rail, truck, or water carriers to transport cargo. **Air freight forwarders** use regular airline service to ship their loads. Most freight forwarders offer free pick-up and delivery. This door-to-door service is very convenient to the customer.

Small Package Express Service. Have you ever taken an express bus or train to another city? *Express* means *direct*. An express bus goes directly from one city to another. It does not stop at little towns in-between. Therefore, it gets you to your destination faster.

Now, when we say *small package express*, we are talking about delivering packages in the fastest and most direct way possible. Many companies provide express delivery of small packages, or **parcels**. One well-known company is the United Parcel Service (UPS). This company has a **fleet** (group) of their own trucks and aircraft. The vehicles deliver parcels quickly to any place in the country.

Small package delivery is a big business, and many companies are "getting into the act." The United States Postal Service has always provided package delivery service. However, it now also offers special overnight delivery. Federal Express is another company that provides this service. See Fig. 9-20. Most express deliveries involve intermodal transportation.

ON-SITE TRANSPORTATION

Moving things and people over long distances is not the only kind of transportation. Things and people are also commonly moved over short distances. This movement usually takes place inside a building. It is called **on-site transportation**. (*Site* means *place* or *location*. Therefore, on-site transportation always happens within one particular area.)

There are two basic kinds of on-site transportation: materials handling and people moving.

Materials-Handling Equipment

There are three basic steps or stages in manufacturing. First you have the material needed to make something. This is called the **raw material**. To make something from raw material, you must change, or **process**, it. Then you have a **finished product**.

Raw materials can be transported to a factory by truck, train, ship, or barge. Then, at the factory, the materials have to be moved around to different areas for processing. After processing, the finished products also have to be moved. They are moved into storage to wait for shipment. All of this movement inside the factory is called **materials handling**.

There are three basic types of materials-handling equipment: conveyors, cranes and hoists, and industrial trucks.

Fig. 9-20 Small package express companies can deliver packages quickly almost anywhere in the country by using intermodal transportation.

Fig. 9-21 **Moving boxes from one place to another is easy using a roller conveyor.**

Conveyors. Conveyors move materials constantly over a fixed path. There are several different kinds of conveyors used for different purposes.

Belt conveyors and **roller conveyors** usually carry things at or close to ground level. See Fig. 9-21. **Trolley conveyors** usually carry parts overhead. The parts hang from hooks.

Bucket conveyors carry loose materials like sand and gravel. The buckets travel from one level to another.

A **chute** is a simple type of conveyor. It works by gravity. The material simply slides down the chute.

Cranes and Hoists. Cranes and hoists are overhead devices. They are used to lift and carry heavy things from one place to another. A **hoist** is a lifting mechanism. Usually it consists of a motorized drum with steel cable wrapped around it. As the drum rotates, it winds up the cable. This lifts the load attached to the cable.

A **crane** is a moveable hoist. It can lift *and* move things. There are several kinds of cranes. Standard overhead cranes travel on rails mounted close to the ceiling. These cranes can lift and move very heavy loads like engines, steel beams, and rolls of paper. See Fig. 9-22.

Industrial Trucks. Industrial trucks are not used on highways. They are only used on-site, like in a factory or at a loading site. Industrial trucks can be simple hand-powered carts or motorized vehicles.

One type of industrial truck is the **pallet jack**. See Fig. 9-23. The load rests on a low wooden platform called a **pallet**. To move the load, the worker pushes the "fork" of the jack into the pallet. The load can then be jacked up and moved away.

Industrial **fork lifts** can be battery-powered, or they can have an engine. Figure 9-24 shows the most common type of fork lift. The driver drives up to the load and inserts the "fork" into the pallet. A hydraulic system lifts the fork and the load. The driver can then move the load to where it is needed.

Fig. 9-22 **Cranes are used to move heavy loads short distances.**

Fig. 9-23 **The materials-handling device being used here doesn't look like a truck. But that's the proper word for it.**

Fig. 9-24 A fork lift truck is used to load and unload other vehicles.

People-Moving Devices

Several types of on-site transportation devices are used to move people. Some of these devices are very common. Other are used only in certain places, like large airports or amusement parks.

Elevators. Elevators are a common way to move people vertically in a building. The part of the elevator that carries people is called the **elevator car**. The car is hoisted and lowered by either cables or hydraulic cylinders. A computer controls the starting, stopping, and speed of the car. High-speed elevators can travel from the ground floor of a building to the top floor in just seconds.

Elisha Makes His Mark

Sometimes when people plan their careers, they are very sure of what they want to do. They go to a certain school, take a job with the right company, and are satisfied with their work. We say that these people have found their niche ("nitch") in life. However, many people take quite a while to find their niche.

Elisha Otis was one of these people. Among other things, he was a building contractor, carriage maker, mechanic, and saw mill operator. He would try to set up a business, and things would look good for a while. But then he would lose money in the business and have to start all over again.

After *20 years*, Elisha finally developed an idea that made his name known around the world. Around 1850, elevators were being used to lift freight in warehouses. The trouble was that there were no safety devices on the elevators. If the lifting cable broke, the elevator car would fall and the cargo would be smashed. Elisha changed all this by designing the first **safety elevator**.

The Otis elevator had a device that kept the car from falling if the cable broke. In 1854 Elisha demonstrated his invention at a special exhibition. He stood on the platform of the elevator as it was raised high above the spectators. Then he cut the lifting cable. The elevator didn't fall! Everyone was impressed.

Elisha Otis died a famous man. People called him "Otis the Safety Elevator Man." Look for his name the next time you use an elevator. And remember the story of the man who finally made his mark on the world — after years of looking for a way to make it.

Escalators. An escalator is really a moving stairway. It moves people from one floor of a building to another. See Fig. 9-25. The escalator is always moving. Therefore, it can move a lot of people. Usually, there are two escalators used. One moves up, and the other moves down.

Moving Walkways. Moving walkways are also called **moving sidewalks**. They are really a conveyor that moves people. You just step onto the walkway and let it move you. Or, you can walk on the moving sidewalk and actually be moving twice as fast as normal. See Fig. 9-26.

Monorails. A monorail is a kind of train that moves on one track. (*Mono* means *one*.) Usually the track is a concrete beam, and the train cars have rubber tires. There are usually several cars connected together. You'll see monorails at World's Fairs and at some amusement parks. See Fig. 9-27.

Louisiana World Exposition

Fig. 9-27 A monorail rides on a concrete rail above the ground. This allows the space below it to be used for other purposes.

United Technologies/Otis

Fig. 9-25 Escalators are moving stairways.

Fig. 9-26 A moving walkway is a conveyor used for transporting people.

Fig. 9-28 A cableway system transports people across the Mississippi River at New Orleans, Louisiana.

Mississippi Aerial River Transit

United Technologies/Otis

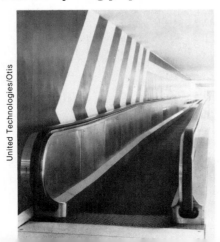

Cableways. A cableway system uses special cars that hang from a moving cable. The cars move people over the tops of buildings, ski slopes, and other areas. See Fig. 9-28. Sometimes a cableway is called a **tram** or **tramway**.

AGT. AGT stands for **Automated Guideway Transit.** AGT systems are sometimes called **people movers.** These systems use vehicles that are like buses. The vehicles run in a special track called a **guideway.**

All the vehicles in the system are controlled by a computer. Large airports like the ones at Atlanta, Dallas, and Houston use AGT for moving people from one terminal to another. See Fig. 9-29. Walt Disney World in Florida also uses an AGT system.

Westinghouse Transportation Division

Fig. 9-29 **This AGT system uses computer-controlled cars that travel on special tracks.**

STUDY QUESTIONS

1. What is transportation?
2. Name six things you can see around you that were transported.
3. What two words are included in the word *transport,* and what do they mean?
4. Fill in the blank: People who are transported are called _____.
5. What are the two basic types of cargo?
6. What is the difference between the two types of cargo?
7. What is place utility? Give two examples.
8. What are the three basic types of transportation? Give an example of each.
9. What are the five modes of transportation?
10. What is intermodal transportation?
11. Give an example of intermodal cargo transportation.
12. What is the proper name for "piggyback" transportation?
13. What is the main benefit of containerization?
14. Who are the two main users of intermodal transportation?
15. What is on-site transportation?
16. What are the three basic types of materials-handling equipment?
17. Name two ways of moving people from one floor of a building to another.
18. What kinds of on-site transportation might you find at a large airport?

ACTIVITIES

1. What different kinds of city, state, and federal government vehicles are used in your community? As you move around the area during the next week, keep an eye out for these vehicles. Make a list of what you see.
2. Using your local telephone book, find and list the different small package express companies that serve your area. (Hint: Look under several different headings, such as "freight," "air," and "delivery.")
3. Write a description of an intermodal transport vehicle that you have seen. Present your description to the class. You could also include drawings or pictures with your report.
4. With a friend, go to a store or other large building that has an escalator. Have one person walk up and down the stairs while the other rides the escalator. Compare the time and energy it takes the two of you to get from one level to another.

The Roadeo

Roadeo — the word looks misspelled, doesn't it? But unlike the word "rodeo," this word does not refer to a contest for bronc riders and calf ropers. It is a contest that gives truck drivers a chance to prove their driving skills.

The Roadeo rules and regulations are set by the American Trucking Associations, Inc. (ATA). In order to qualify, drivers must have at least one year of accident-free driving. Each year, contests are held on two levels — state and national. Drivers must compete and win on the state level in order to be eligible for the National Truck Roadeo.

State and national contests are conducted in much the same way. Contestants are scored on four different aspects of good driving.

First is a *written examination*. This exam tests knowledge of the trucking industry, safe driving rules, first aid, and fire fighting.

Second is a *personal interview*. Judges rate each contestant on personality and professional attitude.

Next is a *pre-trip inspection test*. Within a specified time, drivers must find the defects planted on a test vehicle. The type of vehicle inspected depends on the category the driver chose when entering the contest. There are seven vehicle categories. These include the most commonly used vehicles — straight trucks, tractor semi-trailers, and twin trailers.

The most important test of all is last — the *field test course*. Contestants drive a vehicle through a special course. The course presents problem situations that drivers must cope with in everyday operations. Contestants are judged on how skillfully they handle their vehicles and how accurately they judge distances and clearance (clear space around the vehicle).

Scores on all four tests are added to determine the winners. Prizes are awarded to the champion and to other top scorers — trophies, plaques, gold and silver belt buckles, and cash. But perhaps we are all winners. As the ATA says, "Truck Roadeo competition is a constructive program emphasizing skill, professionalism, and safe handling of motor vehicles. No person can take part in competition without becoming a better driver for it."

Contestants must:

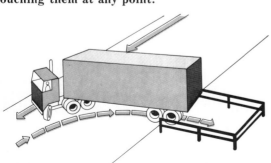

Drive through barricades without stopping or touching them at any point.

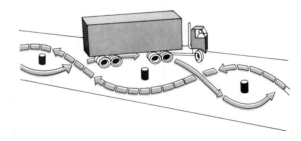

Drive through this course backward and forward without stopping or touching the barrels or curblines.

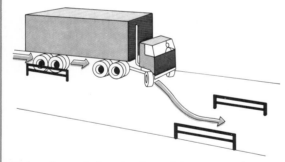

Back to within 6 inches of the line without touching the barricades.

Keep their right wheels between the balls without stopping or moving the balls.

172

Principles of Transportation

As you know, a **system** is an organized way of doing something. A system is also a set of things that are related to each other, or **interconnected**. You have heard of the solar system, the telephone system, and other systems. In our bodies we have the circulatory system, the nervous system, and others. Each system is made up of separate things that work together. The transportation system is like that, too.

As you read on, keep in mind that we can talk about transportation in two ways. First, we can refer to *all* the transportation that goes on as one big system. At the same time, a transportation system can also be a single short-distance trip. On whatever scale, transportation is simply a *systematic* way of getting passengers and cargo from one place to another.

THE TRANSPORTATION SYSTEM

There are many different ways, or modes, of transporting passengers and cargo. However, transportation by any mode is still a system. Like other systems, transportation has three main parts: inputs, processes, and output.

Inputs

Inputs are the things that we "put into" the system. Vehicles, workers, fuel, equipment — these are just a few of the inputs needed for transportation. Actually, there are three main kinds of inputs: capital, people, and energy.

Capital. *Capital* is another name for *money*. Are you ready for another division of terms? Try this one: There are two basic types of capital: working capital and fixed capital.

Working capital is the money that transportation companies use to buy things and pay employees (workers). This type of capital is needed to keep the system operating. Where does working capital come from? Well, the company owners can provide it themselves, or they can get it in the form of a loan from a bank. Working capital can also come from people who own stock, or **shares**, in the company. See Fig. 10-1.

So you see, working capital is basically money. It's a lot different from **fixed capital**, which is all the "real" objects needed in the transportation system. Here are a few examples of fixed capital: airplanes, trucks, computers, radios, and warehouses. See Fig. 10-2. Can you think of more examples?

Consolidated Freightways, Inc.

Fig. 10-1 A transportation system runs on capital (money) in one form or another. Working capital is basically the paper money needed to buy equipment and pay employees.

Fig. 10-2 Another form of capital is fixed capital. Buildings and vehicles are examples of fixed capital.

Fig. 10-3 No matter what kind of work they do, people are the most important input to a transportation system.

Fig. 10-4 **A pilot is the person who sets the controls of an airplane.**

People. No transportation system could operate without people. Companies need employees to fly airplanes, load and unload cargo, serve passengers, repair engines, and do many other jobs. Some jobs require more physical work, like loading furniture into moving vans or driving a fork-lift truck. Other jobs require more mental work. For example, people who work with satellite charts and computers to predict the weather really have to exercise their brains. Other people schedule truck traffic or even air traffic, as you'll see later in this chapter.

Both physical and mental work are important and necessary in transportation. People are the most important input in any technological system. See Figs. 10-3 and 10-4.

Energy. Another very important input is the energy source used to power vehicles. The source is usually a petroleum product, like gasoline, diesel fuel, or kerosene.

Other forms of energy are also used a great deal. One major example is electrical energy. Electricity powers subway trains and is used in diesel-electric locomotives. Electricity is also used in practically all modes of transportation to power lights, air conditioners, computers, and other equipment.

Processes

Processes are the actions that cause the movement of passengers and cargo. First you must have inputs. Then, you can *do* something with your inputs. Or let's put it this way: *Processes are carried out with the use of the inputs.*

Controlling the speed of a ship is one example of a process. Let's take a look at the inputs involved in this process:

First, there are *people* — like the captain and the engineer. There is *working capital* (the money to pay the captain), and there is *fixed capital* (the ship). There is also an *energy source* (for example, fuel oil).

When the captain wants to speed up the ship, he or she notifies the engineer, "Full speed ahead!" The engineer adjusts the engines so they will produce more power and move the fixed capital to its destination faster.

There are two basic types of processes in transportation: management processes and production processes.

Management Processes. To work properly, every part of the transportation system has to be *managed*. A railroad company must have people to manage the activities of the company. Buslines, airlines, and steamship companies also need managers. To be efficient, all kinds and modes of transportation require management. There are three basic types of management activities: planning, organizing, and controlling. See Fig. 10-5.

Planning means thinking about how you want to do something before you do it. In transportation, planning the move can make everything work more smoothly.

Organizing means getting all the necessary things together. Filling the gas tank in your car before you start a long trip is an example of organizing.

Controlling means keeping track of things. Following a map is an example of controlling where you are as you travel.

Fig. 10-5 **Making sure a vehicle has everything it needs is a management activity.**

Production Processes. In transportation, **production** involves all the things that actually cause the movement of passengers and cargo. All of the management activities are very important. However, they do not actually produce movement. There are three basic types of production activities: preparing to move, moving, and completing the move.

Preparing to move can include packing cargo into crates, helping passengers into their seats, loading cargo into containers, or fastening the hatches on a barge. See Fig. 10-6.

"Pick a point — any point": Computer Planning of Truck Routes

A trucking company will often have many trucks that are delivering cargo to many different destinations. Getting a certain truck to a certain destination takes a lot of planning. This is the job of the **dispatcher**. He or she must look at the possible routes for a delivery and figure out the most efficient route.

Dispatching can be a very demanding job. However, now there is a new system that can make the dispatcher's job much easier. This system uses a special computer program to plan the best route. Here's how it works:

The dispatcher asks the computer for a general map of the area in which the delivery is to be made. The map flashes on the computer's display screen. Then the dispatcher touches the departure point and the destination point on the screen.

Next, the computer displays street maps on the screen. Flashing arrows point out the most efficient route for the truck to follow. Pretty simple, isn't it?

But that's not all. The computer also tells the dispatcher if there are any problem areas in the route, like road construction. It estimates the amount of traffic build-up in these areas, along with any other possible delays. Then it calculates the amount of travel time based on the time of day that the route will be driven.

And after that? The dispatcher punches a button and the computer prints out the information into a schedule for the driver!

The system pays for itself by reducing the cost of fuel and labor. The trucks use less fuel because they don't waste time on inefficient routes. The company saves labor costs because the drivers don't waste time either. And the computer lets the dispatcher do more work in less time.

This is an example of how technology serves both the company and the customer. It means less expense for the company and better service for the customer!

Trans Tech Services

Cooking with the Sky Chefs

Sky Chefs is a food catering service used by many domestic and international airlines. Each month, Sky Chefs feeds hundreds of thousands of airline passengers. The company was created in 1942 by Newton K. Wilson. He was manager of passenger services for American Airlines. Wilson started Sky Chefs so that the airline could cut the cost of buying meals for its passengers.

There are now 30 Sky Chefs kitchens around the country. The two kitchens in the Dallas-Fort Worth area cover 200,000 square feet. The ice makers at these kitchens produce over two tons of ice per day. The food prepared daily includes 2000 chickens! And during a 24-hour period, the kitchens produce *25,000* meals. This includes everything from snacks to deluxe dinners. In 1983, Sky Chefs prepared over 40 *million* meals!

Moving activities mainly involve vehicle operation and the providing of in-route services. There are many examples: driving an "18-wheeler" truck, serving lunch to passengers on an airline flight, and steering a tow-boat on the Mississippi River.

Completing the move can include unloading cargo, helping passengers get their suitcases, and delivering freight to its final destination.

You will learn more about the various processes as you read more about transportation.

Output

What's an output? Very simply, an output is what you get when you carry out processes by using inputs. In transportation, the output is the movement of passengers and cargo.

For example, suppose you are a citrus grower in Florida who wants to ship oranges to different parts of the country. Using a transportation system, you could send your oranges to New York, or Chicago, or St. Louis. It all depends on inputs (capital, people, and energy) and processes (management and production). Put these things together, and you've got a transportation system! See Fig. 10-7.

Fig. 10-6 **An example of preparing to move. Barges are being loaded into a special barge-carrying ship. The whole ship can be loaded in just a few hours.**

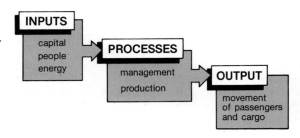

Fig. 10-7 The key ingredients in a transportation system.

BETWEEN DEPARTURE AND DESTINATION

Now that you know a little about the transportation system, let's see how the different inputs and processes work together to get an output. We'll look at each of the following processes:

- Planning
- Organizing
- Preparing to move
- Moving
- Completing the move
- Controlling

Planning

There are many different activities involved in transportation. All of these activities must be **planned** — carefully thought out — before they are actually done. For example, a new bus route must be mapped out before the first bus ever leaves the terminal. Another example is the loading of vehicles. Freight deliveries are more efficient if a truck is loaded according to a plan. The first piece of cargo to come off should be the last piece loaded on.

Transportation by ship or airplane depends on careful planning, too. Passengers and cargo must be loaded in just a certain way. Otherwise, the load will be unbalanced, and the trip will not be safe. The fuel used by the vehicle is part of the load, too. Therefore, the planners have to ask questions like "How much fuel does it take to make this trip? How much will the fuel weigh?" Routing, loading, and fuel supply are just a few of the things that have to be planned.

Organizing

Once the planning is over, the organizing process can begin. One part of organizing is deciding who will do what. For example, every airline flight needs a flight crew and a cabin crew. The **flight crew** — the pilot, co-pilot, and flight engineer — are responsible for operating the aircraft. The **cabin crew**, or flight attendants, are responsible for taking care of the passengers. Someone in management must decide which crews will work on which flight.

Organizing is also deciding which aircraft will be used for a particular flight. In addition, organizing involves getting together all the things needed for the trip. For example, someone must order the jet fuel for the flight, along with the meals needed for the passengers.

There is another very important organizing job to do before any transportation begins. This is making sure that the operator and the vehicle meet certain legal requirements. **Legal requirements** are laws set up to make transportation safe for the public. For example, truck drivers, airplane pilots, ship captains, and railroad engineers all must have a **license** or be certified to operate their vehicle. See Fig. 10-8.

Together, the different legal requirements are called **regulations**. Different agencies, like the Federal Aviation Agency and the Interstate Commerce Commission, set up the regulations. They also check to see that transportation companies follow the rules.

Have you heard the word *deregulation* on the TV news, or seen it in the papers? **Deregulation** is the relaxing or removal of some of the rules. Regulations are never totally removed, however. This is because they are needed to make transportation a safe and orderly process. See Fig. 10-9.

Fig. 10-8 Special training programs are available to help future truck drivers learn to safely operate their vehicles.

Duane Zehr

Fig. 10-9 In most cities, heavy trucks must travel only on certain roads. This helps the flow of traffic through the city.

United Airlines

Fig. 10-10 Making passengers feel comfortable is a part of preparing to move.

Preparing to Move

After planning and organizing comes preparations for the move. Several things must be done before any cargo or passengers can be moved. Transport preparations depend on three things: (1) *what* is to be moved, (2) *where* it is going, and (3) *when* it has to be there. Preparing a load of dynamite for shipment from Virginia to Alaska is a lot different from preparing passengers for an airline flight.

For example, airline passengers must first be gathered together in one place. This is usually a lounge or waiting area. Meanwhile, the plane is being prepared for flight. Then, as the passengers board (get on) the plane, their tickets must be checked.

Airline attendants must make sure the passengers are boarding the right plane. They must also be ready to answer any questions the passengers may have. Some passengers may need special help in getting to their seats and getting settled for the flight. See Fig. 10-10.

Preparing cargo is different from preparing passengers. There is no real movement of cargo until the shipper turns the freight over to the carrier. The **shipper** is the person or company that sends the cargo. The cargo could be something as simple as a birthday card, or something

179

Fig. 10-11 **This person is shipping a small package. The carrier accepts responsibility for safe transportation of the package.**

as fragile as a crate of eggs. In either case, the shipper must prepare the cargo. For example, you put the card in an envelope, address it, and put a stamp on it. Or you carefully pack the eggs so they will not be damaged during shipment. Then you turn the prepared cargo over to the carrier.

The **carrier** is the company that does the actual moving. When the shipper gives the cargo to the carrier, the carrier accepts responsibility for the goods. For a certain price, the carrier agrees to transport the freight safely. See Fig. 10-11.

Sometimes the carrier must store the cargo for a short time before transporting it. This, too, is part of preparing to move. The carrier must store the cargo according to its size, shape, and content. For example, meat must be refrigerated, and explosive chemicals must be very carefully stored. The carrier is responsible for any damage to the cargo.

Moving

Once cargo and passengers have been brought to a terminal and loaded aboard a vehicle, they are ready to be moved. There are two processes involved in the actual movement of goods and people: operating vehicles and providing in-route services.

Vehicle Operation. No freight or passengers could be moved if someone wasn't operating the vehicle. Operating basically involves *controlling the speed of the vehicle* and *controlling the direction of the vehicle*. Different modes of transportation require different types of vehicles. But no matter what the mode, the speed and direction of the vehicle must be controlled.

In highway transportation, the operator uses several basic control skills. Controlling speed and direction is much the same whether the driver is operating a car, a truck, or a bus. See Fig. 10-12. In controlling speed, the driver must consider the condition of the road, the weather, the traffic, and other safety factors. The driver controls direction by steering.

Vehicle operation is different in rail transportation. Here the basic route is controlled by the rails. The train goes where the tracks go. Of course, the engineer can control whether the motion is forward or reverse. He or she can also control the speed of the train.

In water transportation, the vehicle operator is called a **pilot**. The pilot controls speed and direction much like a truck driver or bus driver does. However, there are two important differences. First, a ship or boat doesn't have the traction that a highway vehicle has. It doesn't have rubber tires that can "grip" the roadway on a turn. During a turn, a ship or boat pilot must allow for a certain amount of drift.

Fig. 10-12 **A bus driver controls the speed and direction of the bus.**

The second big difference is related to the first. A ship or boat needs more distance to stop in than a highway vehicle does. Because of this, the pilot must constantly think ahead.

In air transportation, the pilot must keep several things in mind. In addition to controlling the plane's direction, the pilot must control the plane's **altitude** (its height above ground). Every plane has a specific altitude range. The pilot has to keep the aircraft within this range to provide for the safety of the passengers and crew.

Speed is a big factor in air transportation, too. Commercial pilots have to keep the plane on schedule. And the plane itself must maintain a certain speed for efficient operation. See Fig. 10-13.

In pipeline transportation, there is no vehicle. The cargo is simply moved through the pipe.

United Airlines

Fig. 10-13 Piloting an aircraft is a complex task. Just look at all of the instruments that the pilot uses to control speed, direction, and altitude!

Directional Control, or "How do you steer this thing?"

Since most transportation devices use engines for power, controlling speed is not much different from one device to another. The same basic principles are involved whether you are accelerating a car or slowing down a turbofan airliner. However, there *is* a big difference in controlling the direction. Basically, a moving vehicle can be controlled in one of three ways:

- The operator can control the direction by steering the wheels, which are in contact with the roadway. The vehicle goes in whatever direction the wheels are turned.

- A set of rails or a guideway provides a *fixed* route. Trains and monorails use this type of directional control.

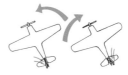

- The vehicle is steered by a **rudder**. Airplanes and ships use this method. When the pilot turns the rudder, it presses against the air or water that is passing by it. This pushes on the rear end of the vehicle with a sideways motion. As a result, the front of the vehicle turns in the same direction that the rudder is turned. For example, if you turn the rudder to the right, the vehicle turns to the right.

Got it? Now, how do you steer this thing?

However, controlling the speed and forward direction of the cargo is still important. Pumps provide the movement and speed, while valves control the direction of the cargo. Of course, *people* are needed to control the pumps and valves.

In-Route Services. Providing in-route services is an important part of the moving process. One of the most common examples is the job of the airline flight attendant. This person serves meals to passengers, instructs them in emergency procedures, and provides for their comfort throughout the flight. See Fig. 10-14.

In-route services apply to both passengers and cargo. For example, maintaining the cabin temperature in an airliner is a service for the passengers. Maintaining the temperature of a refrigerated truck or railroad car is a service

United Airlines

Fig. 10-14 **On an airline flight, flight attendants serve meals and assist passengers.**

Tender Loving Care in the Air

During the first years of commercial air travel, the airlines had a concern about their passengers. They thought the passengers would soon become bored as the novelty of flying wore off. So a European airline hired young boys as attendants. The boys served food and drink and made the flights more interesting. These "cabin boys" wore uniforms like those of hotel bellboys.

In America the airline executive S.A. Simpson suggested adding a registered nurse to each airplane crew. So, in 1930, nurses were recruited to be the first "stewardesses." They were all single and under 25 years of age. They wore their nurse's uniforms in flight.

On May 3, 1933, American Airlines (then American Airways), flew its first trip with a stewardess. The flight was aboard the brand-new Curtis Condor airplane. The stewardess was Velma Maul, from Burlington, Iowa. She served coffee and sandwiches to the passengers.

The duties of the first stewardesses weren't very glamorous. Before each flight, they were expected to bolt the seats to the floor. After takeoff, the stewardesses had to offer cotton to the passengers

for their ears. (The cotton was for muffling the noise of the engines.)

Today both men and women work as flight attendants. They don't have to be nurses. But they *do* receive special training. And nearly every passenger airline hires flight attendants to serve the needs of their passengers.

American Airlines

done for the cargo. A steady, cool temperature makes sure that bananas won't spoil on a trip from New Orleans to Minneapolis.

Completing the Move

At the end of the trip, certain activities are done to complete the move. These activities are much like the things that were done before the move. But this time they are done in reverse order. For example, both cargo and passengers must be unloaded in an orderly way. With cargo, the first item that was loaded is the last item to be unloaded.

Cargo may also have to be stored and protected before the move is completed. The cargo carrier doesn't give up responsibility until the goods are actually delivered to the contracted receiver. The receiver may be another shipper, or it may be the actual **consignee** (the person or company that ordered the goods). See Fig. 10-15.

Fig. 10-15 **The last step in transporting a small package is making sure it is delivered to the proper place.**

Federal Aviation Administration

Fig. 10-16 **These air traffic controllers are responsible for keeping aircraft on certain courses.**

Controlling the System

By now you should have an idea of the different processes that are involved in moving something from its departure point to its destination. However, there is another important process that applies to the whole move. This is called **system control**.

System control involves four basic activities: directing, monitoring, reporting, and correcting. **Directing** is telling the people in the different parts of the system what to do. It is also telling them how and when to do it. **Monitoring** means keeping track of what is happening in the system. **Reporting** is telling what is happening or what has already happened. Reports can be verbal (spoken) or they can be in writing. Finally, **correcting** is giving new orders or directions based on the information that is reported.

Air traffic control is one of the best examples of transportation system control. **Air traffic controllers** are people who direct air traffic so that it flows smoothly and efficiently. At the departure point, controllers give the pilot take-off instructions. They also provide weather information and describe the positions of other aircraft. During the flight, other controllers use radar to monitor the airplane's location. See Fig. 10-16. The controllers keep in radio contact with the pilot and report on any major changes in the weather. They also keep track of other planes flying near the airliner. If two aircraft get too close to each other, the controllers give new directions to one pilot to correct the plane's course.

STUDY QUESTIONS

1. What is a system?
2. What are the three main parts of a transportation system?
3. What is the difference between fixed capital and working capital?
4. Could a transportation system operate without people? Explain.
5. What are the two basic types of processes in transportation?
6. List the three major types of management activities.
7. List the three major types of production activities.
8. Why are regulations needed in transportation?
9. What is the difference between a shipper and a carrier?
10. What two things must be controlled in vehicle operation?
11. What are the three ways in which a vehicle's direction can be controlled?
12. What is a consignee?
13. List the four basic activities involved in controlling a system.

ACTIVITIES

1. Visit any type of transportation terminal. Make a sketch that shows the shape and plan of the building and the surrounding area. Label the different areas, such as waiting rooms, storage areas, parking lots, and so on.
2. While visiting a transportation facility, make a list of everything you see going on. Then classify the items on your list according to the three parts of a transportation system (inputs, processes, and outputs).
3. Obtain a transportation schedule for one of the four passenger transportation modes. Also obtain a map of the U.S. and/or Canada. Then plan a major trip from your home to anywhere you'd like to travel to in the U.S. or Canada. Organize your plans in a step-by-step order.

Transportation Industries

Each of the five basic modes of transportation has its own special characteristics, advantages, and disadvantages. In the following chapters, you'll learn all about the features of the different modes. Then you'll be able to *compare* the modes, and find out which mode is best for a certain situation.

Each chapter covers a transportation *industry*. Chapter 11, "Highway Transportation," tells you about the big trucks you see so often. The next chapter covers rail transportation. Railroads are still a vital part of the total transportation system. Air transportation is the subject of another chapter. It is the fastest mode.

Water transportation is a fairly slow way of transporting things. But it is also flexible and inexpensive. And let's not forget pipeline transportation. It is the most unusual mode. The cargo moves while the vehicle stands still!

But there's no time to waste in your journey through technology. Reading is a kind of vehicle, too. So get moving!

Highway Transportation

The movement of passengers and cargo over roads and highways is called *highway transportation*. There are two basic types of highway transportation: inter-city and intra-city.

As you know, *inter* means *between*, so *inter-city* means *between cities*. Inter-city transportation is travel from one city to another. In the trucking business, this type of travel is called **line-haul travel**.

Intra means *within* or *inside*, so *intra-city* means *inside a city*. Intra-city transportation is travel from one part of a city to another part. This type of travel is also called **local travel**.

Streets, roads, and highways are everywhere. Therefore, highway transportation is probably the mode of transportation that is most familiar to you. To get to school this morning, you probably travelled over a city street, or maybe a country blacktop road. While you were going to school, thousands of passengers were being moved from one place to another over the highway system. So were millions of tons of freight!

FEATURES AND BENEFITS

Of all the modes of transportation, highway transportation offers the greatest flexibility. Just think of all the different types of wheeled vehicles! There are buses, cars, and all kinds of trucks. What do they have in common? Well, they all have rubber tires, and they are all *independently controlled*. This means that the driver decides where the vehicle goes.

Of course, the driver is limited by the number of roads that go to any particular location. But there are *millions* of miles of public roadways.

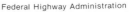

Federal Highway Administration

Fig. 11-1 **We can travel almost anywhere, anytime because of automobiles and highways.**

This makes it possible for a person to travel almost anywhere by highway. In addition, you can leave at any time, and arrive at any time — day or night. See Fig. 11-1.

COMPANIES AND SERVICES

Some highway travel is strictly personal business. For example, driving the family car on a vacation trip is personal transportation. Moving furniture to a new home with a rental truck is another example.

Much highway travel, however, is **commercial transportation**. It is provided by "for-hire" companies. These companies charge a fee for the services they provide.

Some companies offer only freight service. Other companies deal strictly in passenger transportation. A few companies offer both passenger service and freight service.

Commercial Passenger Service

Passenger service is provided by several different types of companies. These companies include taxi and limousine agencies, local and regional bus lines, and large inter-city bus lines.

Taxis provide the fastest service in intra-city travel. Local buses are slower, but they operate on regularly scheduled routes. Buses also provide cheaper transportation.

Regional bus companies provide inter-city transport in only one state or area of the country. Larger bus companies provide both inter-city regional transport and trans-continental service. Bus companies also provide charter service. (**Charter** means to hire a bus for a special trip.)

Trailways

Fig. 11-2 Several companies offer coast-to-coast, border-to-border passenger service.

Commercial Freight Service

The main commercial use of roads and highways is freight service. That's why you see so many "18-wheelers moving down the line." Some manufacturing companies own their own fleet of tractor-trailers. For example, a company that makes TV sets may use their own

The Greyhound Story

In 1914 a Minnesota car salesman named Carl Wickman was trying to sell a seven-passenger car called a Hupmobile. He couldn't make his first sale. So he used the car as a bus instead. He started carrying people on short trips between towns.

The fare was 15 cents one way or 25 cents for a round trip. Passengers were provided with lap robes. They were also given hot bricks for their feet.

The service was so popular that Wickman bought more Hupmobiles. To each one, he added three extra seats. Still, Wickman's buses were often overloaded. On many trips, they chugged along with extra passengers standing on the running boards and clinging to the fenders.

The buses were equipped with tow chains and snow shovels. Muscle was essential for the drivers. They often had to fight for their fares. Few women dared to make the hazardous trip.

The expanded cars were long and sleek and grey in color. Their appearance made someone remark that they looked "just like greyhounds streaking by."

Carl Wickman's "nickel-and-dime" venture was destined to become a national bus system. In 1915

the service was expanded to take in the 90-mile stretch between Hibbing and Duluth. In 1930 the company officially became the Greyhound Corporation. Its trademark became the now-famous running dog. Today Greyhound is one of the largest intercity passenger companies in the world.

Greyhound Lines, Inc.

Fig. 11-3 Motor freight companies transport a wide variety of cargo to many cities across the country.

trucks to deliver TV sets to stores throughout the country.

Another common freight-hauling operation is that of **trucking companies**. Trucking companies can be large, or they can be fairly small. These **motor freight carriers** transport freight to almost any place. See Fig. 11-3. Some carriers run on regularly scheduled routes. Others provide transportation only when needed.

Owner-Operators

Besides trucking companies, there are also many independent owner-operators. An owner-operator is a person who owns a tractor and leases (hires out) his or her driving services. For a certain price, the owner-operator will pull your trailer wherever you want. Some owner-operators even lease their tractor and driving skills to trucking companies on a full-time basis. See Fig. 11-4.

Fig. 11-4 This husband-wife team are owner-operators. They own the tractor and pull a trailer for someone else.

HIGHWAY TRANSPORT VEHICLES

Well, so far you know about commercial passenger transport and commercial freight transport. And let's not forget private highway transport, which is what you do with the family car. There are many different kinds of passenger cars, from small sports cars to large luxury cars. Later on in this book, you'll get a chance to learn about the inner workings of automobile engines. But for right now, let's look at two types of commercial transport vehicles: buses and trucks.

Buses

Private cars are designed to meet individual desires and needs. Buses are also designed to meet different needs. For example, school buses are made basically for short-distance travel. Large **motor coaches** are designed for long-distance travel. Passengers will be on board these buses for long periods of time. Therefore, motor coaches are equipped with comfortable seats and a restroom. Some charter buses also have a stereo system, extra-wide seats, and other special features. See Fig. 11-5.

Trucks

Freight is moved by many different kinds of trucks. However, there are basically two types

Fig. 11-5 Passengers can relax while riding in this bus. It is equipped with a TV, custom-built furniture, and a kitchen.

A. Step van

B. Dump truck

C. Pick-up truck

D. General hauler

Fig. 11-6 **Different kinds of single-unit trucks are used to haul different kinds of cargo.**

or categories of trucks: single-unit trucks and tractor-trailer combinations.

Single-Unit Trucks. Do you remember the two main types of cargo? That's right — bulk cargo and break bulk cargo. Single-unit trucks haul only one type of cargo. For example, dump trucks haul bulk cargo, such as sand or gravel. Step vans, on the other hand, haul many small packages. This is a kind of break bulk cargo. See Fig. 11-6.

Tractors. The tractor is the base unit of the tractor-trailer combination. It has a high-horsepower diesel engine and a 12-speed or more transmission. (You will learn about diesel engines in Chapter 17 and 18. These engines are more fuel-efficient than gasoline engines.) A regular automobile has only a few speeds in its transmission. (The **transmission** is a combination of gears that transmits engine power to the drive wheels.) Tractors must pull very heavy loads. The extra gears in the transmission let the engine apply a lot of torque, or turning

force. This force gets the truck moving. Once it is moving along, the driver can shift into a higher and more efficient gear.

Truck tractors are the perfect vehicle for pulling heavy loads. Why? Because they have a strong engine, a wide-range transmission, and a fairly lightweight frame. See Fig. 11-7.

Fig. 11-7 **An over-the-road tractor such as this one serves as "home" for the driver on long-distance trips.**

Like buses, tractors are used for both short-distance hauling and long-distance hauling. See Fig. 11-8. However, most tractor transport is for long distances.

The tractor has a driver's compartment called a **cab**. Long-distance tractors usually have very nicely equipped cabs. Special seats, air conditioning, and stereo sound systems are common. Drivers may be travelling for several days at a time. Therefore, many tractors have a sleeping compartment.

Sometimes drivers travel in teams, and take turns driving. One driver sleeps in the back while the other one keeps right on driving. Some cabs even have a mini-kitchen. All of these extra features add to the cost of the tractor. Some fancy tractors cost as much as a house!

Trailers. Trailers are the other half of the tractor-trailer combination. There are many different types of trailers. Some are very common. Others are very uncommon and special.

You've probably seen many **dry freight vans**. This type of trailer is made of lightweight aluminum and looks like a large box on wheels. See Fig. 11-9. Dry freight vans are good for carrying break bulk cargo.

Other commonly used trailers include refrigerator vans, flatbeds, tankers, and livestock trailers. See Fig. 11-10. Each trailer has a special purpose.

The Fifth Wheel. The main advantage of tractor-trailer transport is that almost any tractor can pull any trailer. A special hitch called the *fifth wheel* makes this possible.

"Fifth wheel" is really an odd name for the special hitch, because it doesn't look like a wheel and it doesn't ride on the road. However, the term and the hitch itself are used throughout the trucking industry.

One part of the hitch is located on the back of the tractor. This part is a heavy circular metal plate. The other part of the hitch is located on the front of the trailer. This part is a large metal pin called the **kingpin**. See Fig. 11-11.

To hitch a tractor to a trailer, the driver simply backs the tractor up to the trailer. The metal plate on the tractor slants up as it contacts the trailer. This allows the kingpin to slide into the plate. The hitch then automatically locks in place.

The truck is now "ready to roll," except for one very important safety duty. The trailer will be travelling right behind the tractor — obviously. Therefore, it must have the same braking ability as the tractor. Otherwise its weight and momentum wouldn't allow the truck to stop. So before the driver goes anywhere, he or she must be sure to hook up the air brake hoses between the tractor and the trailer. See Fig. 11-12. The driver must also make an electrical hook-up, so that the trailer will have rear lights and turn signals.

PHYSICAL FACILITIES

As you read each chapter in Section IV of this book, you'll learn about the physical facilities for each mode of transportation. Facilities include the following:

Fig. 11-8 **This spotting tractor is used only to move trailers inside a freight terminal yard.**

Fig. 11-9 **The dry freight van is the most common type of trailer.**

Consolidated Freightways

Capacity of Texas, Inc.

Great Dane *A.* Refrigerator van, double shown here

B. Flatbed   Mack Trucks

C. Tanker   Fruehauf

D. Dry bulk tank

Fig. 11-10 There are many special types of trailers used to haul different types of freight.

Fig. 11-11 The fifth wheel allows any trailer to be hooked to the tractor. The kingpin fits into the notch in the plate and is locked in place.

Peterbilt Motor Co.

Fig. 11-12 Hooking up the air brakes is an extremely important safety duty.

Fig. 11-14 At the terminal, freight is unloaded from incoming trailers. Then it is sorted and loaded onto other trailers for delivery.

Consolidated Freightways

Fig. 11-13 The interstate highway system connects major cities and makes travel easier and faster. Yet it is only one part of the whole network of highways.

- The actual physical **route** that the vehicle travels on (for example, a roadway, an air route, or a railroad track).
- The **terminals**, which are buildings used as the beginning and end points of the trip.
- **Intermediate facilities**, which are buildings or stations at different points along the trip.

In this chapter, we'll look at the highway network, passenger and freight terminals, and truck stops.

The Highway Network

Modern technology consists of systems within systems within systems. As you recall, a system is a group of things that are related to each other. Another name for a group of inter-related things is **network**.

Highway transportation requires a good network of roadways. First, there are the local roads and streets through towns and cities. Then there are the county and state roads and highways. Higher up, there is the Interstate Highway System. This system stretches from coast to coast and from border to border. See Fig. 11-13. Do you see how the highway network consists of systems within systems within systems?

The Interstate Highway System is a high-speed roadway. Vehicles must stay at a certain speed limit. However, there are few slow-downs or stops on the interstates. Travel is usually a "straight shot" from one major city to another.

Interstate highways are **limited-access** highways. This means that you can enter and exit (leave) only at certain places called **interchanges**.

The federal government and the state governments work together to construct and maintain the interstate highways. Sometimes the driver must pay a **toll**, or fee, to travel over certain roads. This money is used to maintain the roadway. Highway maintenance may include patching potholes, painting fresh stripes on the road, and cutting weeds along the road.

Most interstates and other highways can be used free of charge. Actually, though, everyone who pays federal or state taxes pays for the upkeep of the roads. Roadway funds also come from taxes on gasoline and diesel fuel, along with vehicle registration fees.

Terminals

A terminal is a building used for loading and unloading passengers and cargo. For example, bus terminals are both starting and ending places for the transport of people and packages.

Freight terminals are very busy places. Many workers are needed to load and unload trucks, and to sort out the cargo. See Fig. 11-14.

Freight terminals have a large paved area for trucks to back in, back out, and move around in. Part of this area is also used to park trailers. The terminal building itself is large and open.

It has many large garage-type doors. There is a raised area at each door. This is the **dock** that the trailers are backed up to for loading and unloading.

Truck Stops

There are many truck stops next to busy highways across the country. These service centers are designed to meet the needs of the thousands of truck drivers that work on the highways. The basic services, of course, are the providing of fuel for the tractor and food for the trucker. See Fig. 11-15. However, many truck stops are like miniature cities. They provide hot showers, motel rooms, clothing, and even banking services!

Fig. 11-15 **At the truck stop, truckers get food and fuel.**

STUDY QUESTIONS

1. What is highway transportation?
2. What makes highway transportation so flexible?
3. Name two kinds of commercial passenger service.
4. What is another name for commercial freight companies?
5. What is an owner-operator?
6. Give two examples of passenger vehicles.
7. What is the main advantage of the tractor-trailer combination?
8. Where are the two parts of the fifth wheel located?
9. What is a limited-access highway?
10. What happens at a freight terminal?

ACTIVITIES

1. Pretend that you're in the market for a good car or truck. Visit a dealer and check out a vehicle you're interested in. Note down important information, such as sticker price, miles per gallon, carrying capacity, and warranty. Then go to another dealer and look at a similar vehicle of a different make (manufacturer). Compare your notes on the two vehicles to figure out which is the best buy.
2. Obtain a local bus schedule. Plan and compare the different routes from a point near your home to a point on the other side of the community. (If your community doesn't have a bus line, find out about the regional bus line that serves your area. What times of the day does the bus pass through town? What cities can you reach using the regional bus line?)
3. With a friend, locate a safe place from which to watch a major highway. Keep a record of the number and make of the different trucks that go by. You can compare other information, too, such as color, type of trailer, and cargo.
4. Make a list of the different interstate, state, and local highways that serve your area. Also list any important lesser roads, such as two-lane blacktops that are used often.
5. Using a telephone book, find out which highway freight companies serve your area.

Here's how it works...

1 A computer manufacturer is going to ship a load of computers from Idaho to California. The shipping clerk is making the arrangements with a trucking company to pick up the load and make the long haul.

2 This truck line offers local pick-up. The cargo is loaded from the shipper's loading dock onto a small truck.

5 Inside, the terminal building looks like a long, narrow warehouse. Some workers are transferring freight from one truck or trailer to another. Others are sorting out freight for local delivery. And some are loading the shipment of computers onto a trailer.

6 The company dispatcher has a big job to do. Before any freight goes out, the dispatcher plans the loading, assigns a tractor and trailer, and determines the driver's route.

3 This truck is not the one that will be used to make the long haul. Small trucks are used within the city because they get around easily in city traffic.

4 The trucking company's local terminal is located close to the city limits, near an interstate "cloverleaf" intersection. The trucking company operates similar terminals in other major cities.

7 The drivers are reporting to work. The dispatcher has assigned two drivers to the same truck. The tractor and trailer are ready to go. The truck will carry many other items besides the computers. All the items need to be delivered in California.

8 The big wheels are rolling! Having the two drivers will make the haul a fast one. The two members of the driving team will take turns sleeping and operating the vehicle.

Photography by David R. Frazier Photo Library

9 There are weigh stations along the highway. State or federal authorities weigh each axle of the tractor and the trailer. Overloading can put too much stress on tires, thus creating a safety hazard. Also, overloaded vehicles can damage highways sometimes. The drivers are asked to show their driving log book (trip record) to make sure they do not drive for overlong periods of time.

10 A tractor may carry over 150 gallons of fuel! However, after about eight hours of driving, time for a break. The drivers stop to have t tractor refueled and serviced. They also ge themselves "refueled" with food and take t to relax before moving "on down the line."

13 A hostler or spotting tractor picks up the trailer and backs it up to the unloading dock. Workers will unload and store the freight until the company's local trucks can make a delivery.

14 The local delivery truck is on its way.

11 Most major highways are paid for by tax money. But users pay some money. Vehicles are charged a toll to use the road. Truckers pay a toll based on the number of axles on the vehicle.

12 California, here we are! At the terminal, the drivers "drop" the trailer. After resting up and getting the tractor serviced, the drivers will be ready for another assignment.

15 The driver unloading the freight notes any shortages or damages. This information must go on the freight bill before the receiver signs it.

16 Hey, is that one of the computers? It's had quite a journey — from an Idaho manufacturing company to California school room. And all this was brought about by something called **highway transportation.**

Rail Transportation

Rail transportation is a system of moving people and goods in vehicles that run on rails. There are several different kinds of rail transportation. For example, **subways** are electrically powered trains. They run on special tracks in tunnels underground. **Elevated trains**, or "Els," run on tracks above the ground. However, the most common kind of rail transportation takes place at ground level, or **on-grade**. Diesel- or coal-powered engines are used to move freight and passengers over steel tracks.

FEATURES AND BENEFITS

One special characteristic of rail transportation is the design of the vehicles and the railway. Steel wheels on steel rails can support

and carry large and heavy loads. See Fig. 12-1. The large load capacity of a single railroad car provides for efficient transportation. Shippers pay a low cost per mile for every ton of freight transported.

Rail transportation provides other special benefits, too. Railroad cars are actually rolling warehouses. The goods are packed and stored just as they would be in a warehouse. But there's one big difference: The "warehouse" moves from one city to another. This characteristic of "rolling storage" is a big benefit for shippers.

One other important benefit of rail transportation is its relatively minor environmental impact. Per mile travelled, trains create very little pollution and cause few traffic jams.

Fig. 12-1 **Rail vehicles have steel wheels that ride on steel rails. This combination is efficient for carrying heavy loads.**

Chessie System Railroads

COMPANIES AND SERVICES

Rail transportation service is provided by two basic types of companies, or **lines**. Some lines provide only passenger service. Others provide only freight service.

Passenger Lines

As in highway transportation, travel can be either intra-city or inter-city. Maybe you've seen or heard advertisements for AMTRAK. The name *AMTRAK* comes from **Am**erican **Tr**avel Tr**ack**. This passenger line provides service between major American cities. See Fig. 12-2.

There is also intra-city rail transport for passengers. For example, some students in New York City go to school on **rapid transit trains**. These trains run only on tracks within the city.

Some large cities also have large **regional** rapid transit trains. The tracks cover not only the city but outlying areas. Examples of regional lines include the BART (Bay Area Rapid Transit) in San Francisco, the Metro system in Washington, D.C., and the MARTA (Metropolitan Atlanta) system. See Fig. 12-3.

Freight Lines

Freight rail lines provide transportation service to more than 50,000 towns and cities in the U.S. and Canada. There are two basic types of freight service: unit train service and regular freight service.

Unit trains carry the same type of freight to the same location, trip after trip. See Fig. 12-4.

Fig. 12-2   AMTRAK provides long-distance rail transportation for passengers.

Fig. 12-3   In several major cities, passengers can travel using regional rapid transit systems.

Fig. 12-4   This coal train is a unit train.

Fig. 12-5 **A regular freight train carries a wide variety of cargo. Notice the different types of cars.**

Fig. 12-6 **Container on Flat Car transport is very flexible. A train provides fast delivery from city to city.**

Railroad workers usually don't have to uncouple or rearrange cars from trip to trip. The same cars carry the same load.

Regular freight service works much like long-distance truck hauling. First a local train picks up loaded rail cars from a shipper's loading dock. The local train then pulls the cars to a large freight yard. There the cars are reorganized with cars from other local trains. Together, the cars and engines make up a long train for long-distance hauling. See Fig. 12-5.

At the destination city, the cars are again put into local trains for delivery. Do you see how this is much like what happens in long-distance trucking? Regular rail freight doesn't have the flexibility of trucking. However, it costs the shipper much less per mile.

Compared to unit train service, regular freight service is slower. However, on the flexibility scale, regular service rates much higher.

Regular freight rail also offers two special services called **TOFC** and **COFC**. These stand for **Trailer On Flat Car** and **Container on Flat Car**. TOFC is also called "piggyback," and it is a very common form of intermodal transportation. Almost any train you see going by has some truck trailers "hitching a ride." Piggyback service has grown tremendously since it was first used in the 1950s.

COFC is just like piggyback service, except that containers are used instead of trailers. (Containers were discussed in Chapter 9, remember?) Using containers gives the shipper an extremely flexible form of transporting his or her goods. Once a container is loaded, it can go from truck to train to ship without any further handling. See Fig. 12-6.

RAIL TRANSPORT VEHICLES

The vehicles used on railroads are called **rolling stock**. Rolling stock includes all the different types of wheeled vehicles used. We'll take a look at three basic types: engines, railroad cars, and maintenance vehicles.

Engines

The steam engines used to pull trains 50 years ago were large, heavy, and powerful. A large coal car behind the engine carried the fuel for the engine. Today, railroad engines are still big and powerful. However, there are two major improvements. The engines are lighter, and they use better power sources.

Another advantage of modern engines is that they have better vehicle control methods. This allows an engineer to operate several engines hooked together from just one set of controls.

Fig. 12-7 This new diesel-electric engine costs over one million dollars ($1,000,000)!

Some railroad engines run on electricity, and a few run on ground-up coal. However, most engines are **diesel-electric locomotives**. See Fig. 12-7. They use diesel oil as the fuel. Diesel engines power generators that supply electricity to traction motors, which drive the steel wheels.

Railroad Cars

Some railroad cars are 90 feet long. However, most train cars are from 50 to 55 feet long. It takes about 100 cars to make up a mile-long train.

The most familiar type of railroad car is the **boxcar**. See Fig. 12-8. It has doors on either side and can carry all kinds of break bulk cargo.

There are also many special types of railroad cars for different purposes. See Fig. 12-9. There

Fig. 12-8 The boxcar is the most common type of railroad car.

A. **High side gondola**

B. **Open hopper**

C. **Flatcar, with bulkheads**

D. **Gondola**

E. **Covered hopper**

Fig. 12-9 Several types of railroad cars.

are flat cars for intermodal transport, livestock cars, refrigerator cars, "hoppers" for hauling bulk cargo, and many many more.

Not all of the railroad cars are owned by railroad lines. Chemical companies may own their own cars for carrying liquid or gaseous chemicals. The railroad line charges the chemical company for moving, maintaining, and storing the cars.

Other railroad cars are owned by large banks, insurance companies, and specialty transportation companies. In fact, these companies and banks own about one-quarter of all the railroad cars in North America. These cars are called **leased** vehicles. Shippers pay the owners for the use of the vehicles.

Maintenance Vehicles

Railroad lines own and operate many vehicles needed to maintain the railways. (You will learn more about the "ways" used in rail transpor-

Fairmont Railway Motors

Fig. 12-10 This "hi-railer" truck can be driven on either a highway or a railroad track.

tation shortly.) Maybe you have seen a railroad truck that has both steel wheels and rubber tires. See Fig. 12-10. This vehicle is called a "hi-railer." It can travel anywhere a truck or train can.

Railroad lines also use a variety of special vehicles and machines. There are vehicles to maintain the track and roadbed. There are also weed and brush cutting machines, cranes, and special snowplows that travel on the tracks.

Have you read a good freight train lately?

Have you ever stood and watched a long freight train go by? There are many names (Sante Fe, Great Northern, Union Pacific, etc.), and many types of cars (boxcars, flat cars, tankers, etc.). How does a freight line keep track of the different cargo and the companies that own it?

Well, the rail lines have devised a system called **Automatic Car Identification**, or **ACI**. In this system, each car has its own identification number. This number is on the side of the car, but it doesn't look like a number. Instead, it is a series of lines or stripes — just like the lines you see on canned goods or magazines.

Many supermarkets now have electronic readers that "read" the lines on groceries. A small screen then displays the price of the item. Freight companies use readers, too. These devices read the lines on rail cars as they go by. The reader can tell who owns the car and what it is carrying. This information is then transmitted to the freight company office. By having readers at different points along the track, the company can tell where the car is at any particular time.

David Frazier

Fig. 12-11   The United States railroad system.

PHYSICAL FACILITIES

The facilities used in rail transportation can be divided into two categories. First, there are the ways and roadbed that the trains travel on. Second, there are the buildings or areas used to classify, schedule, and maintain vehicles.

Ways and Roadbed

Years ago, when the railroads were first being developed, the government granted special sections of land to the railroad lines. These strips of land provided a way for the trains to travel through communities and the countryside. Since the railroad lines had the right to use the land, the land was (and still is) called the railroad **right-of-way**.

There are over 200,000 miles of right-of-ways (or simply "ways") making up the railroad system in the United States. See Fig. 12-11. A single railway has four parts: land, ballast, ties, and rails.

Today, the land is either owned by the railroad line or is leased by them. The **roadbed** consists of the land plus the **ballast** (rock) used to build up the land. **Ties** are lengths of wood or concrete that are settled into the ballast. The steel rails are laid on top of the ties.

All railroad tracks in North America are built to a standard **gauge**, or width apart. See Fig. 12-12. The rails themselves may be different lengths. However, the shape is always the same.

Railway Use and Maintenance. The money needed to build, maintain, and pay taxes on the

Fig. 12-12   A cross-section of a railroad "way."

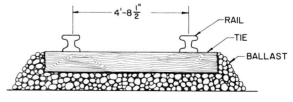

railways is one of railroad lines' major expenses. Most tracks are built to last many years. However, sometimes temporary tracks are laid. These are used to transport single very large pieces of equipment, such as an electrical transformer or a nuclear reactor boiler.

Sometimes one railroad line will pay another to use their tracks. For example, the AMTRAK system operates trains over the entire United States. However, the system actually owns very little track. Instead, AMTRAK pays each line to run their trains over that line's tracks.

As trains roll over the rails, the tracks become thinner and thinner. Federal, state, and local inspectors constantly check the railways. They look for thin or broken rails, rotted or missing ties, and washed-out ballast. These are the major causes of railroad accidents, such as derailed engines and cars.

Union Pacific Railroad

Fig. 12-14 A rail classification yard.

Yards, Shops, and Terminals

The other railroad facilities include classification yards, repair shops, and terminals. Each type of facility is used for a specific purpose.

Classification Yards. You have already read about how local trains carry rail cars to yards for regrouping into larger trains. At these yards, long trains are also "taken apart" and regrouped into several local trains. See Fig. 12-13. Classification yards are also used to switch cars from one railroad line onto another.

Figure 12-14 shows an aerial view of a classification yard. One of the main jobs of workers

in the **switch crew** is to position cars in a train so that the transport will be safe. For example, cars carrying chemicals may cause dangerous reactions if they are placed next to each other. The workers must also arrange the cars so that delivery will be efficient and convenient.

The progress of incoming and outgoing trains is monitored by a **dispatcher** in a control tower. See Fig. 12-15. This person makes sure that trains and cars are switched onto the right tracks.

Repair Shops. Railroad vehicles may require maintenance or repair at any point along the

Fig. 12-13 **At classification Yard 1, incoming cars are assembled into trains for transportation to another city. Yard 2 is at the destination city. The train is disassembled there into several local trains for final delivery.**

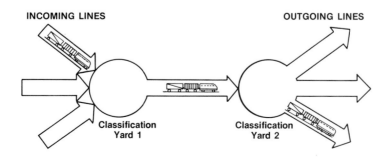

INCOMING LINES

OUTGOING LINES

Classification Yard 1

Classification Yard 2

Southern Pacific

Fig. 12-15 **A dispatcher coordinates train traffic.**

way. Therefore, rail lines operate many repair shops. Some repair shops are portable. If a train breaks down at a remote spot, repair tools and equipment can be brought to the train. This is much easier than uncoupling the train and bringing a broken car or engine to the repair shop.

Major repairs and maintenance work are done in large shops. These are called **engine shops** or **car shops**.

Terminals. Terminals are meeting places for trains and their crews. Often, the business offices of a rail line are located in the terminal building. There are basically two types of terminals: freight and passenger.

At freight terminals, shippers can deliver and pick up freight. At passenger terminals, passengers buy tickets and board (get on) trains. Workers for the line handle the passengers' baggage.

STUDY QUESTIONS

1. What is rail transportation?
2. What characteristic makes rail transport good for large and heavy loads?
3. Give two advantages of rail transportation.
4. What two basic kinds of service do rail lines provide?
5. Arrange the following in order of flexibility, from least flexible to most flexible: regular rail freight service, long-distance trucking, unit train service.
6. What word describes all rail vehicles?
7. What kind of fuel is used in most railroad engines?
8. Name four kinds of railroad cars.
9. Name the four parts of a single railway.
10. What happens at a classification yard?

ACTIVITIES

1. Locate a safe place from which to watch trains go by. For each train you observe, do one or more of the following:
   A. Record the names of the different lines that own the cars.
   B. Identify the different types of cars.
   C. Count the cars. Then estimate the length of the train.
   D. Identify whether the train is a local train or a "long-haul" train. (Hint: Long-haul trains require more horsepower.)
   E. Determine which car is the oldest. (Look for a date listed under "BUILT" on the side of the car.)
   F. Determine what kind of maintenance has been done on the cars lately. (Again, look for information on the side of the car.)
2. Using the local telephone book, list the different railroad lines that serve your area.

Here's how it works...

1 Mr. Gonzales raises oranges in southern Texas. He has sold his entire crop to a buyer in New York City.

2 Here Mr. Gonzales is meeting with a salesperson from a major rail freight line. They will decide on the type of cars needed, the day of departure, the loading area, and other details of shipping. The salesperson answers all of Mr. Gonzales' questions and explains how the rail line will take care of the oranges during shipment. It is important that the produce not be damaged.

5 Once the shipment is packed, Mr. Gonzales' foreman seals the car doors. He notes the number of the car and the number of the seal.

6 A local switch engine arrives to pick up the refrigerator car. The conductor of the local train notes down the car's number again. This information goes into the rail company's computer, and the predetermined route plan goes into action!

4 The rail line has allowed 24 hours to load the car. The loading crew gets to work right away. They pack the crates well so that there will be no bouncing around during shipment.

3 On the departure day, a switch engine spots (positions) a refrigerator car at the community railroad dock. The dock isn't far from Mr. Gonzales' grove and warehouse.

8 The road crew inspects the engines they will use on the trip. Meanwhile, the produce car, along with many other cars, is "built" into a long freight train.

7 The yard operator already has a "hang sheet" on the carload of oranges. This tells him where to "hang," or spot, the car as it enters the classification yard.

9 The engines check out O.K. It's time to hook up to the long train of cars. The trainmaster has decided which outgoing track the train will use. The shipment of oranges will soon be on its way!

10 The engineer pulls the train out of the yard and "stretches it out." Now the train has 176 cars — nearly one-and-one-half miles long! Its fastest speed will be only about 5 miles per hour.

13 The "new" train continues on to New York. By this time, several different crews have operated the train. Each crew runs the train for only certain amounts of time — usually four or five hours. This is the time needed to cover a typical section of the track.

14 The train arrives at a classification yard just outside New York City. This time, the car carrying the oranges will be transferred to a local train for delivery.

208

12 The train must switch onto several different lines during transport. At each transfer point, the engines and caboose are changed. A new road crew gets on to the train further on down the line.

11 A lot of system control goes on during shipment. The engineer receives orders on how fast to move the train. He also learns of track conditions up ahead and the locations of other trains on the same line. While the engineer and the firer watch the front of the train, the conductor and braker watch the rear. The crew stays in contact with each other using portable radios.

15 A "belt line" railroad company moves the car from the large yard to the buyer's loading dock.

16 The shipment has arrived and is inspected! The freight line has done its job well. There is no damage to the oranges. The buyer signs a receipt, and Mr. Gonzales is billed for the freight service. Rail transportation has done it again!

Air Transportation

In air transportation, special vehicles move passengers and cargo through the earth's atmosphere. Different types of aircraft are used. However, they all go through the same basic process. The aircraft takes off from one airport, flies to another, and lands there.

FEATURES AND BENEFITS

Compared to other modes of transportation, air transport is relatively new. The Wright brothers made their first historic flight in 1903. See Fig. 13-1. Since that time, air transport has developed into a very large and important part of the entire transportation system.

Two features make air transportation especially attractive. First, there is very little "lost time" during an air flight. There are no traffic lights, stop signs, or speed limits. Second, planes can fly from one place to another in a straight line. And as we all know, a straight line is the shortest distance between two points.

The above features make air transport the fastest mode of travel. However, compared to the other modes, it is also the most expensive per ton per mile.

COMPANIES AND SERVICES

Like other modes of transportation, air transportation can be used to move both passengers and cargo. Sometimes planes carry only passengers. At other times, planes carry only cargo, as with air freight companies. However, most air transportation is the movement of passengers and their personal baggage.

There are two basic types of carriers in air transportation: private sector and for-hire. Private sector operations are also called **general aviation**. For-hire operations are also called **commercial aviation**. One other category of air transport is military aviation.

General Aviation

Do you know anyone who owns a small plane? Flying a small plane is a part of general aviation. General aviation includes all privately owned aircraft that are used for personal flying.

General aviation also includes some aircraft that are used for business purposes. Crop dusters, medical helicopters, and business jets are a few examples. See Fig. 13-2.

Smithsonian Institution

Fig. 13-1 **We can trace the beginnings of successful air transportation to Orville and Wilbur Wright's first flight at Kitty Hawk, North Carolina.**

Fig. 13-2 Many business executives travel to meetings in small jets like this one.

United Airlines

Fig. 13-3 This airline offers regularly scheduled passenger and freight service to many different cities in the U.S. and around the world.

Commercial Aviation

Commercial aviation is what most people think of when they think of air transportation. Commercial air carriers are in business to make a profit. The many regularly scheduled airline flights are the best example of this business. See Fig. 13-3.

There are two main types of commercial air carriers: scheduled and non-scheduled. **Scheduled airlines** provide regular service to and from towns and cities. The aircraft always fly on a set schedule.

Non-scheduled airlines are also called **charter airlines**. These airlines operate only when a person or group of persons hires them for a special flight.

There are four basic types of scheduled airline services: international, domestic, regional, and commuter. Let's start with the last one — commuter service. **Commuter airlines** generally provide transport to a major airport from several smaller towns or cities. **Regional airlines** fly only in specific regions of the country. They connect several major airports with smaller airports. See Fig. 13-4.

Domestic airlines fly to and from major airports across the United States. **International airlines** offer service between the U.S. and other countries. Of course, these same airlines offer service within the United States.

Fig. 13-4 Regional airlines use smaller airplanes to serve cities in several different states.

Air Wisconsin

AIR TRANSPORT VEHICLES

People may think that there is only one type of vehicle used in air transportation — the airplane. However, there are really two basic types of vehicles. **Aircraft** includes all the vehicles that fly through the air. This would include airplanes, jets, helicopters, and so on. **Ground support vehicles** are land travel vehicles. They are needed to maintain and support the aircraft.

Aircraft

All vehicles that fly are called *aircraft*. Aircraft can be divided into two basic categories: those without fixed wings and those with fixed wings. One example of an aircraft without fixed wings is a **helicopter**. This vehicle uses spinning blades to lift it off the ground and drive it through the air. **Lighter-than-air vehicles** are another type of aircraft that don't have fixed wings. For example, a **blimp** uses helium to lift it off the ground. Motor-driven propellers drive the craft.

Aircraft that have fixed wings are called **airplanes**. Airplanes are heavier than air. However, they are designed to fly when moved forward by an engine.

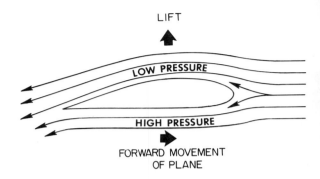

Fig. 13-5 The shape of an airplane wing causes the air traveling over the wing to go faster than the air traveling under the wing. This creates lift.

Igor and the ''Whirlybird''

Igor Sikorsky had a love for flying. He was especially fascinated with helicopter designs. He studied drawings made around 1500 by the artist/inventor Leonardo Da Vinci. These drawings gave him the idea for the world's first practical helicopter.

Sikorsky had many problems in designing his helicopter. His main problem was in coming up with a design that would work well. He also had to teach himself to fly the new aircraft. But the Russian-born inventor was not overwhelmed by these difficulties. He kept on improving his design until it worked. He made 19 major design changes over two years. Finally he developed his original idea into a practical aircraft. People called the new aircraft a ''whirlybird.''

In 1942 a Sikorsky helicopter flew from Connecticut to Ohio. This was the first extended flight of such an aircraft. The flight began a new chapter in military and aviation history.

Helicopters changed the basic tactics of war. They provided new ways to carry troops into battles and to carry the wounded out. Of course, helicopters also became a very useful method of civilian transportation. Sikorsky was especially proud that helicopters served mainly to save lives rather than to destroy them.

Sikorsky's favorite quote hung on the wall of his office. It read: *According to recognized aerotechnical tests, the bumblebee cannot fly because of the shape and weight of its body in relation to its total wing area. The bumblebee doesn't know this, so it goes ahead and flies anyway!* This quote told the story of Igor Sikorsky's lifework. He designed and built a new aircraft. He *knew* it would fly, even though other people said it would never fly.

Figure 13-5 shows a cross-sectional view of an airplane wing. Notice the wing's special shape. As the engine moves the plane down the runway, air rushes both over and under the wings. This creates a high-pressure area under the wings and a low-pressure area over the wings. Pressure always moves from high to low. Therefore, the higher pressure below the wings pushes them up. The upward movement is called **lift**.

There are many different sizes and shapes of airplanes. Basically, we can classify them according to the type of power source used. **Prop planes** are powered by an engine-driven propeller. **Jet planes** are powered by jet engines. See Fig. 13-6. (You'll learn more about engines in Section V of this book.) Airplanes can also be classified by the number of engines used. Some planes use only one engine. Others use two or more engines.

Airplanes have two main parts: the airframe and the powerplant. The **powerplant** is the engine or engines used on the plane. The **airframe** basically consists of the structural frame and the metal "skin" of the plane. See Fig. 13-7.

Fig. 13-6 **Some planes have one or more propellers that "pull" them through the air. Other planes use turbojet engines.**

Fig. 13-7 **An airplane has two basic parts: the powerplant and the airframe. The engines make up the powerplant. The airframe includes the fuselage (main body of the plane), the wings, and the tail.**

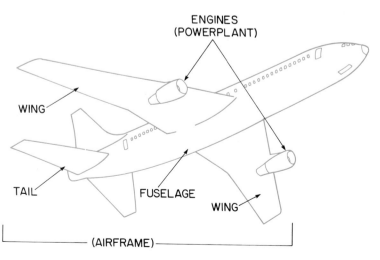

Ground Support Vehicles

The air transport industry requires many different vehicles besides aircraft. These **ground vehicles** support the operation of the aircraft. For example, special trucks are used in the preparation stage of a flight. Each type of truck has a different purpose. See Fig. 13-8.

Tank trucks carry jet fuel to the plane. They have built-in pumps for transferring fuel into the plane's tanks. **Food service trucks** look much like ordinary trucks. However, the bed or box can be raised to the same level as the plane. This way, workers can unload food directly from the truck onto the plane.

The **potable water truck** supplies fresh drinking water to the plane.* The **sanitary service truck** empties waste water from storage tanks on the plane. **Baggage-loading trucks** have built-in conveyors. The conveyors are used to transfer baggage into and out of the airplane's cargo hold.

In addition to trucks, airports also use tugs and special tractors for pushing and pulling. Small tractors are used to pull baggage carts. Larger tractors pull air compressors and other equipment. **Tugs** are heavy-duty ground vehi-

City of Houston, Dept. of Aviation

Fig. 13-9 **A powerful tug is used to push large airplanes back out and away from the loading terminal.**

cles. They have powerful diesel engines and large rubber tires. An airplane can't back up by itself. So, to move a plane, workers hook up a tug to the plane's nose wheel. They can then push the airplane out and away from the terminal building. See Fig. 13-9.

Other ground support vehicles include baggage trailers, air conditioning trailers, and auxiliary power units. All of these are needed to service the airplane properly.

Fig. 13-8 **Many special ground support vehicles are used at every large airport.**

Los Angeles International Airport

A. Tank truck

B. Food service truck

City of Houston, Dept. of Aviation

C. Rescue vehicle

D. Baggage-loading truck

City of Houston, Dept. of Aviation

City of Houston, Dept. of Aviation

PHYSICAL FACILITIES

Like other transportation modes, air transport requires certain physical facilities for operation. The routes that aircraft take in air transportation are called **airways**. The **airport** is a facility that includes several different buildings.

Airways

There are no visible roads or highways in the sky. There is just **airspace**. We can talk about the general airspace above the surface of the earth. Or we can speak of the particular airspace above a certain area of land or water.

In the United States, the Federal Aviation Administration (FAA) has set up a system of airways. The airway system is a series of zones and routes for aircraft to fly in. There are specific routes for private, commercial, and military aircraft.

Planes can fly anywhere in the air. This makes them a very flexible form of transportation. However, for safety, planes must be kept apart from each other while flying. Therefore,

the FAA system uses **air layers** to separate the aircraft. See Fig. 13-10. Here's how the layers work:

- All airspace from the ground level up to 75,000 feet is divided into three main layers.
- Each main layer is sub-divided into smaller layers 1000 feet "thick."
- All the layers are numbered according to their height. For example, an **even-numbered** layer would be 28,000 feet. An **odd-numbered** layer would be 15,000 feet.
- Even-numbered layers are used by aircraft flying in a westerly direction. Odd-numbered layers are used by aircraft flying in an easterly direction. See Fig. 13-10 again.

Besides using the correct air layer, the pilot must also follow certain rules and regulations. For example, aircraft must always be separated by 1000 feet of airspace above and below. There should be no less than 10 minutes of flying time in front of or in back of the aircraft. Furthermore, aircraft must be separated by at least 10 miles on either side.

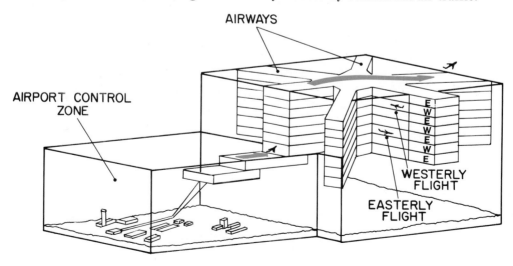

Fig. 13-10 **This drawing shows air layers used by commercial air traffic.**

The Airport

All flights begin and end at an airport. The terminal building and the runways are two of the most visible parts of an airport. But there are many other features that are very important. See Fig. 13-11.

Fig. 13-11 **Large airports, like this one at Chicago, have many features.**

Runways. Airplanes take off and land on runways. Runways are usually made of steel-reinforced concrete several feet thick. The length of the runway determines the size of the plane that can land there. At some airports, runways are almost three miles long.

Runways are used for take-offs and landings. **Taxiways** are the paved "roads" that connect the runways with the rest of the airport area. Planes **taxi,** or travel at slow speeds, between the terminal and the runway. See Fig. 13-12.

Chicago Department of Aviation

Fig. 13-12 **Paved areas of an airport. Notice that the large area between the terminal and the taxiways is called the *apron*.**

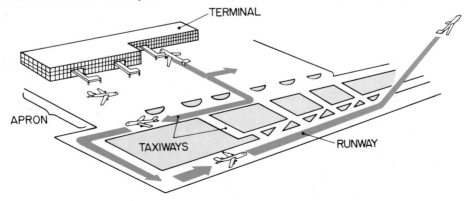

The Terminal. The terminal building generally has more than one level. The upper level houses airline offices, ticket counters, restaurants, and other facilities. See Fig. 13-13. Passengers wait at certain **gates** before boarding their flights. On the lower level, workers sort baggage, load and unload trucks, and do many other things that support the various flights.

Control Tower. The control tower is another very important building. It has a glass-enclosed room at the top. From this room, air traffic controllers monitor arriving and departing flights. They also give directions to the pilots of the various planes. See Fig. 13-14. In another room, usually somewhere below the tower, other controllers watch radar screens. These controllers give directions to pilots flying toward and away from the airport. See Fig. 13-15.

Other Buildings. There are several other buildings at major airports. Hotels, fire stations, and **hangars** (garages for aircraft) are just a few examples. Other facilities, such as radar towers, special lights, and water towers, are needed to support the airport operation.

City of Houston, Dept. of Aviation

Fig. 13-13 The terminal building is the center of airport activity.

Federal Aviation Administration

Fig. 13-14 From the control tower, air traffic controllers can see all the planes and give directions.

Federal Aviation Administration

Fig. 13-15 At the TRA-CON (Traffic Control Center), air traffic controllers watch radar screens to give directions to the many planes flying near the airport.

Here's how it works...

1 Ray Dixon is in the last semester of a two-year drafting program at the local community college. With graduation in sight, Ray has inquired about jobs in the machine tool industry. A company in Chicago has asked Ray to come for an interview. Here Ray is making flight arrangements with a travel agent.

2 Ray arrives at the airport on the day of departure. A "skycap" greets Ray and carries his luggage to the flight desk. Ray's ticket is checked and his luggage is marked for the same flight. The luggage will be transported in a separate section of the plane.

5 The food service truck has pulled up to the plane. Workers unload prepared meals, snacks, and beverages into the galley (kitchen). The hot meals are stored in movable insulated cabinets.

Photography by Tom Cinoman

3 While the plane is being serviced, the cabin crew prepares for the flight. The co-pilot checks the weather report, and the pilot files a flight plan with the airport authority.

4 Several things have to be done before the passengers board the plane. Mechanics check the engines and other parts of the plane. The fuel tanks are filled with jet fuel. And the waste water from the last flight is emptied.

6 Here Ray is finding out at which gate he will board his flight. Any changes in departure time will be posted on the board.

7 Each gate has a check-in counter and a waiting area. After checking in, Ray receives a boarding pass. Soon a flight attendant will open the *jetway* — an enclosed walkway from the gate to the plane.

8 At boarding time, flight attendants greet the passengers at the entrance to the aircraft. The attendants check the passengers' passes and help them find their assigned seats.

9 Ray's seat is near a window. He has already stored his carry-on luggage in the cabinet above his seat. Now, with his seat belt fastened, he is ready for departure.

11 The pilot has received take-off clearance from the control tower. When the plane reaches the correct ground speed, it lifts off. When the plane is off the ground, the pilot will retract the wheels.

12 The flight to Chicago takes several hours — plenty of time for the passengers to get hungry. Here the flight attendants are serving dinner.

10 Some aircraft — such as helicopters — can move in reverse. Airliners, however, must be pushed out of the loading area by a tug.

13 The plane is guided into O'Hare Field in Chicago by air traffic controllers. After landing the aircraft, the pilot receives permission to taxi to the boarding ramp area.

14 Ray has left the plane and is going to pick up his luggage. But wait! There's his potential employer on hand to greet him. This could be the start of a new career for Ray!

STUDY QUESTIONS

1. What two features make air transportation the fastest mode of travel?
2. Name the two basic categories of air transportation.
3. What are the two categories of commercial aviation?
4. Name the two basic types of vehicles used in air transportation.
5. Name one kind of aircraft that doesn't use fixed wings.
6. Name the two main parts of an airplane.
7. What is a powerplant?
8. What does the airframe include?
9. List four kinds of ground support vehicles.
10. Why is the airway system necessary?
11. List some standard rules a pilot must follow.
12. What is the purpose of a runway?
13. Describe what goes on in the control tower.

ACTIVITIES

1. Each commercial airport is registered with the FAA with a proper name and a special three-letter identification code. For example, the main airport in Dallas, Texas, is registered as Dallas-Fort Worth Airport, or DFW. Find the proper name and identification code for the nearest commercial airport in your area. (Hint: A travel agency may be able to help you.)
2. Using the local telephone book, list all the airlines that fly into the nearest airport.
3. Collect and compare different airline advertisements from your local newspaper. Find out which line offers the best fare to a certain place. Also compare the best times for travel. (Hint: Sunday papers are a good source for this type of advertisement.)
4. Keep notes on the different types of aircraft flying overhead. From your notes, try to determine the most common direction of travel.

Water Transportation

Vehicles that travel on water are called **vessels**. Boats, ships, and barges are just a few examples. The movement of cargo and passengers by these vessels is called *water transportation* (or simply **water transport**).

The first European explorers of North America used water transport a great deal. For example, the Frenchmen La Salle and Joliet used canoes in their explorations of the Mississippi Valley. Later, as people moved inland from the east coast, water transport developed into a system. Mules pulled barges along the Illinois-Michigan Canal from Chicago to the Illinois River. Flat-bottom wooden boats hauled freight from the Illinois River to the Mississippi River, and from there to New Orleans. See Fig. 14-1. Paddle-wheel steamboats carried passengers and cargo from New Orleans to Memphis and further north.

Without water transport, it would have taken many more years for the United States to extend its borders westward. Today, we depend on this mode of transportation more than ever.

Missouri Historical Society, Collet, Neg. 190

FEATURES AND BENEFITS

Water transport is an inexpensive and fairly flexible mode of transportation. Ships and barges are slower than vehicles used in some other modes. However, water vessels are more fuel-efficient.

All kinds of loads — including passengers and all types of cargo — can easily be moved on water. Water transport has only one obvious limitation: It can only take place where there are waterways. As you read on, you'll learn about the many international and domestic waterways.

COMPANIES AND SERVICES

There are two main types of water transportation companies: passenger lines and cargo lines. Most water transport involves the shipping of cargo, either within the country or to and from other countries.

Passenger Lines

Passenger lines are either private **luxury lines** or public transportation lines. Luxury lines are also called **cruise ship lines**. These companies offer vacation cruises, such as a cruise through the Caribbean islands.

Fig. 14-1 **Flatboats were used to transport passengers and cargo in the 1800s.**

Public water transport lines are operated by both state and local governments. For example, in some parts of Louisiana, school children ride to school in yellow "bayou buses" instead of regular school buses. See Fig. 14-2. Passenger and vehicle ferry service is another common type of public water transport.

Cargo Lines

A **port** is a place where ships stop, or **dock**. Cargo lines dock at ports all over the world. Cargo ships transport most of the goods that are imported into the U.S. or exported from the U.S. (To *import* something means to bring it into the country. To *export* something means to send it to another country.)

The largest amount of water cargo transport is done by **barge lines**. Some barge lines own several hundred barges. The barges operate on both inland waterways and ocean waterways. This lets the barge lines offer very flexible service.

Types of Services

There are two basic types of water transport services: domestic inland shipping and international merchant shipping. You may remember the word *domestic* from Chapter 13. It refers to things that take place within a country. In this chapter, **domestic inland shipping** refers to all shipping that connects different parts of inland North America. The vessels used in international merchant shipping connect North America with other countries that have seaports. See Fig. 14-3.

Ocean-going vessels can provide two types of service. One type is called **berth**, or **liner**, service. A liner is a vessel that follows the same route on a regular basis, like a regularly scheduled airliner.

The other type of ocean-going service is for-hire, or **tramp**, service. Tramp service is like the charter flights offered by airlines. A tramp vessel will haul cargo to any port the shipper wishes.

Lykes Lines

Fig. 14-2 Do you ride a school bus to school? Some students in Louisiana ride a school *boat*.

Fig. 14-3 Both basic types of water transport services are shown here. The barges are examples of domestic inland shipping. They are loaded onto a barge carrier with several levels, or *decks*. The barge carrier is an example of international merchant shipping.

Fig. 14-4 Barges can be connected to form a *tow*. This tow is carrying a variety of cargo.

WATER TRANSPORT VESSELS

Water transport vessels vary greatly in size. They may be as short as an average car or longer than a football field. In the same way, a vessel may be fairly inexpensive or may cost millions of dollars to build. Generally speaking, though, a water transport vessel makes up for its construction expense with a long and efficient transport life.

Most vessels used for transportation can be classified into three categories. There are (1) barges, (2) ships, and (3) tugboats and towboats.

Barges

The best way to describe a barge is as a "floating box." Some barges have lids, or **hatches**, to cover the cargo. Others don't. A barge can be loaded and sealed in the same way that a container can be loaded and sealed.

Another name for barge is **lighter**. This term comes from the fact that a barge is lighter than a ship.

Barges don't have engines. Therefore, they must be pushed or pulled by an engine-driven vessel. Several barges can be lashed (connected) together to form a **tow**. See Fig. 14-4. Many tows consist of more than 30 barges.

Some barges are built to carry liquids or gases. Other barges carry dry bulk cargo, such as sand and gravel. Still others carry break-bulk cargo. Some special barges even transport the containers that are used in Container On Flat Car shipping.

Ships

There are many different kinds of ships. The ones most common in water transport are general cargo ships, tankers, and containerships. We'll also take a look at two special types of ships: LASH ships and cruise ships.

General Cargo Ships. General cargo ships are the "freight trains" of water transportation. These ships usually have two or three levels, or **decks**. Inside the ship are four or five areas called **holds**. A hold is a large cargo storage area. Each hold opens onto the top deck and can be covered with a hatch.

Most cargo ships have their own cranes for loading and unloading cargo. See Fig. 14-5.

Fig. 14-5 These cargo ships are being loaded with break-bulk cargo. Notice the on-board cranes.

American Petroleum Institute

Fig. 14-6 This natural gas tanker is being prepared for an ocean voyage. The on-shore facility cools the gas to a temperature of −260° F *(−162° C)*. The liquified gas is then pumped into the ship's insulated domes.

Fig. 14-7 This supertanker holds over two million (2,000,000) barrels of oil.

Sea Land Service

Fig. 14-8 This containership can carry many containers. In port, it can be quickly unloaded and loaded.

Cargo ships are also called **break-bulk ships**. This is because the cargo is loaded or unloaded one piece at a time.

Tankers. Tankers are special ships designed to carry liquid cargo. A natural gas tanker is easy to identify because of the large metal spheres used to hold the liquified gas. See Fig. 14-6.

Supertankers are used to transport huge amounts of crude oil. These ships are the largest ships in the world. Some are over 1200 feet *(365 meters)* long! See Fig. 14-7. Supertankers are loaded by pumping the crude oil into several large holds. Partitions between the holds keep the cargo from sloshing too much during transport. At the destination port, the tanker is unloaded with on-board pumps.

Containerships. If you've read Chapters 11 and 12, you know about container transport. Containers can be shipped by highway transport, rail transport, and — you guessed it — water transport. Containerships are vessels that are especially designed to carry containers. See Fig. 14-8.

The advantage of a containership is that it can be loaded and unloaded quickly. General cargo ships are slow-loading because the cargo must be handled one piece at a time. But containerships carry cargo that has already been loaded. The shipper does the packing. Then the containers are sealed and transported by truck, train, or barge to the seaport. There the containers are loaded into containership holds or simply fastened to the top deck.

Barge Ships. There are two basic types of barge ships: LASH ships and SeaBees. Both types are large ships used to carry barges across the ocean.

LASH stands for **L**ighter **A**board **Sh**ip. LASH ships have cranes that pick lighters (barges) out of the water and hoist them onto their top deck. See Fig. 14-9.

On a **SeaBee ship**, the **aft** (rear) of the ship actually opens to receive cargo. First, lighters are floated onto a special elevator in the water near the ship. The elevator lifts them to one of several decks. The lighters are then pushed on rails into the ship and secured in place. (Figure 14-3 shows a SeaBee ship being loaded.)

Fig. 14-9 **This special barge ship can hold many smaller barges for international transport.**

The Floating Railroad

The Alaska Hydro-Train is a different kind of railroad. It operates over the water between Seattle, Washington, and Whittier, Alaska. There is no direct railroad link between the lower 48 states and Alaska. Therefore, the floating railroad provides a valuable service.

Now, how can a railroad float? The Alaska Hydro-Train floats on giant barges that are each larger than a football field. These special barges have railroad tracks on their deck. In Seattle, rail cars are **rolled on** the barges and secured in place. Special wheel locks hold each car in position. Each whopper barge can carry 55 rail cars!

Workers connect two barges with towing chains. Then a powerful tugboat pulls the barge train up the Pacific coast. Can you imagine the weight of 110 railcars, *plus* the weight of the barges themselves? It takes a 9000-horsepower tugboat to pull this kind of tow. The 1400-mile trip takes five-and-a-half days.

On arrival at Whittier, the railcars are **rolled off** the barges onto the Alaska Railroad line. Then the cars are delivered to the different locations by locomotive.

Another name for the service provided by the Alaska Hydro-Train is **RO/RO**, or **Roll On/Roll Off**. Much of the cargo is never rehandled from its departure point to its final destination. Of course, this makes for efficiency and economy. Add to this the overall economy of water transport, and what do you have? . . . An ideal way of connecting the lower 48 states with the great 49th!

Cruise Ships. Cruise ships are large passenger transport vessels that make sea cruises or ocean crossings. They have several decks. One deck may have staterooms (bedrooms) for the passengers. Other decks have restaurants, shops, game rooms, or swimming pools. Cruise ships need many workers to provide services for the passengers. See Fig. 14-10.

Carnival Cruise Lines

Fig. 14-10 **Relax! Take a vacation cruise on a ship like this one.**

Tugboats and Towboats

Tugboats and towboats are small but very powerful. They use diesel engines that can develop more than 5000 horsepower. Many people confuse tugboats and towboats, but there is a clear difference between the two.

Tugboats are used for pulling barges and tows. They are also used to guide large ships into port. See Fig. 14-11. **Towboats**, on the other hand, do not "tug" other vessels. Instead, they have a flat **bow** (front end) for pushing barges and tows. See Fig. 14-12.

Crowley Maritime

Fig. 14-11 **Tugboats are used for pulling barges and other vessels.**

Fig. 14-12 **Towboats have a flat bow for pushing barges.**

PHYSICAL FACILITIES

Like any other transportation mode, water transport requires certain facilities to operate. The waterways themselves are a necessary part of the system. We'll look at the two basic types of waterways: sea-lanes and inland waterways. The other water transport facilities include ports and terminals.

Sea-lanes

A sea-lane is an established shipping route across an ocean. There are sea-lanes on all the oceans of the world. See Fig. 14-13. International merchant ships follow these routes for the safety of the crew, the ship, and the cargo. Can you imagine the confusion if ships traveled anywhere they wanted?

Inland Waterways

There are thousands of miles of inland, or **domestic**, waterways in the United States. See Fig. 14-14. Some of the waterways are used mainly by barges. Other waterways can be used by other vessels, such as ore ships and general cargo ships. The following waterways are the most important:

- The **Atlantic Intracoastal Waterway**, which extends from New Jersey to Florida along the Atlantic coast.
- The **Great Lakes — St. Lawrence Seaway**, which connects the Great Lakes with the Atlantic Ocean. The St. Lawrence Seaway is made up of the St. Lawrence River and a series of man-made canals. It is the result of a joint building effort on the part of the United States and Canada. Ocean-going vessels can enter the Seaway at Montreal, Canada, and travel all the way to Chicago, Illinois.

Fig. 14-13 **Ships that travel on the Atlantic Ocean must follow the sea-lanes shown here.**

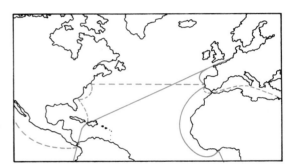

American Waterway Operators

Fig. 14-14 **The inland waterways provide a way to ship cargo from state to state. Also, barges can deliver freight to seaports and pick up freight from international shippers.**

- The **Gulf Intracoastal Waterway**, which runs along the coast of the Gulf of Mexico. It extends from Brownsville, Texas, to northern Florida.
- The **Mississippi River System**, which consists of the Mississippi River and the major rivers that flow into it. These rivers include the Illinois River, the Arkansas, the Missouri, the Ohio, and the Tennessee. Most of the traffic on this system is barge traffic. Barges can carry freight all the way from Minneapolis, Minnesota, to the Gulf of Mexico.
- The **New York State Barge Canal**, which connects the Hudson River in New York with Lake Erie. This waterway was originally the Erie Canal, the first major waterway built in the U.S.

In addition to the waterways listed above, there are three major waterways on the Pacific coast. These are (1) Puget Sound, (2) the Columbia River and the Snake River, and (3) the Sacramento River. See Fig. 14-14 again.

There is also a new major waterway being developed in the southeastern United States. The Tennesee-Tombigbee (TENN-TOM) Water-

way will serve the area between northwestern Mississippi and Mobile, Alabama.

Ports

A port is a place where boats or ships dock to load and unload passengers and cargo. There are many ports located along sea coasts, lake fronts, and rivers. The port at which a vessel stops is called the **port of call**.

Some ports offer a wide variety of services, such as loading and unloading, fueling, and repair. Other ports are very specialized. For example, they may only offer rapid loading and unloading of certain types of cargo. The cargo could be containers, grain, or oil. This service goes on 24 hours a day. See Fig. 14-15.

Some ports are very large. The Gulf ports of Houston and New Orleans are two examples. Other ports are small, like the river port at Cairo, Illinois.

Vessels are usually not docked long in port. The reason for this is that ships and boats make money only when they're moving. Therefore, a port that offers fast service is popular with water transport companies.

Port of Long Beach

Fig. 14-15 **This port specializes in loading and unloading containerships.**

Terminals

There are two basic types of water transport terminals: passenger terminals and cargo terminals.

Passenger Terminals. As in the other transportation modes, water transport companies need facilities to take care of embarking and disembarking passengers. (To *embark* means to get on board. To *disembark* means to leave the vessel.)

Passenger terminals have waiting areas, restaurants, and shops. Here, workers gather the passengers' luggage and load it onto the ship. Food and other ship's supplies are also loaded at the terminal.

Cargo Terminals. The main purpose of a cargo terminal is to provide a place to store incoming and outgoing cargo. At many ports, the actual loading and unloading of cargo can be done in a short time. However, it takes much more time to assemble the outgoing cargo at the port. It also takes time to prepare the unloaded cargo to be forwarded to some other point in its journey.

At cargo terminals, cargo is also transferred from ships and barges to other modes of transportation. These other modes may include highway transport, rail transport, air transport, and even pipeline transport.

After discharging their cargo, ships and barges take on fuel, food, and spare parts. These and other items are needed for the next voyage.

One special kind of terminal is the **barge terminal**. At these places, barge tows are either assembled or "broken apart."

STUDY QUESTIONS

1. What is water transportation?
2. What are two advantages of water transportation?
3. Name a limitation of water transportation.
4. List the two main types of water transportation services.
5. What is liner service?
6. What is tramp service?
7. What is another term for barge?
8. Why is a general cargo ship also called a *break-bulk ship*?
9. What do LASH ships carry?
10. What is a sea-lane?
11. List the major inland waterways in the U.S.
12. Why don't ships stay in port very long?

ACTIVITIES

1. Obtain a map of the U.S. and/or Canada. Using the map and the information in this chapter, determine which inland and international waterways serve your community. Also identify the nearest port for each waterway. (Note: You may need references other than the ones given here. Check out your local library.)
2. Name three major cities served by domestic waterways and three smaller communities served by inland waterways.
3. Locate advertisements for cruise ship tours in the travel section of a Sunday newspaper. List the ports of call that the ships visit.

Here's how it works...

1 Every year, millions of tons of grain are shipped off North American farms and onto inland waterways. The waterways carry the grain to seaports for shipment to other countries. Here a hopper barge is being loaded with grain in Minneapolis, Minnesota. Its final destination? The Soviet Union.

2 When the barge is full, its hatch covers are sealed. This protects the cargo from the weather and thieves. Then the barge is floated away from the loading facility.

5 The tow is on its way down the Mississippi. On its journey, it may pass other tows going upriver.

3 In a tow-building area, the hopper barge is tied, or "rigged," to other barges. Several different types of cargo will be shipped down-river to New Orleans.

4 Once the tow is ready, a large line-haul towboat is rigged to the barges. The boat shown here has over 5,000 horsepower.

6 The pilot of the towboat has a broad view of the tow and the waterway. He or she operates many controls from the console in the pilothouse.

7 While the tow is underway, the crew members work at maintenance and repair. Most large towboats have two crews on board. The crews take turns working and resting in six-hour shifts.

8 The cook on a towing vessel prepares four meals a day and brews gallons of coffee. Food is one of the most important "inputs" in barge transport.

11 The pilot steers the tow by following navigational markers on the land. At night, some of the markers flash colored lights. Others are highly reflective. If the markers are difficult to see, the pilot can also navigate by radar.

12 The tow is nearing New Orleans. Once there the hopper barges will be loaded onto barge ships and shipped to the Soviet Union.

The Waterways Journal, St. Louis, Missouri, provided support services for the preparation of this photo essay.

SCNO Barge Lines, Inc.

9 On its way south, the tow must pass through 26 locks. A **lock** is a kind of basin in which vessels are raised or lowered to a new level by adjusting the level of the water. Some barges must be disassembled from the tow to pass through the lock.

10 There's the St. Louis Arch! Only 1,021 miles to go to New Orleans!

13 For the young deckhand, however, the story is not over. Now its time to clean the towboat and prepare it for the return trip up the mighty Mississippi!

Pipeline Transportation

Pipeline transportation is the movement of cargo through a tube or pipe. Pipelines were used to transport water as early as the Roman Empire. The English first added power to a pipeline system. In 1582, they connected a mechanical pump to the London water supply. Since then, the size and shape of the pipes have changed, and so has technology for moving the cargo.

Pipeline technology really began to advance when oil was discovered in Pennsylvania in 1859. See Fig. 15-1. Between 1865 and 1880, many crude oil and natural gas pipelines were built in New York and Pennsylvania. Pipeline construction also followed the discovery of oil and gas in Ohio, Illinois, Kentucky, California, Louisiana, and Texas. By 1910, there were two parallel eight-inch-wide pipelines going from a large refinery in Baton Rouge, Louisiana, to the east coast.

Fig. 15-1 **The world's first commercial oil well. Production of oil commercially encouraged the development of pipeline technology.**

American Petroleum Institute

FEATURES AND BENEFITS

Pipeline transport is the only transportation mode that does not carry passengers. Some other special characteristics of pipelines are:
- Pipeline transport is the only mode in which cargo moves while the vehicle stands still,
- Pipelines are usually buried in the ground. See Fig. 15-2. Therefore, they are unseen and quiet. They also cause no traffic congestion or accidents.
- Pipelines are laid out in straight lines across the country. This decreases the travel time.
- Theft from pipelines is difficult. Also, damage or contamination of cargo is rare.

Pacific Gas & Electric

Fig. 15-2 **Most pipelines are installed underground. This way, the land can still be used for other purposes.**

Fig. 15-3 **The Trans-Alaska Pipeline is partially above ground.**

Fig. 15-4 **The fuel in these tanks was pumped here through company pipelines.**

Pipelines are expensive to install. However, there is generally not much maintenance required on the installed lines. Along with the characteristics listed above, this makes pipeline transport the cheapest way of moving cargo.

Not all pipelines are underground. Some are above ground. For example, the Trans-Alaska pipeline is above ground for over half of its 800-mile total length. It is used for transporting crude oil. See Fig. 15-3. This major pipeline crosses 20 large rivers, 300 streams, and three mountain ranges. The Trans-Alaska line must constantly have oil moving through it. If the oil ever stopped moving, it would cool and thicken. This would block the line.

COMPANIES AND SERVICES

At different times in U.S. history, the federal government has helped develop transportation industries. During World War II, the government helped develop many petroleum pipelines. More recently, the governments of both the U.S. and Canada joined with several large oil companies to build the Trans-Alaska system. Today, however, most pipeline systems are owned and maintained by private companies only.

Pipeline transportation is provided by a small number of large companies. These companies offer very limited services.

Types of Companies

Most companies that ship material by pipeline own the pipelines themselves. They are in the transportation business to move their own products. These products are generally petroleum and natural gas. See Fig. 15-4. However, some shippers do not own their own pipelines. Instead, they pay the larger companies to move their products.

The larger companies benefit in two ways from leasing their lines to other firms. First, they make a profit by moving someone else's cargo for them. Second, the lines operate more efficiently when they are full. Having many users makes sure that the lines are full.

Types of Services

Pipeline operators offer a very simple service. They simply transport special types of cargo directly through a pipe from one place to another. Pipeline transport is not as flexible as other transportation modes. However, it can be the most efficient way to transport goods. There are only two conditions: (1) the shipper must be located near a pipeline, and (2) the product must be able to be moved through a pipe.

Types of Products. Three types of products can be transported through a pipeline: solids, liquids, and gases. (Of course, these are very broad categories.) Each type of product is pumped through the line in some way. For example, compressors are used to pump natural gas. Compressed air is used to move ground grain.

Other products moved through pipelines include crude oil, coal, wood chips, molasses, gravel, lard, and sulfur.

Some solid products, like coal, are moved through the line in a **slurry**. A slurry is a kind of rough solution. It is made by grinding the solid, then mixing it with a liquid. The product can then be moved through the pipeline. Slurry transport requires a lot of power to keep the cargo moving. Other words for the slurry are **batch** or **mass**. If the slurry stops moving, the ground-up solid settles to the bottom of the line. Starting the mass moving again is very difficult.

System Layout. There are over a million miles of pipelines crossing the U.S. and Canada. These lines range from two inches in diameter to fifteen feet in diameter. About half of the lines carry natural gas. Most of the remaining lines carry crude oil or petroleum products, like gasoline and kerosene. See Fig. 15-5.

PIPELINES

The most important part of a pipeline system is the line itself. A pipeline is made of plastic or steel sections that are joined together. See Fig. 15-6. There are three basic types of pipelines:

Gathering pipelines are used to "gather" together small quantities of unrefined products, such as crude oil. These lines bring the smaller amounts to a central holding tank.

Transmission pipelines are larger than gathering lines. They are used to transport cargo for long distances. **Distribution pipelines** deliver the cargo from a terminal to the customer.

All three types of pipelines are like one-way streets. The cargo can move in only one direction. If the cargo has to be returned, another line must be used.

American Petroleum Institute

Fig. 15-6 Steel pipes must be welded together to form the pipeline before being lowered into the ground.

Fig. 15-5 There are two types of petroleum-related pipelines.

A. Crude oil pipelines transport crude oil from oil field to refineries.

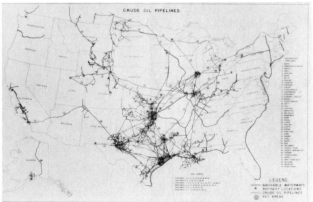

American Petroleum Institute

B. Product pipelines transport refined petroleum products from refineries to users.

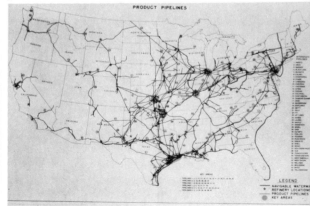

PHYSICAL FACILITIES

Several kinds of facilities are needed in pipeline transport. These facilities include pumping stations, control stations, measuring stations, and exchange stations.

Pumping Stations

As mentioned earlier, pipeline cargo is *pumped* through the line. The pumping stations house the compressors or pumps needed to keep the cargo moving. See Fig. 15-7. The pressure applied to the cargo depends on two things: (1) the type of pipe, and (2) the type of cargo. The pressure may range from 50 pounds per square inch to 2000 pounds per square inch.

The speed of the cargo also varies. Gas or powdered solids travel at about 15 miles per hour. Liquid or slurried cargo travels at about two to five miles per hour. To keep the cargo moving steadily, the pumps are positioned every 30 to 150 miles apart. As the cargo passes through a pumping station, it gets a "boost" to its movement.

System Control

The cargo that passes through the pipeline doesn't always travel at the same speed. Also, a pipeline can transport different types of cargo at different times. This operation requires a **control station**. The operator at the control station can start the cargo moving, stop it, speed it up, or slow it down.

As many as 30 different products can be transported through a pipeline. For example, a shipment of kerosene could be followed by a shipment of diesel oil. The diesel oil could be followed by a shipment of gasoline. All cargo or products shipped by pipeline are divided into batches. The order of the different products is called the **batch sequence**.

The control center operator monitors the flow of products through the pipeline. He or she makes sure the correct sequence is being followed. The operator also keeps in constant touch with other stations along the route. This way, the company will know if the cargo is being transported according to plan.

American Petroleum Institute

Fig. 15-7 **Pumping stations are also called boost stations. They keep the cargo moving steadily through the line.**

One important part of pipeline system control is to watch for leaks in the lines. A drop in pressure along one line may signal a leak. Some companies also use "flying pipeliners" to patrol the line from the air. The pipeliners can spot a leak before it causes any damage to the environment. In case a leak *does* happen, special safety devices shut off the faulty section of pipe from the rest of the line. This keeps the leaking to a minimum. Repair crews can then take care of the problem.

Like the water lines in your home, pipelines can become clogged. To prevent clogging, system workers routinely clean the line with a barrel-shaped brush called a **pig**. The pig is either blown through the line with compressed air or is pushed through by the moving cargo. See Fig. 15-8. This keeps the batch moving efficiently.

Fig. 15-8 **Different types of pigs are used to scrub pipelines clean.**

Exxon Pipeline Co.

Here's how it works...

1 Scott Travis owns several small oil wells. Each well pumps about 20 barrels of crude oil a day into a nearby storage tank. Every four months, the "crude" must be delivered to a refinery 600 miles away.

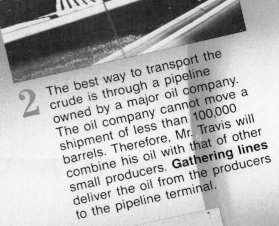

2 The best way to transport the crude is through a pipeline owned by a major oil company. The oil company cannot move a shipment of less than 100,000 barrels. Therefore, Mr. Travis will combine his oil with that of other small producers. **Gathering lines** deliver the oil from the producers to the pipeline terminal.

5 The oil moves through the underground line slowly — about five miles per hour. At this rate, it will take the batch five days to travel the 600 miles to the refinery.

6 About every 100 miles, the crude passes through a boost station. Pumps there keep the speed of the oil constant and the pipe full. If the pipe were to lose pressure, the oil would slow down and back up the entire line.

3 At the pipeline terminal, the oil from the different producers is measured and pumped into a holding tank. When the pipeline is ready to receive the combined batch, the combined number of barrels will be determined. The temperature of the crude will be measured as well.

4 The next step is to pump the batch out of the holding tanks. The oil goes through a boost station and into the main, or **trunk**, line. The journey has begun!

7 The crude oil arrives at a terminal near the refinery. Here the batch is switched from the trunk line into short **transfer lines**. The transfer lines deliver the oil to a "tank farm." When it is time for the oil to be processed, it is pumped into the refinery.

8 In the refining process, the crude is transformed into several different petroleum products. Gasoline, kerosene, and diesel fuel, are examples. The products are shipped to consumers through different modes of transportation. The pipeline company will make a profit, and so will Scott Travis!

American Petroleum Institute

Fig. 15-9 Measuring is an important part of pipeline system operation.

Measuring and Exchange Stations

One way that companies keep track of their cargo is to measure it at different points in its journey. First the batch is measured before being placed in a pipeline. The batch is measured again during its travel. Then it is measured once more at its destination. All of this measuring is done at **measuring stations**. See Fig. 15-9.

As in railroad transport, pipeline cargo travels in a straight line. To change direction, the cargo must be switched onto another line. This switching is done through the use of **crossover valves**. Some cargo is switched from one pipeline to another within the same company's system. Cargo may also be switched from one company's line to another company's line. This second type of switching is done at an **exchange station**.

STUDY QUESTIONS

1. Define pipeline transportation.
2. List four characteristics of pipeline transport.
3. True or false: All pipelines are underground.
4. What two types of companies ship products by pipeline?
5. What two benefits do large companies get from leasing their lines to other firms?
6. What three types of products can be transported through pipelines?
7. What are pipelines made of?
8. List the three basic types of pipelines.
9. How is cargo kept moving at a steady pace through a pipeline?
10. What is the order in which different products pass through a pipeline called?
11. How are pipelines kept from clogging?
12. What device is used to switch pipeline cargo from one line to another?

ACTIVITIES

1. Using your local telephone book, find out if there is a long-distance pipeline company that serves your area. If there is, what products does the company transport?
2. Since many pipelines are underground, they may be hard to locate. However, as you travel around your community, look for signs that tell of a pipeline below the ground. (Hint: You may have to travel on area roads or highways to find a pumping station, valves, or a metal sign.)
3. Local utility companies transport natural gas and water through pipelines. Make a list of the companies in your area that provide these services. Also find out how much the companies charge for their product. (Note: The cost of transportation is included in the rate charged to the customer.)

Engines for Transportation and Industry

Webster's *New World Dictionary* defines **engine** as *any machine that uses energy to develop mechanical power*. Using this definition, we could say that a bicycle is an engine. It converts the rider's energy input into mechanical power. However, when we talk about engines, we usually mean **heat engines**.

Heat engines convert heat energy into mechanical energy. Steam engines, gasoline engines, diesel engines, jet engines, and rocket engines are all heat engines. In this section, you'll see how heat engines can be grouped into two basic categories:
- external-combustion engines
- internal-combustion engines

You'll also learn about the most common heat engine in use today — the automotive engine.

There are two basic uses for heat engines. Many heat engines are used to power **transportation devices**, like cars, trucks, and ships. Other heat engines are used to provide the power needed by industry and society. These engines are called **stationary engines**. Both transportation engines and stationary engines are very important in our technological society.

External-Combustion Engines

The heat used in heat engines is produced by the combustion (burning) of fuel. This heat is used to increase the pressure of gases. The pressurized gases are then used to produce usable motion.

As you know, there are two basic types of heat engines: external-combustion and internal-combustion. The difference between the two types is in *where* the gases are heated and pressurized.

In **external-combustion engines**, the heat energy is produced *outside* of the engine. This is done by burning fuels such as oil or coal, or by producing a controlled nuclear reaction. The heat is used to produce pressurized gas. Then the gas is piped into the engine, where it produces motion.

Internal-combustion engines operate differently from external-combustion engines. In these engines, heat and pressure are produced *inside* the engine. This is done by burning fuels such as gasoline, diesel fuel, and jet fuel. The pressure is converted into motion immediately.

In this chapter, you'll learn all about external-combustion engines. The next chapter will tell you about internal-combustion engines. First, however, you should know what kinds of motion we can get from heat engines.

MOTION

Heat engines can produce three types of motion: reciprocating, rotary, and linear. See Fig. 16-1. **Reciprocating motion** is back-and-forth motion. This is the kind of motion produced in a piston engine. **Rotary motion** is turning motion. Pressurized gases can be used to turn a rotor or turbine. **Linear motion** is motion in a straight line. Jet and rocket engines produce linear motion.

ENGINE EFFICIENCY

If you think back to the material you read in Chapter 2, you will remember that energy cannot be created or destroyed. However, with any energy conversion there is always a loss of energy. This energy loss is usually in the form of friction, or heat.

Heat engines are particularly inefficient when it comes to energy conversion. They are generally much less than 50 percent efficient at converting chemical energy (fuel) into mechanical energy (motion). This means that over half of the energy we put into a heat engine is lost.

As you read on about engines, notice how there is always a concern for efficiency. Early steam engines were less than five percent efficient. Today's heat engines are much more efficient. However, designers are still working to produce better engines.

TYPES OF EXTERNAL-COMBUSTION ENGINES

Most external-combustion engines are **steam engines**. The heat energy produced outside the engine is used to change water into steam. The steam is then used inside the engine to drive pistons or turbines.

Another type of external-combustion engine is the **Stirling Cycle engine**. Its driving force is produced by the alternate expanding and contracting of a special gas.

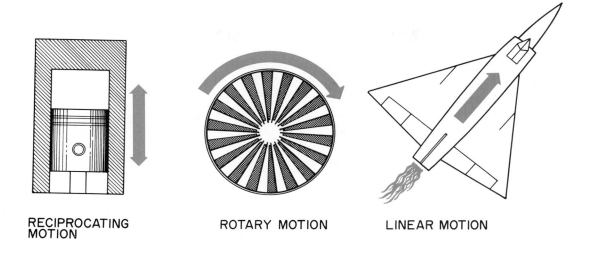

RECIPROCATING
MOTION   ROTARY MOTION   LINEAR MOTION

Fig. 16-1   **Heat engines can produce three types of motion.**

Steam Engines

The **boiler** is an important part of any steam engine. A boiler is a closed container that holds water. A heat source applied to the boiler changes the water into high-pressure steam.

The steam travels through pipes to the engine. Inside the engine, the steam pressure drives pistons or turbines. In this way, heat energy changes to mechanical energy.

The following diagram shows the main steps in steam engine operation:

Steam engines can use any energy source to produce heat. This is because the energy source never enters the engine. All of the fossil fuels (coal, oil, and natural gas) have been used to power steam engines. Other energy sources can also be used, including solar energy, geothermal energy, nuclear energy, and synfuels.

There are many different steam engine designs. Steam engines are made in hundreds of different sizes and shapes to fit special needs. However, all steam engines can be grouped into two classes: reciprocating (piston) steam engines and steam turbines.

Heat Boils water, changing it into steam Steam drives engine Mechanical energy (motion) is produced

Reciprocating (Piston) Steam Engines

Reciprocating steam engines were the first steam engines built for practical use. These engines were first developed to pump water from coal mines. One of the best-known early steam engines was developed by an Englishman, Thomas Newcomen, in 1712. See Fig. 16-2.

Water expands about 1700 times when it is changed into steam. Newcomen's engine made use of this expansion and the force behind it. First the water inside the boiler was heated to produce steam. The expanding steam passed into a cylinder and pushed a piston upward.

When the piston reached the top of the cylinder, an operator closed the **steam valve** between the cylinder and the boiler. See Fig. 16-2 again. The operator also opened a valve that allowed cold water to spray on the inside of the cylinder. The water cooled the steam and condensed it (changed it back into water). As the steam condensed, it produced a partial vacuum in the cylinder. This vacuum caused atmospheric pressure to push the piston back down. See Fig. 16-3.

Auto Specialties Mfg. Co.

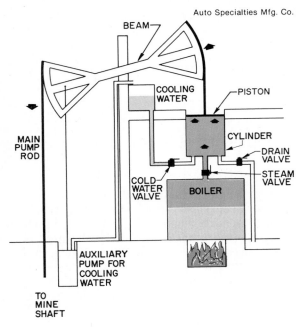

Fig. 16-2 The first practical steam engine was used to pump water from mine shafts. This engine was not very powerful or efficient.

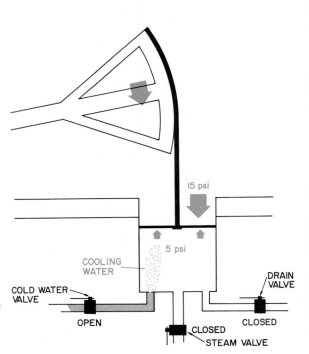

Fig. 16-3 Cold water sprayed into the cylinder condensed the steam. Atmospheric pressure then drove the piston down.

Once the piston was down, the operator turned off the cold water and opened the steam valve. This would start a new cycle. The cooling water and condensed steam were removed from the cylinder through a drain valve.

In the Newcomen engine, the piston was attached to a moveable beam. As the piston moved up and down, it moved the beam up and down. The moving beam operated a pump that removed excess water from the mine.

Another English inventor improved on the steam engine about 50 years after Newcomen's development. James Watt added a special condenser to the Newcomen engine. This condenser was a hollow cylinder connnected to the steam cylinder. The condenser cooled the steam without cooling the cylinder itself. See Fig. 16-4.

This new design improved the engine's efficiency. In the original Newcomen engine, the steam was cooled by cooling the cylinder. The cold cylinder reduced the temperature of the incoming steam. This in turn reduced steam pressure. Watt's design used less fuel and developed more power.

James Watt further experimented to produce the **double-acting steam engine**. In this engine, steam pressure pushed on both sides of the piston alternately. (This was the first actual production of reciprocating motion by steam. In the earlier engines, the steam only produced movement in one direction. The reverse movement was due to atmospheric pressure.) The double-acting design further improved the efficiency of the steam engine.

Soon, inventors made other improvements in the steam engine. They learned to increase the pressure of the steam by heating it in a confined (enclosed) container. Earlier steam engines did not use confined boilers to create steam pressure. Therefore, their steam pressure was limited. With confined boilers, steam engines could produce much more pressure. See Fig. 16-5. The higher the steam's temperature, the higher its pressure. The greater the steam pressure, the more force it could produce.

Fig. 16-5 **Steam heated in a confined boiler produces greater pressure. The greater the heat, the more pressure is produced.**

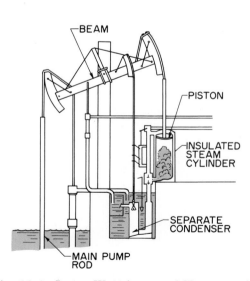

Fig. 16-4 **James Watt improved Newcomen's steam engine by adding a condenser to cool the steam.**

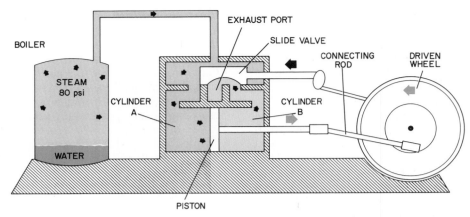

Fig. 16-6 A steam engine converts the force produced by high steam pressure into controlled, usable motion. In this drawing, steam is being supplied to cylinder A. This causes the piston to move to the right.

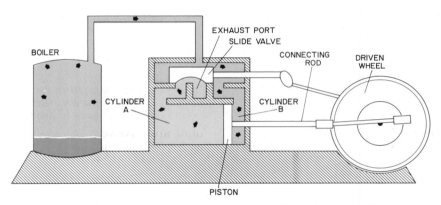

Fig. 16-7 When the slide valve cuts off the steam for cylinder A, the steam enters cylinder B and pushes on the other side of the piston.

Fig. 16-8 This large four-cylinder steam engine provides power for a furniture-manufacturing facility.

Skinner Engine Co.

Figure 16-6 shows a reciprocating steam engine that uses high-pressure steam. Pressurized steam from the boiler flows into cylinder A. The steam pushes the piston to the right. This movement turns a wheel one-half turn. The rotation of the wheel pushes the slide valve to the left. The valve stops the flow of steam into cylinder A.

The steam then flows into cylinder B, as shown in Fig. 16-7. This forces the piston back to the left. The wheel makes another half-turn, completing one rotation. With this half-turn, the sliding valve moves back to the right. The cycle then begins again.

The constant back-and-forth piston movement keeps the wheel turning. The turning wheel produces mechanical energy. This energy can be used to drive generators or operate industrial machinery. See Fig. 16-8.

The force produced by a modern steam engine depends on two things. These are (1) the boiler pressure of the steam and (2) the size of the piston. The force pushing the piston increases as either factor increases. See Fig. 16-9. We can find the amount of force with this equation from Chapter 3:

Force = Pressure × Area

The piston head is round. Therefore, we can use the formula for the area of a circle to find the piston's surface area.

Area of a circle = πr^2

Where:

π = 3.14

r = radius

In Figure 16-9, for example, the radius is one-half of the 14-inch diameter. Therefore, the piston's surface area is 3.14 × 7^2, or 154 square inches.

Example problem:

Find the force exerted by the piston shown in Figure 16-9.

Force = Pressure × Area
Force = 80 psi × 154 sq. in.
Force = 12,320 lbs.

Steam engines brought about great changes in the way people lived. These engines introduced the age of self-propelled vehicles. Steam engines powered automobiles, trains, and ships. The steam engine mechanized agriculture. It provided power for plows, reapers, threshing machines, water pumps, and other farm equipment.

Today, however, reciprocating steam engines are not used much. They have been replaced in most cases by steam turbines and internal-combustion engines. Both steam turbines and internal-combustion engines are more efficient.

Steam Turbines

A Greek scientist named Hero described the first steam turbine in the first century A.D. It consisted of a hollow metal ball mounted on pipes from a steam kettle. See Fig. 16-10. Water in the kettle boiled to produce steam. The steam was allowed to escape from the ball through other pipes. This caused the ball to rotate, producing rotary motion.

Hero's steam turbine was never put to practical use. It was a long time before the steam turbine was truly developed. This did not happen until after Newcomen developed his reciprocating steam engine.

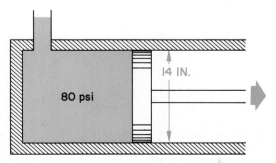

Fig. 16-9 **Steam pressure and piston size are the two factors that determine the force produced by a steam engine.**

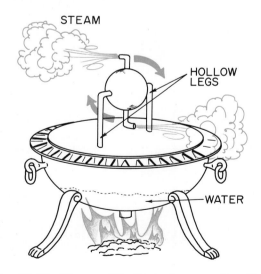

Fig. 16-10 **The world's first steam engine. It was actually a type of steam turbine.**

The American inventor Charles Curtis did important work with steam turbines. In 1903, he developed a steam turbine to run an electrical generator. See Fig. 16-11. The generator produced 5000 kilowatts of electricity (nearly 7000 horsepower).

The turbine took up one-tenth as much space as the steam engine it replaced. It cost one-third as much. It also used less steam. Many steam turbines had been built before the Curtis turbine. However, none of them were as powerful or efficient.

Fig. 16-11   **Charles Curtis developed the first multiple-rotor turbine. A scale model is shown here.**

British Crown Copyright, Science Museum, London

Fig. 16-12 Steam turbines operate on either the reaction principle or the impulse principle. The two principles are shown in practice here.

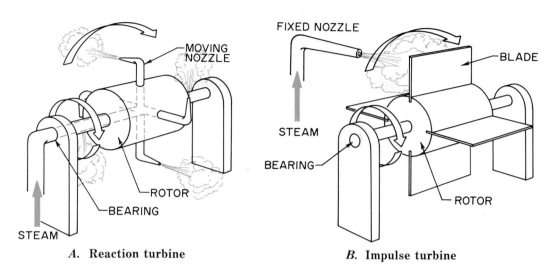

A. Reaction turbine

B. Impulse turbine

Fig. 16-13 This photograph shows the inside of a six-stage (six-rotor) impulse turbine.

Modern steam turbines are among the most powerful engines in the world. They operate on one or both of two different principles: reaction and impulse. See Fig. 16-12.

A **reaction turbine** operates like Hero's turbine. Steam escapes from nozzles on a rotor. It applies a force that produces rotation. The force is opposite the direction of the escaping steam.

An **impulse turbine** has a fixed source of escaping steam. See Fig. 16-12 again. In this engine, steam is directed against the blades of a rotor. The force rotates the rotor in the same direction as the steam.

In both types of turbine, the speed of the steam is important. The faster the steam moves, the greater the speed of rotation. Large steam turbine rotors turn at about 10,000 revolutions per minute. Small turbine rotors turn at much higher speeds. Figure 16-13 shows an inside view of a small six-rotor impulse steam turbine.

In impulse turbines, special nozzles are used to increase the speed of the steam. See Fig. 16-14. The steam's pressure forces it through the narrow part of the nozzle. The volume of the steam decreases. This increases the speed of the steam. (The same principle allows you to squirt water with a hose. Holding your thumb over part of the hose end forces the water to flow faster.)

Fig. 16-14   The nozzle of an impulse turbine increases the speed of the steam by decreasing its volume. The faster-moving steam then increases the speed of the turbine.

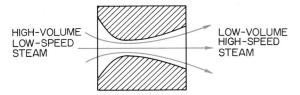

HIGH-VOLUME
LOW-SPEED
STEAM

LOW-VOLUME
HIGH-SPEED
STEAM

Figure 16-15 shows how steam works to produce motion in an impulse turbine. Steam passes into the turbine and through the nozzles. The nozzles increase the speed of the steam and direct it against a set of moveable blades. These blades are on the outside of a wheel that is attached to an output shaft. The steam rotates the blades, which turn the output shaft.

The moveable blades are cup-shaped. Therefore, when the steam hits them, they reverse the steam's direction. The steam then moves against a set of fixed blades. The fixed blades change the steam's direction again and aim it at another set of moving blades. This sequence repeats through more sets of blades until the steam loses most of its speed. In this way, the force of the steam is transferred to the turbine blades. The direction changes are needed to move all the rotating blades in the same direction.

In reaction turbines, high-speed steam also travels between moving and fixed blades. However, the steam increases speed as it travels through the moving blades. See Fig. 16-16. The

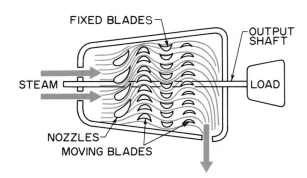

FIXED BLADES

OUTPUT SHAFT

STEAM

LOAD

NOZZLES
MOVING BLADES

Fig. 16-15   In an impulse steam turbine, the nozzles and fixed blades direct high-speed steam to the moving blades. The moving blades are connected to an output shaft.

Fig. 16-16   In a reaction turbine, the steam picks up speed as it passes through the moving blades. The fixed blades direct the steam to other moving blades.

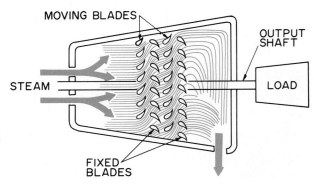

MOVING BLADES

OUTPUT SHAFT

STEAM

LOAD

FIXED BLADES

blades act like rotating nozzles. The high-speed steam applies force to the moving blades as it escapes from them.

The steam then goes through fixed blades. These blades reverse the direction of the steam and direct it to another set of moving blades.

Fig. 16-17 The blades at the output end of the steam turbine are larger in diameter than those at the input end.
Pacific Gas & Electric

This action gives some impulse power to the moving blades. Reaction turbines actually work on a combination of reaction and impulse principles.

Two things happen to steam as it moves through a turbine. First, it loses speed. (In reaction turbines, the steam *does* pick up speed as it passes through. However, the steam gradually slows down.) Second, the steam loses pressure. Both of these changes happen as the steam transfers energy to the turning blades.

The pressure decrease in steam turbines causes an increase in the steam's volume. Therefore, the turbine blades are larger toward the output end. See Fig. 16-17. The larger blades contact more of the steam. This makes up for the steam's lower speed and pressure.

Steam turbines are the most important external-combustion engines. We use them to power generators that produce large amounts of electricity. For example, just one large turbine can supply electricity for a city of 200,000 people.

Steam turbines also power large ocean-going ships. See Fig. 16-18. All nuclear-fueled ships in the United States Navy are powered by steam turbines.

Stirling Cycle Engine

Engineers in government and industry are constantly working to develop more efficient external-combustion engines. They are both redesigning traditional types of engines and studying new types of engines.

Fig. 16-18 The *U.S.S. Enterprise*, a nuclear-powered aircraft carrier, uses steam turbines to generate electricity and propel the ship.

U.S. Navy

Fig. 16-19 The Stirling engine is an external-combustion engine that uses hydrogen or helium instead of steam. The gas expands and contracts to move a piston up and down.

One promising engine under development is the Stirling Cycle engine. It is named after its inventor, the Scottish engineer Robert Stirling. See Fig. 16-19. The Stirling engine principle is not new. It was patented in 1816. Like the Newcomen engine, the Stirling engine was first used to pump water. Today's Stirling engines have many possible uses, including electrical power generation and power for transportation.

The Stirling engine works somewhat like a reciprocating steam engine. Both engines use a gas to move a piston. And both engines use an external energy source to heat or produce the gas. In a steam engine, however, the gas (steam) is piped into a cylinder to move the piston. In a Stirling engine, the gas — usually hydrogen or helium — passes back and forth between *two* cylinders.

HEATER TUBES

COMBUSTION CHAMBER

REGENERATOR

GAS COOLER

COMPRESSION PISTON

EXPANSION PISTON

CRANKSHAFT

Fig. 16-20 As the gas moves back and forth from the compression cylinder to the expansion cylinder, it produces reciprocating motion. This motion is changed to rotary motion with a crankshaft.

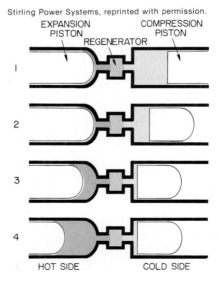

EXPANSION PISTON

REGENERATOR

COMPRESSION PISTON

1

2

3

4

HOT SIDE   COLD SIDE

Fig. 16-21 The stages in the Stirling cycle:
1-2: Compression piston compresses gas on cold side.
2-3: Gas moves from cold side to hot side.
3-4: Heated gas expands to move expansion piston.
4-1: Gas moves from hot side to cold side.

The constant back-and-forth piston movement keeps the wheel turning. The turning wheel produces mechanical energy. This energy can be used to drive generators or operate industrial machinery. See Fig. 16-8.

The force produced by a modern steam engine depends on two things. These are (1) the boiler pressure of the steam and (2) the size of the piston. The force pushing the piston increases as either factor increases. See Fig. 16-9. We can find the amount of force with this equation from Chapter 3:

Force = Pressure × Area

The piston head is round. Therefore, we can use the formula for the area of a circle to find the piston's surface area.

Area of a circle = πr^2

Where:

π = 3.14
r = radius

In Figure 16-9, for example, the radius is one-half of the 14-inch diameter. Therefore, the piston's surface area is 3.14 × 7^2, or 154 square inches.

Example problem:

Find the force exerted by the piston shown in Figure 16-9.

Force = Pressure × Area
Force = 80 psi × 154 sq. in.
Force = 12,320 lbs.

Steam engines brought about great changes in the way people lived. These engines introduced the age of self-propelled vehicles. Steam engines powered automobiles, trains, and ships. The steam engine mechanized agriculture. It provided power for plows, reapers, threshing machines, water pumps, and other farm equipment.

Today, however, reciprocating steam engines are not used much. They have been replaced in most cases by steam turbines and internal-combustion engines. Both steam turbines and internal-combustion engines are more efficient.

Steam Turbines

A Greek scientist named Hero described the first steam turbine in the first century A.D. It consisted of a hollow metal ball mounted on pipes from a steam kettle. See Fig. 16-10. Water in the kettle boiled to produce steam. The steam was allowed to escape from the ball through other pipes. This caused the ball to rotate, producing rotary motion.

Hero's steam turbine was never put to practical use. It was a long time before the steam turbine was truly developed. This did not happen until after Newcomen developed his reciprocating steam engine.

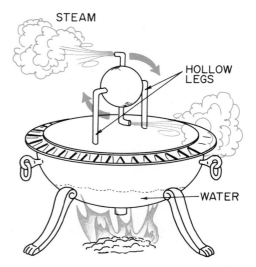

Fig. 16-9 **Steam pressure and piston size are the two factors that determine the force produced by a steam engine.**

Fig. 16-10 **The world's first steam engine. It was actually a type of steam turbine.**

The American inventor Charles Curtis did important work with steam turbines. In 1903, he developed a steam turbine to run an electrical generator. See Fig. 16-11. The generator produced 5000 kilowatts of electricity (nearly 7000 horsepower).

The turbine took up one-tenth as much space as the steam engine it replaced. It cost one-third as much. It also used less steam. Many steam turbines had been built before the Curtis turbine. However, none of them were as powerful or efficient.

Fig. 16-11 **Charles Curtis developed the first multiple-rotor turbine. A scale model is shown here.**

Fig. 16-12 Steam turbines operate on either the reaction principle or the impulse principle. The two principles are shown in practice here.

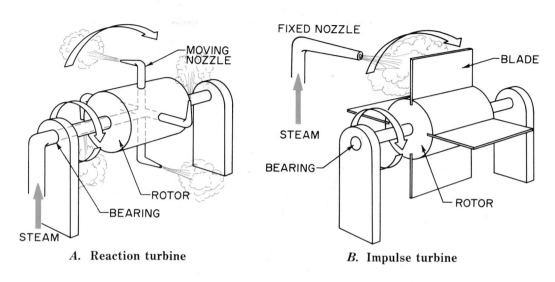

A. Reaction turbine   B. Impulse turbine

Transamerica Delaval Inc.

Fig. 16-13 This photograph shows the inside of a six-stage (six-rotor) impulse turbine.

Modern steam turbines are among the most powerful engines in the world. They operate on one or both of two different principles: reaction and impulse. See Fig. 16-12.

A **reaction turbine** operates like Hero's turbine. Steam escapes from nozzles on a rotor. It applies a force that produces rotation. The force is opposite the direction of the escaping steam.

An **impulse turbine** has a fixed source of escaping steam. See Fig. 16-12 again. In this engine, steam is directed against the blades of a rotor. The force rotates the rotor in the same direction as the steam.

In both types of turbine, the speed of the steam is important. The faster the steam moves, the greater the speed of rotation. Large steam turbine rotors turn at about 10,000 revolutions per minute. Small turbine rotors turn at much higher speeds. Figure 16-13 shows an inside view of a small six-rotor impulse steam turbine.

In impulse turbines, special nozzles are used to increase the speed of the steam. See Fig. 16-14. The steam's pressure forces it through the narrow part of the nozzle. The volume of the steam decreases. This increases the speed of the steam. (The same principle allows you to squirt water with a hose. Holding your thumb over part of the hose end forces the water to flow faster.)

Fig. 16-14 The nozzle of an impulse turbine increases the speed of the steam by decreasing its volume. The faster-moving steam then increases the speed of the turbine.

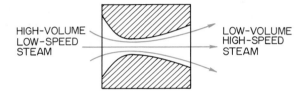

Figure 16-15 shows how steam works to produce motion in an impulse turbine. Steam passes into the turbine and through the nozzles. The nozzles increase the speed of the steam and direct it against a set of moveable blades. These blades are on the outside of a wheel that is attached to an output shaft. The steam rotates the blades, which turn the output shaft.

The moveable blades are cup-shaped. Therefore, when the steam hits them, they reverse the steam's direction. The steam then moves against a set of fixed blades. The fixed blades change the steam's direction again and aim it at another set of moving blades. This sequence repeats through more sets of blades until the steam loses most of its speed. In this way, the force of the steam is transferred to the turbine blades. The direction changes are needed to move all the rotating blades in the same direction.

In reaction turbines, high-speed steam also travels between moving and fixed blades. However, the steam increases speed as it travels through the moving blades. See Fig. 16-16. The

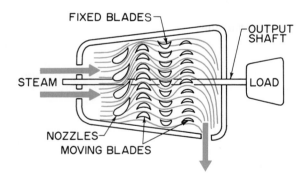

Fig. 16-15 In an impulse steam turbine, the nozzles and fixed blades direct high-speed steam to the moving blades. The moving blades are connected to an output shaft.

Fig. 16-16 In a reaction turbine, the steam picks up speed as it passes through the moving blades. The fixed blades direct the steam to other moving blades.

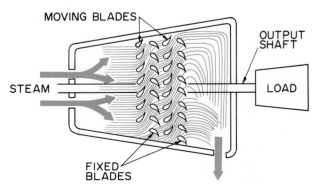

blades act like rotating nozzles. The high-speed steam applies force to the moving blades as it escapes from them.

The steam then goes through fixed blades. These blades reverse the direction of the steam and direct it to another set of moving blades.

Fig. 16-17 **The blades at the output end of the steam turbine are larger in diameter than those at the input end.**   Pacific Gas & Electric

This action gives some impulse power to the moving blades. Reaction turbines actually work on a combination of reaction and impulse principles.

Two things happen to steam as it moves through a turbine. First, it loses speed. (In reaction turbines, the steam *does* pick up speed as it passes through. However, the steam gradually slows down.) Second, the steam loses pressure. Both of these changes happen as the steam transfers energy to the turning blades.

The pressure decrease in steam turbines causes an increase in the steam's volume. Therefore, the turbine blades are larger toward the output end. See Fig. 16-17. The larger blades contact more of the steam. This makes up for the steam's lower speed and pressure.

Steam turbines are the most important external-combustion engines. We use them to power generators that produce large amounts of electricity. For example, just one large turbine can supply electricity for a city of 200,000 people.

Steam turbines also power large ocean-going ships. See Fig. 16-18. All nuclear-fueled ships in the United States Navy are powered by steam turbines.

Stirling Cycle Engine

Engineers in government and industry are constantly working to develop more efficient external-combustion engines. They are both redesigning traditional types of engines and studying new types of engines.

Fig. 16-18 **The *U.S.S. Enterprise*, a nuclear-powered aircraft carrier, uses steam turbines to generate electricity and propel the ship.**

U.S. Navy

Stirling Power Systems

Fig. 16-19 The Stirling engine is an external-combustion engine that uses hydrogen or helium instead of steam. The gas expands and contracts to move a piston up and down.

One promising engine under development is the Stirling Cycle engine. It is named after its inventor, the Scottish engineer Robert Stirling. See Fig. 16-19. The Stirling engine principle is not new. It was patented in 1816. Like the Newcomen engine, the Stirling engine was first used to pump water. Today's Stirling engines have many possible uses, including electrical power generation and power for transportation.

The Stirling engine works somewhat like a reciprocating steam engine. Both engines use a gas to move a piston. And both engines use an external energy source to heat or produce the gas. In a steam engine, however, the gas (steam) is piped into a cylinder to move the piston. In a Stirling engine, the gas — usually hydrogen or helium — passes back and forth between *two* cylinders.

Stirling Power Systems, reprinted with permission.

HEATER TUBES

COMBUSTION CHAMBER

REGENERATOR

GAS COOLER

COMPRESSION PISTON

EXPANSION PISTON

CRANKSHAFT

Fig. 16-20 As the gas moves back and forth from the compression cylinder to the expansion cylinder, it produces reciprocating motion. This motion is changed to rotary motion with a crankshaft.

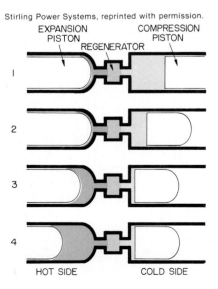

Stirling Power Systems, reprinted with permission.

EXPANSION PISTON

COMPRESSION PISTON

REGENERATOR

1

2

3

4

HOT SIDE   COLD SIDE

Fig. 16-21 The stages in the Stirling cycle:
1-2: Compression piston compresses gas on cold side.
2-3: Gas moves from cold side to hot side.
3-4: Heated gas expands to move expansion piston.
4-1: Gas moves from hot side to cold side.

The gas used in a Stirling engine is called a **captive working gas**. It does not leave the engine after doing work. Instead, the gas is recycled to do more work.

A Stirling engine has a "hot side" and a "cold side." The hot side contains an **expansion piston** in a cylinder. The cold side contains a **compression piston** in another cylinder. Both pistons are connected to a crankshaft. This means that the movement of one piston will affect the movement of the other piston. See Fig. 16-20.

A Stirling engine is started just like an automobile engine. A starting motor turns a flywheel on the engine's crankshaft. At the same time, fuel is ignited in the combustion chamber. The heat of combustion then expands the gas contained in **heater tubes** in the chamber. The pistons go through the stages shown in Fig. 16-21 to produce power.

Driven by the crankshaft, the compression piston moves up to compress the working gas on the cold side (Steps 1-2). The gas then moves

Robert and James Stirling: Brothers of Invention

Most of us have heard of the Wright brothers. Where would the world be without their famous first airplane flight at Kitty Hawk? But have you ever heard of the Stirling brothers? Someday, Robert and James Stirling may be as well-known as Orville and Wilbur Wright.

The Stirling brothers lived in Scotland in the nineteenth century. They came from a long line of mechanically talented Stirlings, both male and female. But there were not only talented engineers in the family. There were also talented ministers. Robert Stirling's uncle was a minister, as was his cousin. Robert himself studied engineering, but became a minister in the Presbyterian Church of Scotland. On the other hand, James Stirling studied for the ministry, but became a professional engineer.

Robert Stirling is generally credited with inventing the Stirling engine, or "hot-air engine." However, it may be more correct to say that he discovered or invented the Stirling *principle*. The Stirling engine of today is very different from the engine that Robert patented in 1816. However, both engines use the same operating principle.

When Robert Stirling wasn't busy at his church duties, he spent time in his own blacksmith shop. He loved to work on mechanical things, and it was in the blacksmith shop that he worked on the hot-air engine.

Some people think, however, that James Stirling should be given equal credit for inventing the hot-air engine. James had more training in engineering than Robert. The brothers patented at least two hot-air designs together. James also had the precise mechanical drawing skills needed to describe the engines on paper. In addition, James had access to bigger and better facilities for manufacturing the engines.

There is an important difference between getting a good idea and actually making something of it. Robert Stirling had enough knowledge of engineering to "dream up" the principle for a new type of external-combustion engine. But he might not have put the principle into practice without the help of his brother James.

Just who did what is still a mystery. But the story of the Stirling brothers emphasizes one thing: Both creativity and application are necessary "ingredients" in technological advance.

through a **regenerator**. The regenerator heats the gas as it moves from the cold side to the hot side (Steps 2-3). The heated gas expands to drive the expansion piston down (Steps 3-4). The turning crankshaft then forces the expansion piston up to move the gas back to the cold side (Steps 4-1).

As the gas moves through the regenerator, the regenerator removes heat from it. Heat is also removed by a liquid cooling system. The gas then contracts, and it is ready to be compressed again.

Stirling engines have the potential to be more fuel-efficient and powerful than internal-combustion engines. They also operate much more quietly and produce far less pollution. Stirling engines have a **multifuel** capability. This means that they can use almost any fuel, including gasoline, diesel fuel, and natural gas. Other possible fuels include methanol, coal, wood, and society's burnable waste products. Stirling engines are already being used to generate electricity. Someday they may power many cars and trucks.

STUDY QUESTIONS

1. What are heat engines?
2. Identify the two broad categories of heat engines. Describe the main difference between the two.
3. Name the three types of motion possible with heat engines.
4. Name two types of external-combustion engines.
5. True or false: Heat engines are generally more than 50 percent efficient in converting energy into motion.
6. Briefly describe the main steps in steam engine operation.
7. Why can steam engines use any energy source to produce heat?
8. Name the two classes of steam engines.
9. What was the first practical use of reciprocating steam engines?
10. How did James Watt improve on Newcomen's steam engine?
11. What advantage did the double-acting steam engine have over earlier steam engines?
12. How did confined boilers improve steam engines?
13. The force of a modern steam engine depends on two things. What are they?
14. Why are reciprocating steam engines seldom used today?
15. Identify the two principles on which steam turbines can operate.
16. Name two important uses of steam turbines.
17. Name two gases used in Stirling engines.
18. Why is the gas in a Stirling engine called a *captive working gas*?

ACTIVITIES

1. Conduct a study on the use of steam locomotives in the United States from 1825 to the present. Identify such important events as the first successful steam locomotive, mail delivery, and connecting the Atlantic and Pacific coasts by rail. Use the encyclopedia and other books in your school library.

2. Do a report on old-time steam locomotives still operating in the United States. Most of these engines are maintained for either tourism or limited passenger and freight transportation. The school encyclopedia and other library books will help you with this activity.

Internal-Combustion Engines

The internal-combustion engine was invented in 1876. It gradually replaced the steam engines that were used to power trains, farm equipment, and even automobiles.

Today, except for steam-powered ships, most transportation devices are powered by internal-combustion engines. These lightweight and powerful engines have made air travel possible. The rocket engine, a very powerful type of internal-combustion engine, has even propelled humans into outer space.

Internal-combustion engines also have important uses as stationary (fixed) engines. They run generators to produce electrical power. They also produce mechanical power for industrial machinery and equipment.

Like steam engines and Stirling engines, internal-combustion engines are *heat engines*. They burn fuel to produce pressurized gases. However, there is a big difference. In internal-combustion engines, the pressurized gases are produced *inside* the engine. The gases don't have to be piped in, as with a steam engine.

Do you remember the three kinds of motion that are possible with a heat engine? They are reciprocating motion, rotary motion, and linear motion (see Fig. 16-1). Different kinds of internal-combustion engines produce different kinds of motion. Gasoline and diesel piston engines produce reciprocating motion. Rotary engines produce rotary motion. Jet engines and rocket engines produce linear motion.

GASOLINE PISTON ENGINES

Wherever you happen to be at this moment, chances are that a gasoline piston engine isn't too far away. We are surrounded by gasoline piston engines used to power cars. See Fig. 17-1. In addition, many small gasoline piston engines are used to power chain saws, small generators, weed cutters, and garden tillers. See Fig. 17-2. You will learn more about small gas engines in Section VI of this book.

Fig. 17-2 This lawn mower is powered by a gasoline piston engine.

John Deere

Chrysler-Plymouth

Fig. 17-1 This sporty car uses a fuel-efficient gasoline piston engine.

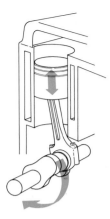

Fig. 17-3 A gasoline piston engine uses a crankshaft to change the piston's up-and-down motion into rotary motion.

Fig. 17-4 Charles and Frank Duryea built one of the first gasoline-powered cars in 1893. Do you see why people called it a "horseless carriage"?

Gasoline piston engines produce reciprocating (back-and-forth) motion. A crankshaft in the engine changes the reciprocating motion into rotary motion. See Fig. 17-3.

There are two basic types of gasoline piston engines: the **four-stroke cycle engine** and the **two-stroke cycle engine**. Both types operate with a piston moving up and down in a cylinder. The difference is in the number of strokes that the piston makes per engine cycle. A **stroke** is the movement of the piston from one end of the cylinder to the other. A **cycle** is a complete set of piston movements. It includes all of the strokes needed to produce a power stroke.

Many times the word *stroke* or *cycle* is left out of a description of the engine. For example, *four-stroke engine* and *four-cycle engine* mean the same thing: a four-stroke cycle engine.

Four-Stroke Cycle Engines

Most modern piston engines operate on the principle of the four-stroke cycle. This principle was developed in 1862 by Beau de Rochas of France. Fourteen years later, a German named Nicholas Otto built the first four-stroke engine. Otto's name was given to the four-stroke principle. We still refer to it as the **Otto cycle**.

Otto's engine burned natural gas. In 1885 Gottlieb Daimler, who worked with Otto, built an engine that used gasoline. This engine worked much better than the Otto engine. Progress continued, and in 1893 two American brothers named Duryea built and operated a successful gasoline automobile. See Fig. 17-4. Today's car engines still operate on the same four-stroke cycle principle used by the Duryea brothers in 1893.

In four-stroke engines there are four separate piston strokes: intake, compression, power, and exhaust. See Fig. 17-5. The function of each stroke is explained in the following paragraphs.

Fig. 17-5 The operation of a four-stroke cycle gasoline engine.

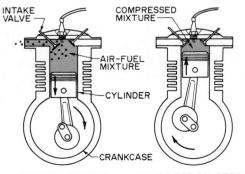

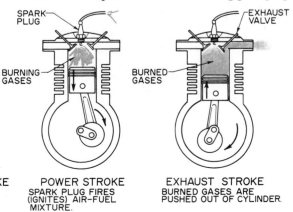

INTAKE STROKE
AIR-FUEL MIXTURE IS PUSHED INTO CYLINDER.

COMPRESSION STROKE
AIR-FUEL MIXTURE IS COMPRESSED.

POWER STROKE
SPARK PLUG FIRES (IGNITES) AIR-FUEL MIXTURE.

EXHAUST STROKE
BURNED GASES ARE PUSHED OUT OF CYLINDER.

On the **intake stroke**, the piston moves down in the cylinder. This creates a partial vacuum at the top of the cylinder. The intake valve is open at this time. Atmospheric pressure pushes a mixture of gasoline and air into the cylinder.

As the piston starts up on the **compression stroke**, the intake valve closes. This seals the cylinder. The piston can then compress the air-fuel mixture into a small area at the top of the cylinder. This area is called the **combustion chamber**.

During the compression stroke, the piston compresses the air-fuel mixture to about one-eighth of its original volume. The comparison between the original volume and the compressed volume is called the **compression ratio**. A compression ratio of 8:1 is enough to produce high pressure in the combustion chamber. Squeezing the air-fuel mixture into a small space heats it up and makes it easier to ignite (set fire to).

The **power stroke** begins as a spark plug ignites the compressed air-fuel mixture. The burning air-fuel mixture raises the temperature inside the cylinder. This further increases the pressure in the combustion chamber. The pressure goes up to about seven times what it was before ignition. The increased pressure drives the piston down. It is the force of this downward movement that produces mechanical power.

The final stroke in the cycle is the **exhaust stroke**. Near the end of the power stroke, the exhaust valve opens. Then, as the piston moves up, it pushes the burned gases out of the cylinder.

At the end of the exhaust stroke, the piston reaches the top of the cylinder. The piston then starts down on another intake stroke. This begins a new cycle.

Two-Stroke Cycle Engines

Two-stroke cycle engines have fewer moving parts than four-stroke engines. This makes them simpler, easier to build, and much lighter. However, two-stroke engines are much less fuel-efficient. They also give off more pollution than four-stroke engines. Therefore, two-stroke engines are generally used when light weight is more important than good fuel efficiency. The

most common uses are for lawn mowers, outboard motors, and small motorcycles. See Fig. 17-6.

A two-stroke engine has only two strokes per cycle: compression and power. As in the four-stroke engine, there are four actions — intake, compression, power, and exhaust. However, these actions all take place in just two strokes. See Fig. 17-7.

Most two-stroke engines do not have intake and exhaust valves. Instead, they have intake and exhaust **ports** (holes) in the cylinder wall. Most of these engines also have a **reed valve** at the bottom of the **crankcase** (the part of the engine that contains the crankshaft). The air-fuel mixture enters the engine through the reed valve.

As the piston moves upward on the **compression stroke**, two things happen. The piston blocks off the intake and exhaust ports. This seals in the air-fuel mixture so that the piston can compress it in the combustion chamber. At the same time, new air-fuel mixture enters the crankcase through the reed valve. This is caused by the partial vacuum created in the crankcase by the upward movement of the piston.

Fig. 17-6 **The two-stroke engine used on this motorcycle can produce a great deal of power for its weight.**

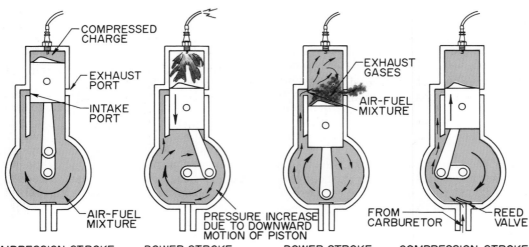

COMPRESSED CHARGE

EXHAUST PORT

INTAKE PORT

AIR-FUEL MIXTURE

PRESSURE INCREASE DUE TO DOWNWARD MOTION OF PISTON

EXHAUST GASES

AIR-FUEL MIXTURE

FROM CARBURETOR

REED VALVE

COMPRESSION STROKE
AIR-FUEL MIXTURE IS COMPRESSED. VALVES ARE SEALED.

POWER STROKE
AIR-FUEL MIXTURE IS IGNITED. PISTON IS FORCED DOWN.

POWER STROKE
EXHAUST GASES LEAVE CYLINDER. AIR-FUEL MIXTURE ENTERS CYLINDER.

COMPRESSION STROKE
PISTON BEGINS COMPRESSION. AIR-FUEL MIXTURE ENTERS ENTERS CRANKCASE.

Fig. 17-7 The operation of a two-stroke cycle engine. The four diagrams show the position of the piston during a single cycle.

Fig. 17-8 A modern four-cylinder diesel engine.

Volkswagen

When compression is complete, the spark plug ignites the compressed air-fuel mixture. The hot expanding gases push the piston down on the **power stroke**. As the piston passes the exhaust port, the burned gases begin to leave the engine. The piston then passes the intake port. New air-fuel mixture pushes up into the combustion chamber. The crankcase pressure built up by the downward movement of the piston causes this push.

As the new mixture enters the combustion chamber, it helps to push out the burned gases. The piston then starts back up on the next compression stroke. This begins a new cycle.

DIESEL ENGINES

The diesel engine was invented by a German mechanical engineer named Rudolph Diesel. He patented his design in 1892. However, it was not until 1897 that he built a working diesel engine.

Today's diesel engines provide power for trains and buses. They also power cars, trucks, boats, ships, and most construction equipment. See Fig. 17-8. As stationary engines, they power generators that produce electricity.

Diesel engines operate on the same two-stroke and four-stroke principles used in gasoline piston engines. However, diesels do not use a spark to ignite the fuel. Instead, the fuel is ignited by the intense heat of compression. This type of ignition is called **compression ignition**.

Compression ignition requires an air and fuel supply system different from the kind used in gasoline engines. In diesels, air and fuel do not enter the cylinder together. Only air is admitted on the intake stroke. See Fig. 17-9.

During the compression stroke, the air is compressed to 1/16 to 1/23 its original volume. That is, the compression ratio is from 16:1 to 23:1. This is more than twice the compression produced on the compression stroke in a gasoline

engine. This high compression ratio raises the pressure and temperature inside the cylinder. The temperature inside the cylinder reaches about 1000° F (*538° C*).

At the end of the compression stroke, **diesel fuel** is sprayed into the cylinder. As the fuel enters the cylinder, it is ignited by the high temperature of the compressed air.

As the fuel burns, it produces very hot gases. The heated gases further raise the pressure in the cylinder. The increased pressure forces the piston down, producing the power stroke. There are some major differences in design between diesel engines and gasoline engines. You will learn more about both types of engines in Chapter 18.

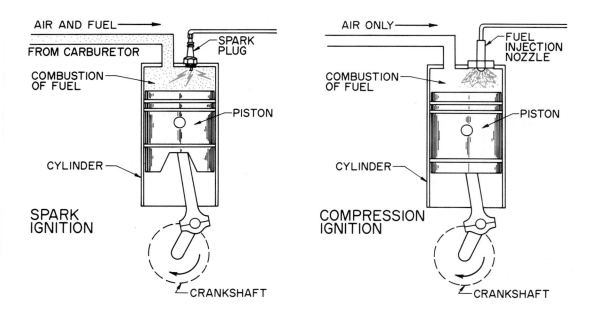

GASOLINE ENGINE   **DIESEL ENGINE**

Fig. 17-9 **Air and fuel are supplied differently in a diesel engine than they are in a gasoline engine. The diesel's method of ignition is also different.**

ROTARY ENGINES

The rotary engine is often called a **Wankel engine**. See Fig. 17-10. This engine was designed by a German scientist, Felix Wankel, in 1958. Today Wankel engines are manufactured in a variety of sizes. They can be used as substitutes for most piston engines. Wankel engines are compact, have very little vibration, and are simple.

The Wankel engine combines features of piston engines and turbine engines. Like the piston engines, Wankel engines have four different actions — intake, compression, power, and exhaust. However, Wankel engines do not produce reciprocating motion. Instead, they produce direct rotary motion. In this way, Wankel engines are like turbine engines.

Wankel engines are more fuel-efficient than turbine engines. However, Wankel engines are less fuel-efficient and more polluting than piston engines. It is also harder to seal the combustion chamber in Wankel engines than in piston engines. This can increase the need for major repairs.

Figure 17-11 shows the main parts of a Wankel engine. The triangular **rotor** plays a central part in producing the intake, compression, power, and exhaust actions. The rotor is sandwiched between the front and rear covers. It spins inside the **rotor housing**.

The rotor has an internal gear that meshes with a stationary gear. These gears keep the rotor properly positioned in the rotor housing at all times.

The rotor has a combustion chamber on each of its three sides. There are **apex seals** at the tips of the rotor. As the rotor turns, the seals are pressed against the rotor housing. This separates the combustion chambers from each other.

Mazda

Fig. 17-10 **A rotary engine performs more smoothly and quietly than a piston engine.**

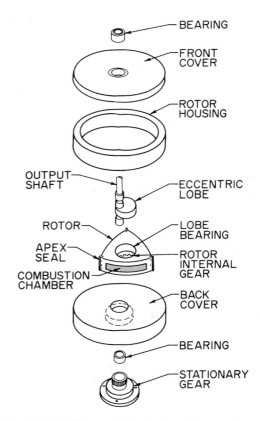

Fig. 17-11 **This exploded view gives a close-up of the rotor and housing assembly on a rotary engine. Each side of the rotor has a combustion chamber.**

The **eccentric** (off-center) **lobe** of the output shaft fits inside the rotor. As the rotor rotates, it drives the eccentric lobe to turn the output shaft. The output shaft is supported by two bearings and spins in a fixed position.

Figure 17-12 shows how a Wankel engine works. Notice that one apex of the rotor has been labeled *X*. On the intake stroke, apex X travels from point A to point B. The intake port is open. Atmospheric pressure pushes an air-fuel mixture into the combustion chamber.

As the apex goes from point B to point C, the rotor covers the intake port. The rotor also starts to compress the air-fuel mixture.

When apex X gets to point D, the air-fuel mixture is fully compressed. At this time, a spark ignites the fuel. The expanding gases produce a power stroke that forces the rotor to continue rotating.

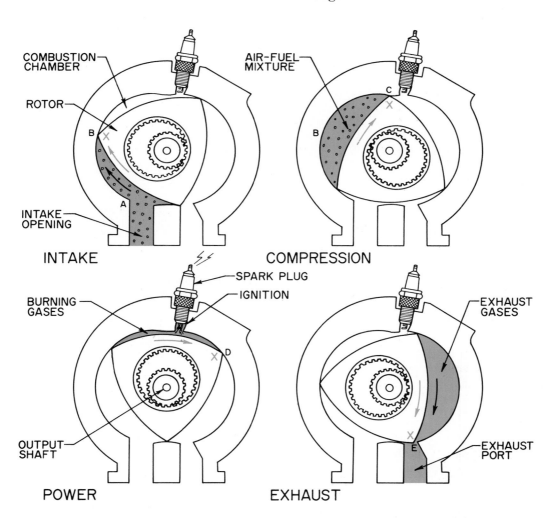

Fig. 17-12 **The operation of the Wankel engine. Note that this diagram follows the action of only one combustion chamber. The other two are also operating at the same time.**

Internal Combustion Meets the Electric Motor: Diesel-Electric Locomotives

An important use of large diesel engines is to power **locomotives**. Locomotives are what people usually call the "engine" that pulls a train. Actually, most trains use a combination of diesel engines and electric motors. The diesels provide the power to turn generators. The electricity produced by the generators is used to run electric motors. The motors then turn the locomotive's driving wheels.

A typical diesel-electric locomotive has two sets of four or six drive wheels. Each set is powered by an electric motor. Most locomotives produce from 1500 to 2500 horsepower!

It takes a lot of power to pull a long and heavy train up a slope. Therefore, some trains combine locomotives to produce more power. Three locomotives of 2000 horsepower each are often used to pull heavy loads.

The next time you see a locomotive go by, listen for the whine of the electric generators. Remember, it's not just an engine that powers the locomotive. It's a combination of two powerful advances in technology.

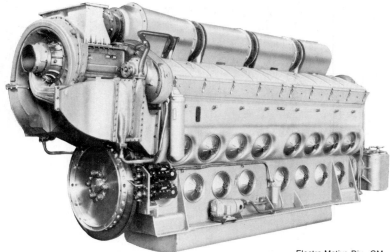

Electro-Motive Div., GM

When the apex reaches point E, the exhaust port is uncovered. The burned gases begin to leave the engine. As apex X rotates back towards point A, the rotor pushes the gases out of the engine. This completes one cycle for one combustion chamber of the rotor.

Remember, all three chambers are working at the same time. While one chamber is on intake and compression, the second chamber is on the power stroke. The third chamber is on the exhaust stroke. This means that the Wankel produces three power strokes for each revolution of the rotor.

Much research and development has been done on the Wankel engine. Engines vary in size from giant industrial engines to miniature engines. See Fig. 17-13. Wankel engines have been made with both single and multiple rotors. They have used fuels as different as methane and diesel fuel. They have powered boats, cars, and airplanes.

Despite its low weight and simple design, the Wankel engine has an uncertain future. It has serious disadvantages: low fuel efficiency, short apex seal life, and the need for more pollution control. These disadvantages may keep the Wankel from competing seriously with piston engines.

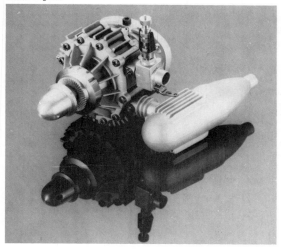

Fig. 17-14 Jumbo jet airliners can carry 429 passengers at a time. They generally fly 40,000 feet above the ground, at a speed of 550 miles per hour.

Fig. 17-13 This miniature Wankel engine is used as a power source for a model aircraft.

JET ENGINES

Jet engines can develop huge amounts of power for their size and weight. They often propel aircraft at speeds of over 750 miles (*1200 km*) per hour. See Fig. 17-14.

To understand how a jet engine works, we must review what happens to gases under pressure. Pressurized gases in a closed container exert pressure equally in all directions. The forces of this pressure are in balance. For example, the air inside an expanded balloon pushes out on all surfaces with the same force. See Fig. 17-15A.

When the person lets go of the balloon, the air starts to escape. This causes the forces inside the balloon to become imbalanced. More force is applied to the surface opposite the stem. This imbalance of force causes the balloon to be "jet-propelled" around the room. See Fig. 17-15B.

Fig. 17-15 A simple demonstration of jet propulsion.

A. When the stem is held closed, the air pressure inside the balloon presses equally on all sides.

B. When the stem is allowed to open, the pressure near the stem is released. The pressure on the opposite side pushes the balloon forward.

265

Isaac Newton described the principle of jet propulsion in his third law of motion. This law states, *To every action there is an equal and opposite reaction.*

Figure 17-16 gives two examples of Newton's third law in practice. A certain amount of force is carrying the boy toward the dock. This force is equal and opposite to the force pushing the boat away from the dock. In the rocket, a certain amount of force is pushing the rocket upward. This force is equal and opposite to the force of the escaping gas.

Both jet engines and rocket engines work on the principle of equal and opposite reaction. In fact, the first jet engines were called *reaction engines.*

Let's look now at several types of jet engines. Later, you'll see how rocket engines work.

Turbojet Engines

The turbojet is a common type of jet engine. It is used on both military and commercial aircraft. See Fig. 17-17.

Like most modern jet engines, the turbojet is shaped basically like a tube or cylinder. This tube is open at both ends. See Fig. 17-18. Air is brought into the front of the tube by an **air compressor** and the forward motion of the air-

Walt Weible

Fig. 17-17 **This F15 Eagle fighter aircraft is powered by a turbojet engine.**

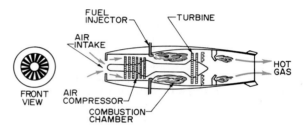

Fig. 17-18 **The operation of a turbojet engine.**

Fig. 17-16 *"For every action there is an equal and opposite reaction."* **This principle includes all motion, whether produced by mechanical force or by pressure.**

craft. The compressor then forces the air into **combustion chambers**. Here the fuel — high-grade kerosene — is added, and the air-fuel mixture is ignited.

The burning mixture creates high temperature and pressure inside the combustion chambers. The pressure tries to exert itself in all directions equally. However, air is rushing in the front of the engine. This keeps the pressurized gases from escaping out the front. Therefore, the gases push out the rear of the engine.

As the gases push toward the rear, they spin a large **turbine**. As the turbine spins, it drives the compressor.

The escaping gases also cause an imbalance of forces inside the engine. The greater force at the front of the engine pushes it forward. This push is called **thrust**. We measure thrust in pounds or in **newtons**. Thrust is the forward force the engine is producing at any given time.

Just like a car engine, the turbojet engine needs a separate source of power to get it started. This power source spins the turbine, which in turn spins the compressor to draw in and compress air. A spark plug ignites the air-fuel mixture. The engine then keeps running until it is shut off.

The action in the jet engine is continuous. Air intake, fuel addition, ignition, and thrust are all maintained on a constant basis.

Ramjet Engines

The ramjet is a very simple kind of jet engine. See Fig. 17-19. Basically it isn't much more than a hollow tube. The ramjet engine has no moving parts. Ramjets are used mainly to power guided missiles.

The forward motion of the ramjet brings large amounts of air into the combustion chamber. See Fig. 17-20. The incoming air compresses the air already in the chamber. The special shape of the chamber causes this compression. Fuel addition and ignition are the same as with the turbojet. Thrust is produced the same way. The incoming air prevents the forward exhaust of burned fuel.

At high speeds, ramjets are more efficient and more trouble-free than turbojets. However, they cannot operate while at rest, during take-off, or at low speeds. Missiles equipped with ramjets must use a rocket engine until a certain speed is reached. At that speed, the ramjet is ignited to produce a great deal of added thrust and speed.

In the future, planes may use **turbo-ramjet** engines. They would start and operate like turbojets. When a certain speed had been reached, the engines would be shifted over to ramjet operation. The ramjets would increase both the speed and the efficiency.

Fig. 17-19 This U.S. Navy target drone is powered by two ramjet engines. These engines can propel the drone faster than the speed of sound.

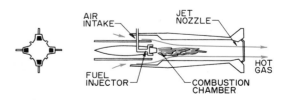

Fig. 17-20 The operation of a ramjet engine.

Fig. 17-21 Many jet airliners use turbofan engines. The larger fan provides additional thrust.

Turbofan Engines

The turbofan is similar to the turbojet engine. However, aircraft that use turbofans have an additional source of propulsion. A large fan is mounted on the front of the engine. See Fig. 17-21.

Figure 17-22 shows how a turbofan engine works. A **fan turbine** at the rear of the engine turns the fan. A **compressor turbine** turns the compressor. The fan forces huges amounts of air both into and around the engine. The air that enters the engine is compressed, mixed with fuel, and ignited.

A turbofan has two sources of thrust. One is the combustion of fuel and the resulting pressurized gases. The other is the air forced around the engine.

The fan supplies a great deal of air to the engine. Combustion can take place more efficiently with more air. Therefore, turbofans are more powerful and efficient at low speeds than turbojets. This is why turbofans are used to power many commercial airliners.

Turboprop Engines

The turboprop is very similar to the turbofan engine. The main difference is that turboprop engines have a propeller instead of a fan. See Fig. 17-23.

As with turbojets and turbofans, the incoming air is compressed by a turbine-driven compressor. The fuel is then mixed and ignited. The resulting pressurized gases pass over a power turbine at the rear of the engine. This turbine drives the propeller.

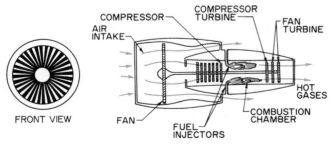

Fig. 17-22 The operation of a turbofan jet engine.

Fig. 17-23 The operation of a turboprop jet engine.

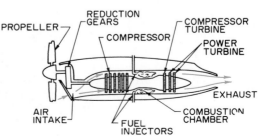

The propeller is the main source of thrust in the turboprop engine. The compressor turbine and power turbine use almost all of the energy produced by the burning fuel. They transfer this energy to the propeller. Very little thrust is provided by the engine exhaust.

Turboprop engines can produce more than twice the power of piston engines the same size. They are used mainly in small business airplanes.

Gas Turbine Engines

The gas turbine is basically a turboprop engine without a propeller. Instead of turning a propeller, the power turbine rotates an output shaft. See Fig. 17-24. Gas turbines are used to power generators, ships, trains, and experimental cars and trucks. See Fig. 17-25.

Many uses of the gas turbines are experimental. The turbines in gas turbine engines rotate at high speeds. They are also exposed to high temperatures. These conditions require the use of expensive materials and manufacturing procedures. Gas turbine engines are also less fuel-efficient than piston engines. These factors make the gas turbine a poor substitute for most piston engines.

Fig. 17-24 **An automotive gas turbine engine operates on the same principle as a turboprop engine. The power turbine turns the drive shaft. A gear system transfers power to the rear wheels.**

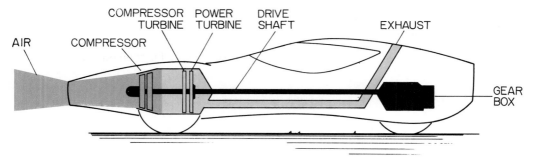

Fig. 17-25 **This experimental automotive turbine engine is fueled by powdered coal. The fuel tank is located in the engine compartment.**

General Motors Corp.

ROCKET ENGINES

Rocket engines are the most powerful internal-combustion engines. A rocket engine the same size as a 100-horsepower piston engine can produce over 300,000 horsepower. Rocket engines that provide the power for space exploration are even more powerful than this. For example, the space shuttle develops about 6.5 million pounds of thrust (*29 million newtons*) during take-off. See Fig. 17-26. Rockets also power guided missiles for national defense.

To produce their tremendous power, rocket engines use up huge amounts of fuel. A large rocket can use up over 500,000 gallons (*1.9 million liters*) of fuel within three minutes after take-off.

Fig. 17-26 The *Columbia* spacecraft is powered by rocket engines. A payload totaling 65,000 pounds (29,000 kg) can be carried in the bay.

NASA

To the Moon with Robert Goddard

In the early part of the century, Robert H. Goddard was both a student and a physics professor at Clark University in Massachusetts. At this time, Goddard was one of only a few people interested in developing a practical rocket. Rockets were used in fireworks and weapons. But no one thought they could be used for transportation.

In 1919 Professor Goddard wrote a paper on rocket theory. He showed mathematically that a rocket could reach the moon. When people heard about this idea, they were excited. They wanted to see a moon launch right away. Of course, there is a lot of work involved in turning an exciting idea into a practical reality. So people were disappointed when the flight didn't happen immediately.

However, Goddard did get to work on practical rockets. In 1926 he launched the world's first liquid-fueled rocket. He fired the rocket himself by igniting the fuel with a blow torch. The rocket flew for only 2-1/2 seconds and traveled only 220 feet (*67 meters*). However, this small rocket was the ancestor of today's mightiest rockets.

In 1930 Goddard began larger-scale experiments in New Mexico. He developed rockets that were 10 to 22 feet long and which could reach altitudes of 8000 to 9000 feet. This is over a mile-and-a-half high. The engines were so loud that people could hear them eight miles away!

Today a mile-and-a-half is tiny compared to the distance that manned space vehicles have flown. Three American astronauts were the first to reach the moon in 1969. This was a distance of more than 200,000 miles!

Robert Goddard died in 1945, so he did not see his dream become a reality. But it was Goddard's pioneering work and spirit that made space travel possible. He is rightly called "the father of modern rocketry."

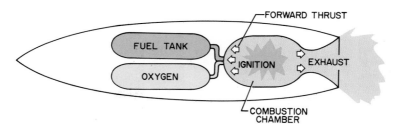

Fig. 17-27 Rocket engines carry not only their own fuel but also their own oxygen. Thrust is produced by mixing and igniting the fuel and oxygen in the combustion chamber.

Like jet engines, rocket engines work on the reaction principle. The main difference is in the supply of oxygen to burn the fuel. Jet engines use oxygen from the air. Rocket engines carry their own oxygen supply, as shown in Fig. 17-27. This makes it possible for rocket engines to operate in space, where there is no oxygen.

The operation of the rocket engine is very simple. Fuel and oxygen are mixed and ignited in the combustion chamber. The burning gases expand. They move rapidly out the exhaust port at the rear of the engine. The opposite reaction pushes the engine ahead, just as in jet engines.

Rocket engines can be divided into two types. One type uses liquid propellant (fuel). The other uses solid propellant.

Liquid-Propellant Rocket Engines

Liquid-propellant rocket engines use liquids for both the fuel and the source of oxygen. Commonly used fuels are kerosene, alcohol, and liquid hydrogen. Liquid oxygen is often used as the oxygen source. The fuel and oxygen are mixed and ignited as they enter the combustion chamber. See Fig. 17-28.

All of the United States' early space exploration efforts used liquid-fueled rockets. The Saturn V, used in 1969 for the *Apollo* moon flights, was a liquid-fueled rocket. See Fig. 19-29. This rocket had three stages (sections), plus a spacecraft carrying the astronauts. As the fuel in each stage was burned up, the engine was cast off. Only the *Apollo* spacecraft continued on to the moon.

Fig. 17-28 A liquid-propellant rocket uses both liquid fuel and liquid oxygen for combustion.

Fig. 17-29 The Saturn V was a three-stage liquid-propellant rocket.

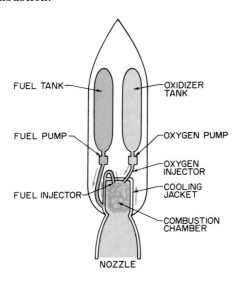

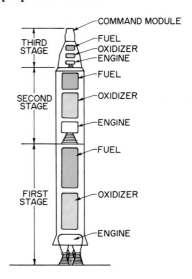

Solid-Propellant Rocket Engines

Since the launching of the Saturn V rockets, much research and development has gone into solid-propellant rocket engines. In these engines, the fuel and oxygen are combined in a solid form. See Fig. 17-30. Solid-propellant engines are much simpler than liquid-propellant engines. The fuel does not have to be pumped to a combustion chamber. It is simply ignited and allowed to burn. This provides thrust, just as in liquid-propellant engines. Solid-fuel engines are so simple that they are used on a small scale to power model rockets. See Fig. 17-31.

One kind of solid propellant is a mixture of a hydrocarbon fuel and a chemical with a high percentage of oxygen. Another solid propellant is a combination of nitroglycerin and nitrocellulose. This propellant also contains a high percentage of oxygen.

The design of a solid-propellant engine determines its power output and duration. Once the engine is made and its fuel is ignited, its power output cannot be regulated. Power output *can* be regulated in liquid-propellant engines. This is one advantage they have over solid-fuel engines.

Fig. 17-30 Solid-propellant rockets burn a solid chemical that contains both the fuel and the necessary oxygen.

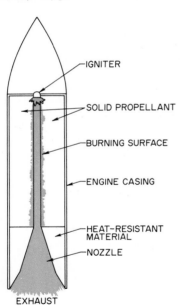

IGNITER

SOLID PROPELLANT

BURNING SURFACE

ENGINE CASING

HEAT-RESISTANT MATERIAL

NOZZLE

EXHAUST

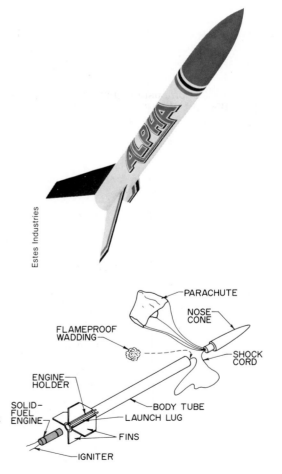

Estes Industries

PARACHUTE

NOSE CONE

FLAMEPROOF WADDING

SHOCK CORD

ENGINE HOLDER

SOLID-FUEL ENGINE

BODY TUBE

LAUNCH LUG

FINS

IGNITER

Fig. 17-31 Model rockets use solid-propellant engines. In addition to propelling the rocket, the engine ejects a parachute at the right time. This provides for a safe landing.

STUDY QUESTIONS

1. Name the two basic types of gasoline piston engines.
2. What is meant by the words *stroke* and *cycle* when referring to engines?
3. Name (in order) the four strokes of the four-stroke engine.
4. In your own words, describe what happens on each of the four strokes of the four-stroke cycle.
5. Name three advantages two-stroke engines have over four-stroke engines.
6. Name two disadvantages of two-stroke engines as compared to four-stroke engines.
7. In your own words, describe what happens on each of the two strokes of the two-stroke cycle.
8. What is the main difference in operation between diesel engines and gasoline engines?
9. Describe how the fuel is ignited in a diesel engine.
10. How many combustion chambers are there in a Wankel rotor?
11. How many power strokes happen during a single rotation of the Wankel rotor?
12. What is Newton's third law of motion?
13. Name five types of jet engines.
14. In your own words, describe how a turbojet engine operates.
15. What is the purpose of the turbine in a turbojet engine?
16. What is meant by *thrust*?
17. Which jet engine has no moving parts?
18. How are turbofan engines different from turbojet engines?
19. What type of jet engine is used to power many commercial airliners?
20. What is the difference between a gas turbine engine and a turboprop engine?
21. What is the main difference between a rocket engine and a jet engine?
22. Explain how a rocket engine works.
23. What are the two types of propellant used in rocket engines?
24. What advantage do solid-propellant engines have over liquid-propellant engines?

ACTIVITIES

1. Find out if it is legal to fire model rockets in your area. Some communities or states restrict rockets to certain areas, or permit their use only under the supervision of teachers or other public officials. You can find out the rules for your area by contacting your local fire department.
2. If rocketry is legal in your area, obtain a model rocket kit. Assemble it, then launch the rocket under the supervision of your teacher. Before the launch, prepare a list of safety precautions for firing. You can get this information from the rocket manufacturer.

Automotive Engines

There are over 100 million automotive engines in daily use. These are all internal-combustion engines. A few cars use rotary engines or two-stroke engines. However, nearly all automotive engines are **four-stroke multi-cylinder engines**. That is, they operate on the four-stroke principle, and they have more than one cylinder. Most car engines have four, six, or eight cylinders.

Most automotive engines also use gasoline as the fuel. However, some engines use diesel fuel. In this chapter, you'll learn about both gasoline engines and diesel engines.

AUTO ENGINE CLASSIFICATION

As you know, cars come in many different sizes and shapes. Car makers also use different engine designs for their products. These designs may change from year to year. However, there are certain basic ways to identify most automotive engines. You have already learned one way: the number of cylinders in the engine. Engines can also be classified according to the arrangement of their cylinders, the type of valve system, and several other features.

Cylinder Arrangement

There are several types of cylinder arrangement possible. However, the most common types are the in-line arrangement and the "V" arrangement. See Fig. 18-1.

An engine that has its cylinders arranged **in-line** is also called a **straight** engine. A *straight four* or *in-line four* engine has four cylinders placed one after another in a straight line. See Fig. 18-2.

A **V-shaped** engine has its cylinders arranged at 45-degree angles. The most common V-type engines are the V-8 and the V-6. A V-8 engine has two banks (sets) of four cylinders arranged in a V. A V-6 engine has two banks of three cylinders. See Fig. 18-3.

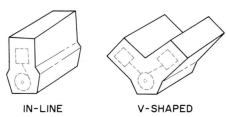

IN-LINE   V-SHAPED

Fig. 18-1 **One way to classify engines is by the cylinder arrangement.**

Fig. 18-2 **This cutaway view of a four-cylinder engine shows its in-line cylinder arrangement.**

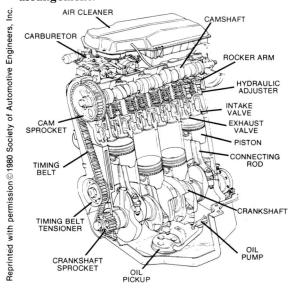

Reprinted with permission © 1980 Society of Automotive Engineers, Inc.

AIR CLEANER — CAMSHAFT

CARBURETOR — ROCKER ARM

HYDRAULIC ADJUSTER

INTAKE VALVE

CAM SPROCKET — EXHAUST VALVE

PISTON

CONNECTING ROD

TIMING BELT

CRANKSHAFT

TIMING BELT TENSIONER

CRANKSHAFT SPROCKET — OIL PICKUP — OIL PUMP

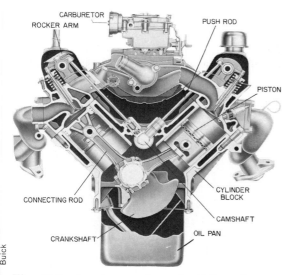

CARBURETOR

ROCKER ARM

PUSH ROD

PISTON

CONNECTING ROD

CYLINDER BLOCK

CAMSHAFT

CRANKSHAFT

OIL PAN

Buick

Fig. 18-3 **A cutaway view of a V-8 engine.**

Valve System

There are two basic types of valve systems used on modern car engines: the overhead-valve system and the overhead-camshaft system.

In overhead-valve engines, the valves are located in the **cylinder head** (the top part of the engine). The valves are directly above the pistons. The V-6 engine in Fig. 18-3 is an example of an overhead-valve engine.

The movement of the overhead valves is controlled by a **camshaft**. The camshaft is a rotating shaft that is located in the **cylinder block** (the part of the engine that contains the cylinders). A series of parts called the **valve train** connects the camshaft to the valves.

In **overhead-camshaft** engines, both the valves and the camshaft are in the cylinder head. The engine in Fig. 18-2 is an example of this type of engine. The camshaft operates the valves directly. This eliminates the need for a valve train. Most modern four-cylinder engines use the overhead-camshaft design.

Fuel and Cooling System

Two other ways to classify auto engines are by (1) the type of fuel they use, and (2) the type of cooling system they use. Most modern car engines are liquid-cooled. Gasoline and diesel fuel are the most commonly used fuels.

Part 1 of this chapter describes the gasoline auto engine and its systems. Part 2 describes diesel auto engines.

PART 1: GASOLINE AUTO ENGINES

If you think about what a car does, you'll see that the engine is part of a mechanical power system. (Figure 5-17 in Chapter 5 shows this system.) The engine provides the mechanical input for the system. However, the engine itself is made up of other smaller systems and assemblies.

Each engine system has its own job. Each must work properly for the engine to work properly. A breakdown in one system will cause problems in the other systems. The result will be an inefficient engine.

Every gasoline auto engine has the following systems:

- Mechanical system
- Lubrication system
- Cooling system
- Fuel system
- Exhaust system

- Ignition system
- Starting system
- Charging system

The diesel engine uses many, but not all, of the same systems. Later on you'll see how a diesel differs from a gasoline engine.

MECHANICAL SYSTEM

The main job of the mechanical system is to convert heat energy into mechanical motion. It first uses heat to produce reciprocating motion. Then it changes this motion into rotary motion. The mechanical system consists of the following:

- Cylinder block and head assembly
- Piston and crankshaft assembly
- Valve assembly

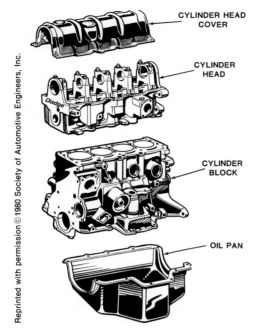

CYLINDER HEAD COVER

CYLINDER HEAD

CYLINDER BLOCK

OIL PAN

Fig. 18-4   **The cylinder block and head assembly has four major parts.**

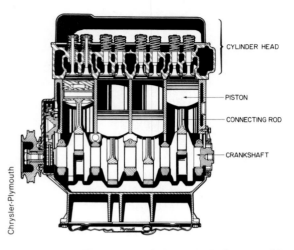

Chrysler-Plymouth

CYLINDER HEAD

PISTON

CONNECTING ROD

CRANKSHAFT

Fig. 18-5   **The piston-and-crankshaft assembly is contained in the cylinder block. Each piston connects to the crankshaft through a connecting rod.**

Cylinder Block and Head Assembly

The backbone of the engine is the **cylinder block**. See Fig. 18-4. An **oil pan** closes the block at the bottom. A **cylinder head** seals it at the top.

All of the mechanical parts and systems needed for engine operation are mounted in or on the cylinder block and head. The block contains the cylinders, pistons, crankshaft, and sometimes the camshaft. The cylinder head contains all or most of the valve assembly.

Piston-and-Crankshaft Assembly

The piston-and-crankshaft assembly is a very important part of the engine. Figure 18-5 shows the piston-and-crankshaft assembly for a four-cylinder engine.

The **crankshaft** receives power from the pistons as they move down in the cylinders. Figure 18-6 shows the crankshaft out of the engine. If you compare this figure with Fig. 18-5, you will see that the pistons do not move down at

the same time. Two pistons are down, and two are up. If all the pistons were down at the same time, the crankshaft couldn't move.

Of the two pistons that *are* down, one is at the end of the intake stroke. The other is at the end of the power stroke. Of the pistons that are up, one is at the end of the exhaust stroke. The other is at the end of the compression stroke. All of the pistons work together to produce a constantly repeating four-stroke cycle.

Fig. 18-6   **This is the same crankshaft used in the engine in Fig. 18-5. The colored areas show where the connecting rods attach.**

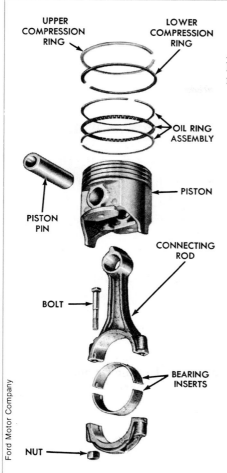

UPPER COMPRESSION RING

LOWER COMPRESSION RING

OIL RING ASSEMBLY

PISTON

PISTON PIN

CONNECTING ROD

BOLT

BEARING INSERTS

NUT

Ford Motor Company

Fig. 18-7 **The piston-and-rod assembly is the link between combustion pressure and the movement of the crankshaft.**

The piston-and-crankshaft assembly contains a smaller assembly called the **piston-and-rod assembly**. See Fig. 18-7. The upper end of the **connecting rod** attaches to the **piston** through the **piston pin**. The lower end of the connecting rod attaches to the crankshaft. The **bearing inserts** provide a low-friction surface for the rotating crankshaft. The different **piston rings** fit around the top of the piston and press against the cylinder wall. They help keep all the combustion pressure contained above the piston.

Valve Assembly

In most modern auto engines, the valve assembly is mounted in the cylinder head. The valves are driven by the camshaft. As you know, the camshaft is located either in the cylinder block or the cylinder head.

Figure 18-8 shows the basic parts of an overhead-valve assembly. Let's take a look at each of the parts:

Camshaft. The camshaft has raised **cams**, or **lobes**, that move the valves as it rotates. Other parts are needed to actually transfer the cam lobe movement to the valves. Figure 18-9 shows how the camshaft operates the valves on the intake and compression strokes.

Hydraulic Valve Lifter, Push Rod, and Rocker Arm. These are the parts that transfer the cam lobe motion to the valves. The hydraulic valve lifter is an oil-controlled device that rides against the cam lobe. It transfers motion to the push rod, which in turn transfers the motion to the rocker arm. The rocker arm then pivots to apply force to the end of the valve. An upward push on the rocker arm changes into a downward push on the valve.

Valve Spring. After the cam lobe rotates past the valve lifter, the valve spring pushes the valve closed. See Fig. 18-9B again. The spring tension keeps the valve closed until the cam lobe raises the lifter again.

Camshaft Drive. The crankshaft drives the camshaft through sprockets and a chain. See Fig. 18-8 again. Notice that the camshaft

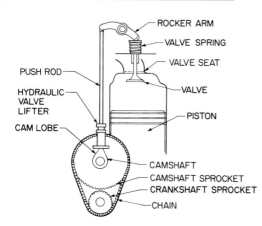

ROCKER ARM

VALVE SPRING

VALVE SEAT

PUSH ROD

VALVE

HYDRAULIC VALVE LIFTER

PISTON

CAM LOBE

CAMSHAFT

CAMSHAFT SPROCKET

CRANKSHAFT SPROCKET

CHAIN

Fig. 18-8 **In an overhead-valve system, the valve assembly is driven by a camshaft located in the cylinder block.**

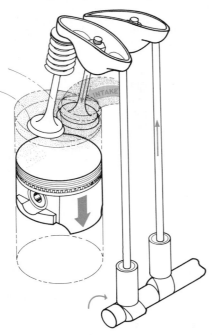 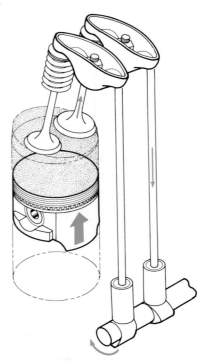

A. Intake stroke: The cam lobe pushes up on other valve assembly parts to open the intake valve.

B. Compression stroke: As the lobe rotates farther, it allows a spring to close the valve.

Fig. 18-9   **The rotating camshaft controls the opening and closing of the valves.**

sprocket is twice the size of the crankshaft sprocket. This means that the camshaft will rotate at one-half the speed of the crankshaft. The crankshaft must make two revolutions to open and close each valve once.

Many engines use an **overhead-camshaft** valve assembly. In this type of assembly, the camshaft operates the valves more directly. The cam lobes move the valves by pushing down on one end of the rocker arm. See Fig. 18-10. Overhead-camshaft assemblies can operate more efficiently at higher engine speeds than overhead-valve assemblies.

LUBRICATION SYSTEM

Large gasoline engines use a **pressure lubrication system**. See Fig. 18-11. The camshaft drives an **oil pump** that is located in the

crankcase. The engine block, crankshaft, and camshaft all have oil passages drilled into them. The oil from the pump passes through an **oil filter** and into the passages. On most engines, all crankshaft and camshaft bearings, hydraulic valve lifters, and rocker arms are pressure-

Fig. 18-10   **An overhead-camshaft valve assembly. The hydraulic adjuster automatically adjusts the assembly for efficient operation.**

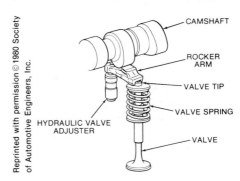

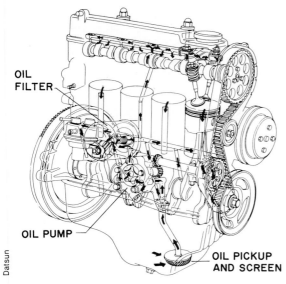

OIL FILTER

OIL PUMP

OIL PICKUP AND SCREEN

Datsun

Fig. 18-11 Automobile engines use an oil pump to pick up and drive oil through the engine under pressure.

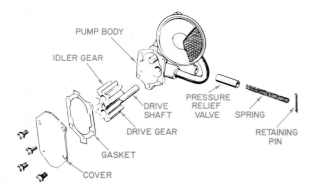

PUMP BODY

IDLER GEAR

DRIVE SHAFT

PRESSURE RELIEF VALVE

SPRING

DRIVE GEAR

RETAINING PIN

GASKET

COVER

Fig. 18-12 An exploded view of an automotive oil pump. The rotating camshaft drives the drive gear. In turn, the drive gear drives the idler gear.

lubricated. Pistons, piston pins, cylinder walls, and piston rings are lubricated by a mist of oil thrown from the crankshaft.

Lubricating oil has important functions in the engine. Its main job is to reduce friction. It also forms a seal between the piston rings and the cylinder wall. The oil also carries heat from the piston to the oil pan. As the car moves, air passing over the outside of the pan cools the oil in the pan.

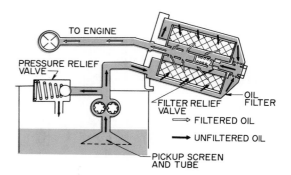

TO ENGINE

PRESSURE RELIEF VALVE

FILTER RELIEF VALVE

OIL FILTER

⇨ FILTERED OIL

➡ UNFILTERED OIL

PICKUP SCREEN AND TUBE

Fig. 18-13 Oil enters through the pick-up screen in the crankcase. It travels through the oil pump and filter before reaching the lubrication points.

In many auto engines, the oil pump is a gear-type pump. See Fig. 18-12. As the pump gears rotate, they force oil through the pump and into the oil filter. See Fig. 18-13. A blocked oil line or very high engine speeds may cause a pressure build-up. A **pressure relief valve** opens if the pressure gets too high.

The filter removes dirt and waste materials from the oil. If the filter is clogged, the oil goes through the **filter relief valve**. See Fig. 18-13 again. Then it passes through the system. In this way, the engine can get lubricated even if the filter is completely plugged with dirt.

COOLING SYSTEM

Some auto engines use an **air cooling system**. In this type of system, air is simply passed over the engine to remove heat. However, most engines use a **liquid cooling system**. The liquid, or **coolant**, is usually water mixed with antifreeze. **Antifreeze** is a liquid that lowers both the boiling point and the freezing point of the coolant.

Figure 18-14 shows the major parts of a liquid cooling system. A **water pump** drives the coolant from the bottom of the **radiator** into the engine block. The coolant travels in passages around the cylinders and through the cylinder head. These passages are called **water jackets**. The coolant removes heat as it travels through the engine. Then it goes to the top of the radiator. The hot coolant cools off as it flows down through the radiator.

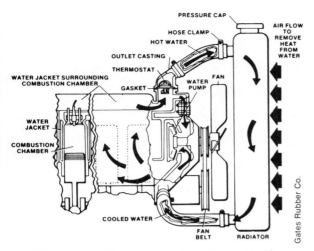

Fig. 18-14 **The arrows show the path of coolant through an engine.**

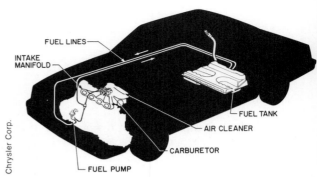

Fig. 18-15 **An automotive fuel system.**

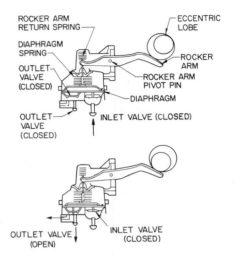

Fig. 18-16 **Automotive fuel pumps are usually driven by a lobe on the camshaft. The inlet and outlet valves are both check valves (one-way valves). Therefore, the fuel can only flow in one direction.**

A **thermostat** between the engine and the radiator controls the flow of coolant. If the engine is cool, the thermostat stays closed. This keeps the radiator from cooling the engine any further. Once the engine reaches its correct operating temperature (usually between 160-210° F), the thermostat opens. In this way, the thermostat keeps the engine at the right operating temperature.

FUEL SYSTEM

An automotive fuel system must provide a mixture of gasoline and air to the cylinders. Figure 18-15 shows the main parts of the system. The **fuel pump** pumps fuel from the fuel tank to the carburetor. The **carburetor** mixes the fuel with clean air from the **air cleaner**. The air-fuel mixture then passes into the **intake manifold**. The intake manifold connects the carburetor to each of the engine's cylinders. When the piston moves down on the intake stroke, a vacuum is formed in the cylinder. Atmospheric pressure pushes the air-fuel mixture into the cylinder to fill the vacuum.

Fuel Pump

Some cars use an electric fuel pump to deliver gas from the tank to the carburetor. However, most cars use a **mechanical fuel pump**. See Fig. 18-16.

The rocker arm of the fuel pump is moved up and down by an eccentric lobe on the camshaft. As the rocker arm moves up, it raises a diaphragm in the pump (Fig. 17-28A). The resulting partial vacuum pulls fuel into the pump. When the arm moves down, the inlet valve closes. The diaphragm spring expands, and the diaphragm pushes fuel out of the pump (Fig. 17-28B).

Carburetor

As you learned earlier, pistons create a partial vacuum in the cylinders when they move down on the intake stroke. This allows atmospheric pressure to push a mixture of air and fuel into the cylinders for combustion. The carburetor, mounted on top of the engine, is the device that mixes the right amount of air with the right amount of fuel.

Figure 18-17 shows a simple carburetor. It looks a little complicated, but to start with just look at the right side of the drawing. A carburetor is basically a hollow tube in which fuel mixes with air. The narrow part of the tube is called the **venturi**. When air rushes through the venturi, it speeds up. (This is just like the water from a garden hose speeding up after you put your thumb over the opening.)

Besides speeding up, the air also creates a low-pressure area (partial vacuum) in the venturi. This vacuum allows atmospheric pressure to push on the fuel in the **float bowl**. The fuel flows from the bowl through the **main fuel jet** and **nozzle** into the venturi.

The fuel nozzle has a small opening. When the fuel rushes through this opening, it separates into many small droplets and mixes with the rushing air. This air-fuel mixture then flows past the **throttle plate** and into the intake manifold and cylinders.

The driver can speed up the air flow further by stepping on the gas pedal. This opens the

throttle plate further and allows more air and fuel to go to the cylinders.

Carburetors must provide different air-fuel mixture for different operating conditions. They use the following parts and circuits to do this:

Float Assembly. Fuel from the fuel pump enters the carburetor past the **float needle**. The float maintains a constant level of fuel inside the bowl. When the carburetor needs more fuel, the float tips down to pull the float needle from its seat. Fuel then flows into the bowl and brings the fuel level back up. The float then tips up to shut the needle valve.

Idle Fuel Circuit. Figure 18-17 shows the fuel flow in the **main fuel circuit**. This is how the carburetor works much of the time. However, when the engine is at **idle** (not moving the car), not too much air flows through the venturi. This air cannot create a high-enough vacuum to pull fuel from the fuel nozzle. Therefore, the fuel must have another way to get into the flow of air.

To provide for fuel flow at idle, the carburetor has an **idle fuel passage**. See Fig. 18-17 again. The vacuum below the throttle plate is enough to pull fuel out of the float bowl and through

Fig. 18-17   The carburetor's job is to mix fuel with air. This mixing takes place in the right side of the carburetor shown here. The fuel supply is in the left side of the carburetor.

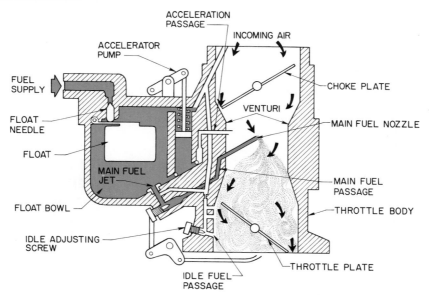

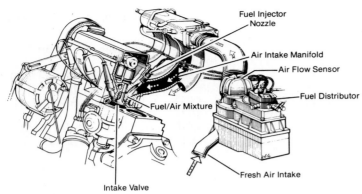

Fig. 18-18 **One type of modern fuel-injection system.**

Fuel Injector Nozzle

Air Intake Manifold

Air Flow Sensor

Fuel Distributor

Fuel/Air Mixture

Fresh Air Intake

Intake Valve

Volkswagen

this passage. The fuel can then mix with the incoming air and travel to the cylinders.

The **idle adjusting screw** controls how much fuel flows through the idle fuel passage. By turning this screw, a mechanic can make the air-fuel mixture **richer** (more fuel) or **leaner** (less fuel).

Choke Circuit. During starting, the engine needs a rich air-fuel mixture. To supply this mixture, the **choke plate** closes partially. This restricts the entry of air. The vacuum in the cylinders then pulls in a greater amount of fuel.

Acceleration Circuit. Sometimes the engine needs a big "boost" of fuel so that the driver can pass other cars. When the driver "floors it," a plunger in the accelerator pump pushes down on the fuel below it. This action squirts extra gasoline into the venturi.

Fuel-Injection Fuel System

Many cars have a fuel-injection system instead of a carburetor. The fuel-injection system does not depend on venturi action to bring fuel into the air stream. Instead, **fuel injectors** spray a measured amount of fuel into the air stream. Fuel injection controls the amount of fuel used more accurately than carburetion does. This improves both fuel economy and pollution control.

Figure 18-18 shows a typical fuel-injection system. Injection takes place just before the air-fuel mixture enters the combustion chamber. The system is controlled as follows:

- Gasoline is pumped from the fuel tank to the **fuel distributor**. The distributor regulates the amount of fuel to be injected. It also sends the fuel to the proper injector just as the intake valve begins to open.

- A throttle plate in the intake manifold controls the air flow into the cylinder. The driver controls the throttle plate with the accelerator pedal.
- When the engine is cold, a heat sensor in the engine block water jacket activates a **cold-start injector**. This injector adds extra fuel to the intake manifold. The cold-start injector serves the same purpose as the carburetor choke.

Fuel-injection systems cost more than carburetion systems. However, they control the fuel more precisely. This increases fuel efficiency as well as engine performance.

Turbochargers

An increasing number of modern cars use turbochargers to increase engine power. See Fig. 18-19. A turbocharger consists of an **air**

Fig. 18-19 **Turbochargers are powered by exhaust gases. When the exhaust turbine spins, the air impellor forces extra air into the cylinder.**

Schwitzer Div., Wallace-Murray Corp.

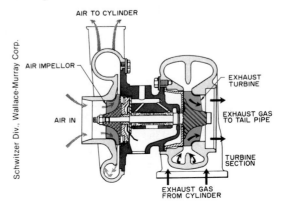

AIR TO CYLINDER

AIR IMPELLOR

AIR IN

EXHAUST TURBINE

EXHAUST GAS TO TAIL PIPE

TURBINE SECTION

EXHAUST GAS FROM CYLINDER

282

impellor and an **exhaust turbine** connected to a shaft. High-pressure exhaust gases leaving the engine drive the turbine. The turbine, in turn, drives the air impellor. The air impellor forces air into the combustion chamber at a pressure higher than atmospheric pressure. This action puts more air into the cylinder. This provides higher compression for greater power output.

Turbochargers improve the performance of all engines, especially small low-powered engines. Today, small engines are popular for reasons of fuel economy. However, they do not have the power that older and larger engines had. Adding a turbocharger to a small engine improves its overall performance, especially during passing at highway speeds.

EXHAUST SYSTEM — EMISSION CONTROL

Since the 1950s, car makers have tried to make automotive emissions (exhaust) less harmful. They have added many controls to auto engines. These controls reduce the amounts of harmful pollutants. The main pollutants are hydrocarbons (unburned fuel), carbon monoxide, and nitrogen oxides.

Before the 1950s, the exhaust system consisted only of an exhaust manifold, a muffler, and a tail pipe. The **exhaust manifold** collects the exhaust gas from the combustion chambers. The **muffler** dampens exhaust noises. The tail pipe is connected to the muffler. It carries the exhaust gas to the rear of the car.

Figure 18-20 shows the emission control devices now used on most cars. Each device has an important function.

IGNITION SYSTEM

Once the air-fuel mixture is in the cylinder and is compressed by the piston, it must be ignited. High voltage must reach the spark plug and produce a spark. Delivering high voltage to the spark plug is the job of the automotive ignition system.

There are three basic types of automotive ignition systems: the breaker-point system, the electronic ignition, and the computerized ignition system.

Fig. 18-20 The emission control system includes special devices added to existing parts of the automobile engine and exhaust system.

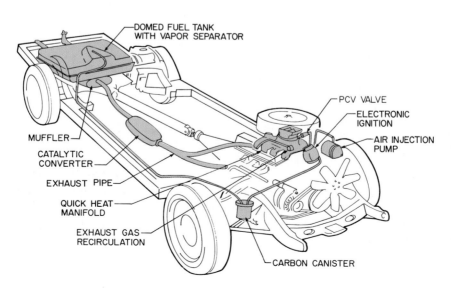

DOMED FUEL TANK WITH VAPOR SEPARATOR

PCV VALVE
ELECTRONIC IGNITION
AIR INJECTION PUMP

MUFFLER
CATALYTIC CONVERTER
EXHAUST PIPE
QUICK HEAT MANIFOLD
EXHAUST GAS RECIRCULATION
CARBON CANISTER

Breaker-Point Ignition

Figure 18-21 shows a typical breaker-point ignition system. The system consists of two separate electrical circuits: the primary circuit and the secondary circuit.

The **primary circuit** carries only low-voltage current. This circuit consists of the battery, coil, breaker points, and condenser.

The **secondary circuit** carries high-voltage current. It consists of the coil, distributor cap and rotor, spark plug wires, and spark plugs.

Notice that the coil is a part of both the primary circuit and the secondary circuit. The **primary windings** in the coil are part of the primary circuit. The **secondary windings** are part of the secondary circuit.

The breaker points are an important part of the whole ignition system. They act as a switch to turn the primary current on and off. The breaker points themselves are controlled by the **distributor cam**, which is turned by the camshaft. Figure 18-21 shows a six-sided distributor cam. The number of sides on the cam is always the same as the number of spark plugs in the system.

When the breaker points are closed, low-voltage current (12 volts) flows from the battery to the primary coil windings. It continues to flow across the points and to the electrical ground of the car. The **ground** is the metal of the cylinder block or the car frame. This metal conducts the current back to the battery to complete the circuit. The symbol for ground is:

The primary current builds up a magnetic field in the coil. When the distributor cam rotates and opens the breaker points, the primary circuit is broken. This causes the magnetic field in the coil to collapse. The collapsing field induces a high-voltage current in the secondary windings. (This is much like the action of a step-up transformer, as was described in Chapter 7.)

Note: When the primary field collapses, the **condenser** keeps current from jumping the space between the points. This prevents point burn-out.

Next, the high-voltage secondary current (over 20,000 volts) travels to the distributor cap. The **rotor** in the cap rotates with the distributor cam. It directs the secondary current to the proper spark plug wire. The current travels through the wire to reach the spark plug. Each spark plug has two **electrodes**, or contacts. The high-voltage current creates a spark when it jumps

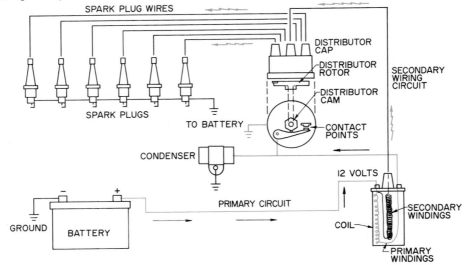

Fig. 18-21 **The breaker-point ignition system is made up of two electrical circuits. The primary circuit is shown here in color.**

from one electrode to the other. The current then flows through the electrical ground and back to the battery.

The ignition system in Fig. 18-21 is for a six-cylinder engine. The distributor cam opens and closes the breaker points six times each time it rotates once. This produces a spark to ignite the air-fuel mixture in each cylinder.

Electronic Ignition

Breaker-point ignition systems require routine maintenance on the breaker points. The current that passes through the points burns and pits the contact area. The points must be replaced and adjusted once or twice a year.

Many cars use an electronic ignition system, which does not use breaker points. All the switching is done with electronic solid-state devices. This system does away with breaker-point maintenance. It also provides a stronger spark at the spark plugs.

Figure 18-22 shows a typical electronic ignition system. The main parts are the impulse generator, reluctor, and transistorized control unit. These parts do the same job as the breaker points, condenser, and distributor cam.

The **reluctor** has iron teeth. There are as many teeth as there are cylinders in the engine. Like the distributor cam, the reluctor is turned by the camshaft. As the reluctor turns, the teeth pass very close to the **pickup unit** of the

impulse generator. As each tooth passes the pickup unit, it produces a small electrical charge in the impulse generator. This charge triggers the **transistorized control unit**. The control unit stops the flow of current in the primary coil windings. The collapse of the primary magnetic field generates high voltage in the secondary circuit.

One of the big advantages of the electronic ignition system is that it can control the ignition timing very precisely. **Ignition timing** refers to the time at which the spark plug fires in relation to the position of the piston in the cylinder. Ignition timing is very important in emission control. An electronic ignition system can help reduce emissions of nitrogen oxides and other pollutants.

Computerized Ignition and Fuel Control

Some car makers are adding small computers to control both the ignition system and the fuel system. In this type of system, there are many sensing devices attached to the engine. These **sensors** measure such things as the exhaust gas temperature, air flow, and engine operating temperature. This information is sent to the computer. The computer "reads" the information. Then it adjusts the fuel system and ignition timing to fit the engine conditions. This type of system reduces pollution and improves the fuel economy.

Fig. 18-22 One type of electronic ignition system.

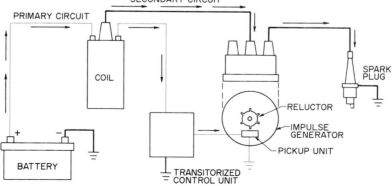

STARTING SYSTEM

Car engines start with a starting motor, or **starter**. The starter starts the pistons moving up and down. This way, they can draw in an air-fuel mixture. The ignition system then provides the sparks to ignite the mixture.

The starter is mounted low on the engine, toward the rear of the crankshaft. As you remember, a heavy flywheel is attached to the end of the crankshaft. The flywheel has a large **ring gear** around its outer edge.

The starter is basically an electric motor with a **drive mechanism** on it. In Fig. 18-23, the drive mechanism consists of a solenoid connected to a shift lever. **Note:** Solenoids and motors are described in detail in Chapter 7.

When the driver turns the ignition key, the motor starts turning. At the same time, the drive mechanism slides a small gear into mesh with the flywheel ring gear. See Fig. 18-23B. The motor then turns the flywheel, which turns the crankshaft and moves the cylinders up and down. When the engine starts, the starter gear disengages from the flywheel gear.

CHARGING SYSTEM

Automobiles need electricity to run the ignition system and accessories during engine operation. The battery also must be kept charged so that it can operate the starting system.

Figure 18-24 shows a charging system. The two main parts are the alternator and the voltage regulator.

The **alternator** is a source of electricity. It is driven by a belt from the engine's crankshaft. The alternator converts rotary motion into electricity, as described in Chapter 7.

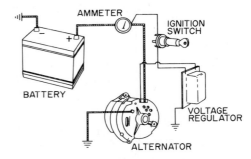

Fig. 18-24   A typical automotive charging circuit.

Fig. 18-23   The operation of one type of automotive starter.

A. **Normally, the starter gear is not engaged with the flywheel gear.**

B. **When the driver turns the ignition key, the starter gear meshes with the flywheel gear and "cranks" the engine.**

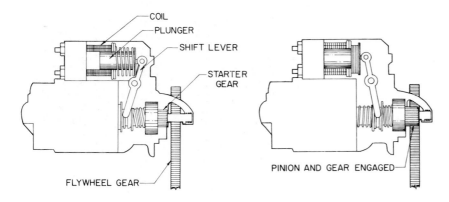

The **voltage regulator** controls the electrical output of the alternator. It maintains a constant voltage output. For the typical 12-volt system, the regulator maintains a voltage of just over 14 volts. This is the voltage needed to keep the battery fully charged and all the electrical parts working.

POWER TRAIN

Automobiles must have a way of delivering engine power to the wheels. This is done with a power train. The power train usually includes a clutch, transmission, drive shaft, and rear end assembly. See Fig. 18-25.

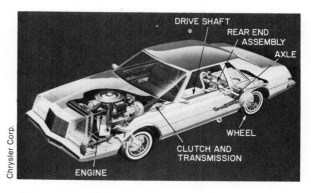

Chrysler Corp.

DRIVE SHAFT
REAR END ASSEMBLY
AXLE
WHEEL
CLUTCH AND TRANSMISSION
ENGINE

Fig. 18-25 **This type of power train is used on many automobiles.**

PART II: DIESEL ENGINES

As you learned in Chapter 17, diesel engines provide power for both transportation devices (like the diesel-electric locomotive) and stationary uses (like electrical power generation). During the past several years, diesels have become a more common power source for cars. See Fig. 18-26. Their popularity is due to their good fuel economy. Diesel-powered cars average about 25 percent more miles per gallon than gasoline-powered cars.

Fig. 18-26 **Mid-size and larger automobiles use large diesel engines, like this V-6 model. Smaller cars use four-cylinder engines.**

However, diesel engines *do* have some disadvantages. They must be made larger, stronger, and heavier than gasoline engines. This is because the pressure of combustion is much higher in diesel engines. The added weight slows down the acceleration, or "pick-up," of diesel-powered cars. Engineers are now designing new cars that will have better performance.

Oldsmobile

DIESEL ENGINE OPERATION

There are two basic types of diesel engines: two-stroke and four-stroke. Most diesel engines used in cars are four-stroke engines. Diesel engines operate very much like gasoline engines. However, there are two major differences. Let's look at Fig. 18-27.

As you can see, a diesel four-stroke cycle is much like a gasoline four-stroke cycle. The piston travels from one end of the cylinder to the other four times during the cycle. The fuel is ignited on every third stroke to produce a power stroke.

But take a close look at the intake stroke. Notice that when the intake valve is open, only *air* goes into the cylinder, not an air-fuel mixture. The air is compressed on the next stroke. Then the fuel is *injected* into the combustion chamber. The compressed air is very hot and ignites the fuel on contact. The result is the same as in a spark-ignited gasoline engine — a power stroke.

Do you see the differences now? First, the diesel engine uses a different method of delivering air and fuel to the cylinder. Second, the diesel uses **compression ignition**, not spark ignition.

COMPRESSION HEAT AND DIESEL FUEL

Diesel engines can use compression heat to ignite the air-fuel mixture because they have much higher compression ratios than gasoline engines. If you recall from Chapter 17, the **compression ratio** is a comparison of the cylinder volume at the bottom of the piston stroke and the cylinder volume at the top of the stroke. Gasoline engines have a compression ratio of about 8:1. Diesel engines have compression ratios ranging from 16:1 to 23:1.

The larger the compression ratio, the greater the heat and pressure built up in the cylinder on the compression stroke. In diesel engines, the temperature reaches to about 1000° F (538° C).

Diesel fuel is a type of oil that is not as highly refined as gasoline. It's thicker than gasoline and has a greater energy content. At the end of the compression stroke, the fuel is sprayed into the cylinder through tiny holes in a **fuel injector**. The fuel separates into a mist or fog of fuel particles. As soon as the mist hits the hot air, it ignites. The resulting sudden rise in pressure forces the piston down in the cylinder.

Fig. 18-27 **The diesel four-stroke cycle has the same four actions as the gasoline four- stroke cycle.**

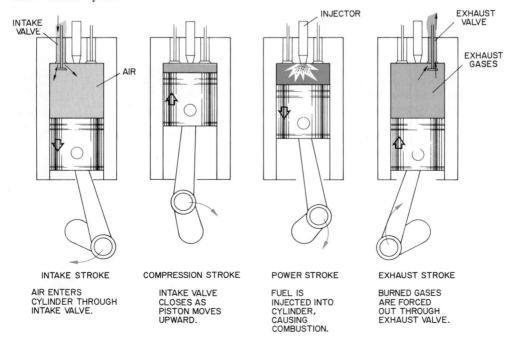

INTAKE STROKE	COMPRESSION STROKE	POWER STROKE	EXHAUST STROKE
AIR ENTERS CYLINDER THROUGH INTAKE VALVE.	INTAKE VALVE CLOSES AS PISTON MOVES UPWARD.	FUEL IS INJECTED INTO CYLINDER, CAUSING COMBUSTION.	BURNED GASES ARE FORCED OUT THROUGH EXHAUST VALVE.

Diesel auto engines look very much like gasoline auto engines. Both types of engines have a piston-and-crankshaft assembly, a valve assembly, lubrication system, and cooling system. But if you're trying to identify the type of engine, look for a carburetor and an ignition system. You won't find them on a diesel engine. The parts of the diesel fuel system do the jobs of both the carburetor and the ignition system.

In gasoline engines, the engine power is controlled by regulating the amount of air-fuel mixture entering the engine. This is the job of the carburetor. In diesel engines, engine power is controlled by regulating the amount of fuel injected into the cylinder. This is the job of the **fuel-injection pump** and the **fuel injectors**. See Fig. 18-28.

Because of the pressure built up in the cylinder on the compression stroke, the fuel that is sprayed into the cylinder must itself be under high pressure. Otherwise, it would not be able to enter the cylinder. The fuel-injection pump provides the necessary pressure. It pushes fuel to an injector in each cylinder at a pressure over 1000 pounds per square inch.

Figure 18-29 shows a typical fuel injector. The injector opens as soon as the injection pump raises the fuel pressure. The high pressure raises the **needle valve** from its seat. Fuel sprays into the cylinder. As soon as the pressure from the injection pump stops, the spring forces the needle against the valve seat. This stops the flow of fuel.

In the injector, a small amount of fuel lubricates the needle valve movement. This fuel returns to the fuel tank through the **leak-off connection**.

The fuel-injection pump has one other important function. Besides controlling the *amount* of fuel that enters the cylinder, it controls the *time* at which the fuel is injected. The injection pump makes sure that the injectors spray fuel just before the end of the compression stroke. This timing provides for the most efficient use of the fuel.

Robert Bosch GmbH

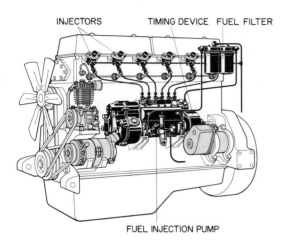

Fig. 18-28 In a diesel engine, the injection pump supplies high-pressure fuel to an injector in each cylinder.

Fig. 18-29 In many diesel engines, each cylinder gets its fuel through an injector like the one shown here.

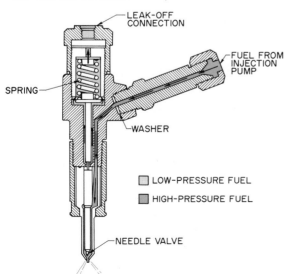

Glow Plugs

Because there is no spark to ignite the fuel, diesel engines can be hard to start. Heat has to build up in the cylinders before the air-fuel mixture will ignite. This is especially true in cold weather. For improved starting, diesels are often equipped with glow plugs.

Glow plugs have a small wire that gets red-hot when electrical current flows through it. There is a glow plug for each cylinder. When the car is to be started, current from the battery heats the glow plugs. The glow plugs then heat the air as it enters the cylinders. This pre-heating improves starting.

STUDY QUESTIONS

1. Name three ways in which automotive engines are classified.
2. Where is the camshaft located in an overhead-valve system?
3. Where is the camshaft located in an overhead-camshaft system?
4. List the eight systems used in gasoline automotive engines.
5. What is the "backbone" of the engine?
6. What is the purpose of the rocker arm?
7. What is the purpose of the valve spring?
8. What is the main function of engine lubricating oil?
9. How does the thermostat work to keep the engine at the right operating temperature?
10. What engine part operates the mechanical fuel pump?
11. What is the narrow part of the carburetor called?
12. What is the purpose of the float in a carburetor float assembly?
13. What do fuel-injection systems have to replace fuel delivery by venturi action?
14. What device is used to force air into the combustion chamber to increase performance?
15. Why do cars have emission control systems?
16. Name the three basic types of automotive ignition systems.
17. What is the purpose of the ignition coil?
18. What are two advantages of computerized ignition and fuel control?
19. What engine part does the starter turn?
20. What is the purpose of the alternator?
21. Name the four parts of the automotive power train.
22. What are the two major differences between diesel engines and gasoline engines?
23. What is a compression ratio?
24. How is diesel fuel different from gasoline?
25. What diesel engine parts regulate the amount of fuel injected into the cylinder?

ACTIVITIES

1. Make a list of the advantages and disadvantages of small cars as compared to large cars. You could compare such things as safety, cost of repairs, and life expectancy. Ask your parents and friends for information. A visit to a new car showroom would be helpful, too.
2. There are several fuels besides gasoline and diesel fuel that can be used to run car engines. For example, LP gas, methanol, and gasohol are all being used. Find out which alternative fuels are used in your community. Gas station attendants and auto repair shop workers would be good sources of information.
3. Find someone who owns a car with a diesel engine. If you can't find someone, visit a car dealer who sells diesel-powered cars. Find out the advantages and disadvantages of diesel-engine cars as compared to gasoline-engine cars. Make up a chart that shows your findings and display it in the classroom.

Small Engines

In Section V, you learned that heat engines are very important sources of power in our society. You also saw that there are two basic types of heat engines: the external-combustion engine and the internal-combustion engine. Of these two types, the internal-combustion engine is the more widely used. For example, practically all cars use an internal-combustion engine.

In Section VI, we will examine another common use of the internal-combustion engine — the small gasoline engine. Small gas engines power many transportation devices such as motorcycles and small boats. They also power small electrical generators. And they are used in lawn mowers, chainsaws, garden tillers, and many other devices.

In general, small gas engines have one cylinder and are air-cooled. They are either two-stroke or four-stroke piston engines, and they produce up to eight horsepower. In this section, you'll get an "in-depth" look at the small gas engine. You'll see how it works and how to take care of it. You'll even learn how to find out what's wrong if there are problems in operation.

Small Engine Operation

19

There are many different types and sizes of small engines. This chapter doesn't explain any one engine in detail. Instead, it gives you the information you need to develop a general understanding of *all* small gas engines. Figure 19-1 shows a cutaway view of a typical small engine.

There's a lot to learn if you really want to know how a small engine works. Basically, a small engine produces rotary motion by burning a mixture of air and gasoline. Now, to understand *how* the engine does this, you have to learn about the parts of the engine. See Fig. 19-2. You have to learn what each part looks

like and what its name is. You also have to find out how the parts fit in the engine. Most important, you should know the **function**, or purpose, of each part.

To make it easier to learn about small engine operation, this chapter is divided into six sections. Each section describes one of the major small engine systems. Each **system** is a group of parts that work together. Together, the parts perform a certain basic function in the engine. The following are the six systems:

- **Mechanical system** — develops up-and-down motion and converts it into rotary motion.
- **Lubrication system** — lubricates (oils) engine parts to reduce friction and wear.
- **Cooling system** — removes excess heat from the engine.
- **Fuel system** — mixes air and fuel properly and delivers the mixture to the combustion chamber.
- **Ignition system** — produces the spark that ignites the fuel.
- **Starting system** — sets the engine in motion.

Study these systems separately. Once you understand each system, you can "put them together" to see how the whole engine works.

MECHANICAL SYSTEM

The mechanical system develops up-and-down motion in the engine. It also converts this reciprocating motion into useful rotary motion. Burning fuel in a sealed chamber produces high-pressure gases. As the gases expand, they push down on a piston. The piston is attached to a crankshaft. When the piston moves down, it

Kohler Co.

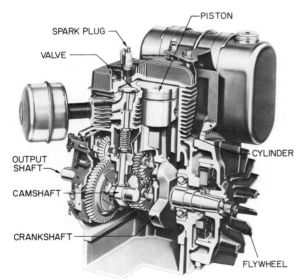

Fig. 19-1 **This small engine has been cut away so that you can see the working parts.**

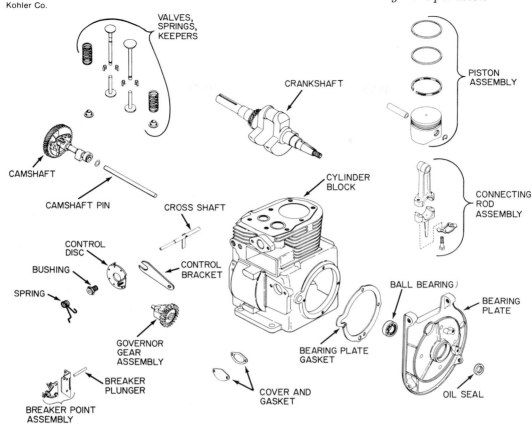

Fig. 19-2 **Each of the many parts in this small engine has an important job in making the engine work. (The top of the engine — the cylinder head — is not shown.)**

turns the crankshaft. The turning crankshaft provides mechanical power. See Fig. 19-3. This power can then be put to work.

The mechanical system operates on either the two-stroke cycle or the four-stroke cycle. (To review these basic cycles, see *GASOLINE PISTON ENGINES* in Chapter 17.)

Two-stroke engines and four-stroke engines basically have the same mechanical parts. The main difference is in the parts used to take in fuel and release burned gases. First, we'll look at all the parts and their functions. Then we'll see how the parts work together in both two-stroke and four-stroke engines.

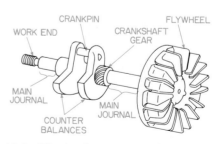

Fig. 19-3 **The basic purpose of the mechanical system is to produce up-and-down motion and then convert it into rotary motion.**

Crankcase, Cylinder, and Cylinder Head

The crankcase, cylinder, and cylinder head make up the framework, or **body**, of the engine. Together, these parts form a sealed unit. The mechanical system works inside this unit. See Fig. 19-4.

Sometimes the parts of the engine body are manufactured separately. Then they are bolted together. Other times the cylinder and the cylinder head are made as one single part. Most of the time, however, the cylinder and the *crankcase* are made as one part. This single part is called the **cylinder block**. See Fig. 19-5.

The **crankcase** is a kind of metal box. It makes up the bottom part of the engine. See Fig. 19-4 again. The crankcase has holes in two sides or ends. These holes support the crankshaft. The crankcase is usually made of aluminum. In four-stroke engines, it acts as a reservoir (holding place) for the engine's oil supply.

The **cylinder** is the middle part of the cylinder block. The piston moves up and down inside it. Some cylinders are made of iron. However, most small engine cylinders are aluminum. Aluminum is lightweight. It also transfers heat quickly. Some engines have an iron sleeve inside the aluminum cylinder. The outside of the cylinder has thin ribs, called **cooling fins**. See Fig. 19-5 again. These fins help cool the engine. (Cooling fins are explained later in this chapter.)

The **cylinder head** is the top part of the engine. It is bolted tightly to the cylinder block. A **head gasket** fits between the cylinder head and the cylinder block. See Fig. 19-5 again.

The head gasket is made of a piece of asbestos sandwiched between two pieces of metal. The gasket is thin and flexible. It produces a tight seal to keep combustion pressure from escaping. Like the cylinder, the cylinder head has fins to help cool the engine. It also has a threaded hole for the spark plug. See Fig. 19-5 again.

Throughout this chapter you will see the term *combustion chamber*. The combustion chamber is not a part, but a space. This space is located under the cylinder head, at the top of the cylinder. See Fig. 19-4 again. The bottom tip of the spark plug extends through the cylinder head into the combustion chamber. The air-fuel mixture is ignited in the combustion chamber.

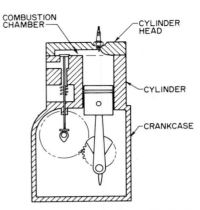

Fig. 19-4   Three parts — the cylinder head, the cylinder, and the crankcase — make up the framework of the engine. Most of the mechanical parts are located inside the framework.

Tecumseh Products Co.

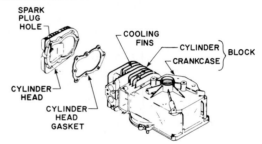

Fig. 19-5   In most small engines, the cylinder and crankcase are made of aluminum. They are usually cast as one part, called the *block*.

Piston

The piston is a hollow metal part that moves up and down in the cylinder. Through the connecting rod, the movement of the piston turns the crankshaft. (The connecting rod is covered later in this chapter.) The piston also works as a moving seal. It keeps the air-fuel mixture in the combustion chamber. It also separates the combustion chamber from the crankcase.

Most small engine pistons are made of aluminum. Figure 19-6 shows the parts of a piston. Notice the hardened steel **piston pin** that fits into the piston. It is also called a **wrist pin**. This hollow pin attaches the piston to the connecting rod.

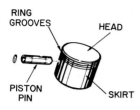

Fig. 19-6 **The main parts of a piston for a four-stroke engine.**

FOUR-STROKE PISTON   TWO-STROKE PISTON

Fig. 19-7 **The heads of pistons used in two-stroke engines and four-stroke engines are shaped differently.**

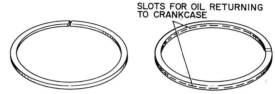

SLOTS FOR OIL RETURNING TO CRANKCASE

A. **Compression ring**   *B.* **Oil control ring**

Fig. 19-8 **Two different kinds of rings are used on small engine pistons.**

The piston is slightly smaller in diameter than the cylinder. This creates a small gap between the piston and the cylinder wall. The top of the piston is made smaller than the bottom. This makes the gap slightly larger at the top of the piston.

Pistons are made the way they are because of the heat produced during combustion. This heat makes the whole piston expand. To keep from scraping against the cylinder wall, the piston must have some space in which to expand. Combustion heat is highest at the top of the piston. Therefore, the piston has to be even smaller at the top. This allows for the extra expansion.

There is a basic difference between two-stroke piston heads and four-stroke piston heads. Four-stroke piston heads are either flat or slightly curved. Most two-stroke pistons have slanted and raised heads. See Fig. 19-7. The raised head directs the incoming air-fuel mixture up into the combustion chamber. The head pushes out burned gases at the same time. It also keeps the new air-fuel mixture from mixing with the burned gases.

Piston Rings

Several grooves are cut around the tops of all pistons. These grooves hold thin metal rings, called *piston rings*. The top rings are called **compression rings**. The bottom rings are called **oil control rings**. See Fig. 19-8.

The main job of the compression rings is to keep the combustion pressure inside the combustion chamber. Without this seal, some pressure would "blow by" the piston. The pressure would not push on the piston. Instead, it would escape down the cylinder wall into the crankcase. This would cause a loss of power.

The oil control rings keep the engine oil in the crankcase, away from the combustion chamber.

They do this by scraping excess oil away from the cylinder wall. Oil must not get into the combustion chamber. Otherwise, the oil would burn and the engine would constantly need more oil. Too much oil in the combustion chamber would also keep the spark plug from firing properly.

All pistons have either one or two compression rings. Four-stroke pistons have either one or two oil rings. Two-stroke pistons usually don't have any oil rings. You'll learn why when you study the *LUBRICATION SYSTEM* section.

To provide a tight seal, the piston rings must press against the cylinder wall. Engine builders provide for this by compressing the rings into the ring grooves as they put the piston into the cylinder. Once inside the cylinder, the rings push out with a spring tension. This tension holds them tightly against the cylinder wall and provides the all-important seal.

Piston rings are usually made of cast iron. Many rings are coated with **chrome**. Chrome is a hard metal that wears slowly. It increases the life of the ring and reduces wear on the cylinder wall.

Connecting Rod

The connecting rod connects the piston to the crankshaft. Figure 19-9 shows how the connection is made. Notice that the connecting rod has a small hole at the top and a larger hole at the bottom.

A **piston pin** fits through the top hole and joins the rod to the piston. The bottom connecting rod hole has a removeable **rod cap**. This makes it possible to connect the rod to the crankshaft. You simply place the rod on the crankshaft and attach the cap with bolts and nuts. This arrangement also makes it possible to remove the rod from the crankshaft.

The holes in the connecting rod are made very precisely. This is so the crankshaft and piston pin will fit well.

During engine operation, the connecting rod is affected by a lot of pushing and pulling. To take all this stress, the rod must be very strong. Therefore, many connecting rods are made of strong forged steel. Some rods are made of cast steel or forged aluminum. However, these rods can only be used in engines with light output loads.

Some connecting rods don't have bearings. Other connecting rods have a bearing at each end. **Bearings** are cylindrical metal parts that support moving engine parts. They are made of low-friction material that wears faster than the moving parts. This keeps the parts from being damaged or wearing out quickly.

Sleeve bearings are the most common bearings used in connecting rod holes. A one-piece sleeve bearing is usually called a **bushing**. It fits into the top hole. A two-piece sleeve bearing is called an **insert bearing**. It fits into the two halves of the bottom hole. See Fig. 19-10. Some small engines use **anti-friction** bearings in the bottom hole. (Anti-friction bearings are covered under *Crankshaft Bearings* later in this chapter.)

Crankshaft

The crankshaft is the heart of the engine's operation. Along with the connecting rod, it changes the piston's reciprocating motion into rotary motion. See Fig. 19-3 again. The rotating crankshaft provides the mechanical power to run the engine. It also provides the power output that is used to do work.

You'll often hear people talk about "turning over" an engine. This is a reference to the turning or rotation of the crankshaft. Another expression is "rpm's." This is a reference to the crankshaft's **revolutions per minute**.

Small engine crankshafts are made of either cast iron or cast steel. Figure 19-11 shows a crankshaft for a one-cylinder four-stroke engine. The gear on the crankshaft is used to drive another engine part called the **camshaft**. Crankshafts in small two-stroke engines do not have a camshaft drive gear. You will learn why later in this chapter.

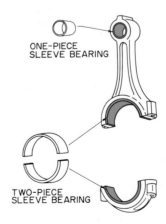

Fig. 19-10 Sleeve bearings are often inserted into the connecting rod holes. This reduces friction and wear on the connecting rod, piston pin, and crankshaft.

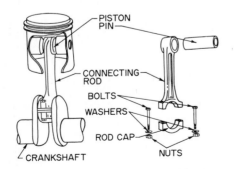

Fig. 19-9 The piston has been cut away so you can see how the connecting rod attaches to the piston.

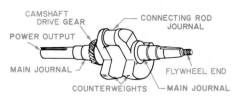

CAMSHAFT
DRIVE GEAR ———   ——— CONNECTING ROD
                        JOURNAL
POWER OUTPUT ———

MAIN JOURNAL ———   ——— FLYWHEEL END
     COUNTERWEIGHTS ——— MAIN JOURNAL

Fig. 19-11 **A crankshaft has one connecting rod journal for each cylinder of the engine.**

Notice the two **main journals**. A **journal** is a part that "rides" inside a bearing. As you know, the crankcase has two holes. The main journals fit through these holes and support the crankshaft in the engine block.

The **connecting rod journal** is the part to which the connecting rod attaches. Sometimes it is called the **crank pin**. All of the journals are precisely made so that the crankshaft will turn smoothly.

Notice that the crank pin is not in line with the main journals. To change reciprocating motion into rotary motion, it has to be set to one side, or *offset*. For the crankshaft to rotate evenly, the crank pin must be balanced by another weight. This is the job of the **counterweights**.

Both ends of the crankshaft are carefully machined (shaped and sized). One end attaches to the flywheel. The other end — the **power**

output — attaches to whatever moving part is used to do work. For example, on a lawn mower the power output would be attached to a cutting blade.

The crankshaft can fit into the crankcase either vertically (up and down) or horizontally (across). The placement depends on what the engine is used for. For example, a lawn mower would have a vertical crankshaft. This is so you can attach the cutting blade in a horizontal position. Figure 19-12 shows how crankshaft placement affects an engine's overall design.

Crankshaft Bearings

As you know, some connecting rods use bearings to reduce friction and wear. The crankshaft main journals also need some kind of bearing surface.

Many different types of bearing surfaces can be used. In some engines designed for light loads, the bearing surfaces are simply the holes in the crankcase. On heavier-duty engines, the bearing surfaces are provided by **main bearings**. These bearings fit into the crankcase holes. Figure 19-13 shows both types of bearing surfaces.

If bearings are used, they are either sleeve bearings or anti-friction bearings. The **sleeve**

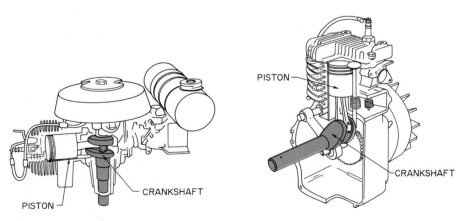

PISTON

PISTON —   ——— CRANKSHAFT

——— CRANKSHAFT

Fig. 19-12 **Lawn mower engines and outboard motors usually have vertical crankshafts, such as you see on the left. In stationary engines (such as the one on the right), the crankshaft is usually horizontal.**

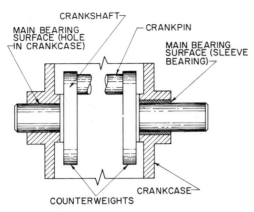

Fig. 19-13 **This diagram shows two types of main bearing surfaces used to support the crankshaft.**

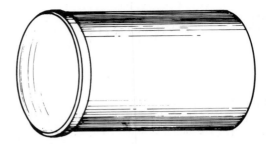

Fig. 19-14 **One-piece sleeve bearings are also called *bushings*.**

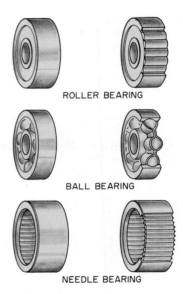

Fig. 19-15 **There are many different types of antifriction bearings. The roller bearing shown here is sealed with its own lubrication. The ball bearing and needle bearing must be lubricated from another source.**

bearings used will be one-piece bearings. They are pressed into place in the crankcase holes. See Fig. 19-14.

Anti-friction bearings consist of steel balls or rollers inside a metal housing. This type of bearing is press-fitted onto the crankshaft journal. As the crankshaft rotates, the inner part of the bearing rotates with it. Figure 19-15 shows three types of anti-friction bearings.

Anti-friction bearings can stand heavy loads and shock. They cost more than sleeve bearings. However, they usually provide smoother performance and longer-lasting service.

Both sleeve bearings and anti-friction bearings require some kind of lubrication. Some anti-friction bearings have their own sealed-in lubrication. Sleeve bearings and other types of anti-friction bearings need another source of lubrication. The engine's lubrication system is discussed later in this chapter.

Flywheel

Take a moment to look at Fig. 19-3 again. Notice the flywheel, which is attached to one end of the crankshaft. The flywheel is basically a heavy metal wheel that helps turn the crankshaft between power strokes.

To understand how the flywheel works, you have to think about how moving objects behave.

For example, if you swing a hammer to hit a nail, the weight of the hammer head adds force to the motion of the hammer. Basically, once you start the swing, the weight will carry the hammer to the nail. The force of the swing, along with the weight of the hammer head, gives the hammer **momentum**.

A flywheel works in the same way. As the piston comes down on the power stroke, it turns the crankshaft and flywheel. The momentum of the heavy flywheel then keeps the crankshaft turning until the next power stroke.

You will learn more about the flywheel in the *COOLING SYSTEM* and *IGNITION SYSTEM* sections of this chapter.

Four-Stroke Cycle Operation

In Chapter 17, you learned how the basic four-stroke cycle works. This cycle is so important that we are going to review it here. See Fig. 19-16.

Notice the positions of the **intake valve** and the **exhaust valve**. On the intake stroke, the intake valve is open to allow the air-fuel mixture to enter the cylinder. On the compression and power strokes, both valves are closed. On the exhaust stroke, the exhaust valve is open to allow the burned gases to escape.

For the engine to work properly, the valves must open and close at the right time, every time. There is a whole group of parts that control the opening and closing of the valves. Together, these parts are called the **valve train**. Besides the valves, the valve train includes the valve guides, tappets, valve springs, camshaft drive gears, and camshaft. See Fig. 19-17.

First we'll look at how the valve train works. Then we'll take a close look at the valves themselves.

Valve Train Operation. The crankshaft drives the valve train and controls its speed. It does this through the use of the **timing gears** and the **camshaft**. See Fig. 19-17 again. The camshaft gear is twice the size of the crankshaft gear. It also has twice as many teeth. This means that the camshaft rotates only once for every two revolutions of the crankshaft.

The camshaft has raised areas called **cams**, or **cam lobes**. The cams operate the valves by

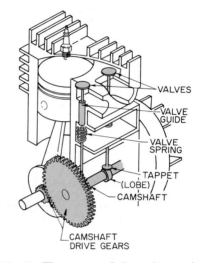

Fig. 19-17 The parts of the valve train work together to control intake and exhaust in four-stroke engines.

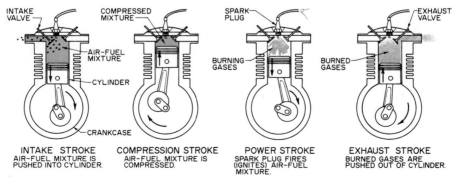

INTAKE STROKE
AIR-FUEL MIXTURE IS PUSHED INTO CYLINDER.

COMPRESSION STROKE
AIR-FUEL MIXTURE IS COMPRESSED.

POWER STROKE
SPARK PLUG FIRES (IGNITES) AIR-FUEL MIXTURE.

EXHAUST STROKE
BURNED GASES ARE PUSHED OUT OF CYLINDER.

Fig. 19-16 The operation of a four-stroke cycle gasoline engine.

Ole and His Outboard

Ole Evinrude was the son of an immigrant Norwegian farmer. Growing up in Wisconsin, Ole often used a rowboat to get from one place to another. One day he rowed two-and-a-half miles to get some ice cream for his girlfriend. It was then that he decided "there must be an easier way."

Ole's easier way lay in the internal-combustion engine. He had been fascinated with gasoline engines, and he thought that he could build an engine that would drive a boat. This "outboard motor" would drive a propeller, which in turn would push the boat through the water.

Ole succeeded in making his idea a reality. He formed a company that produced the first outboard motor in 1909. Over the years, Ole and his son Ralph worked to improve their small engines. They devel-

oped a lightweight two-cylinder two-stroke engine. The engine was quiet, too.

Have you noticed that this story uses both *motor* and *engine* in its description of Ole's invention? An outboard motor is really an outboard *engine*. A *motor* is something that converts electricity into rotary motion. Whatever the name, though, we can thank Ole Evinrude's girlfriend for an invention that has really proved its worth.

---

pushing up on them. There is one cam for each valve. Figure 19-18 shows how the camshaft operates the valves.

As the camshaft rotates, the cam lobe pushes up on the **tappet**. The tappet, or **lifter**, pushes the valve up from the valve seat. As the tappet moves up, it compresses a **valve spring** against the engine block.

As the camshaft continues to rotate, the cam lobe moves away from the tappet. Tension from

the valve spring forces the valve back down. The spring tension closes the valve tightly. The valve stays closed until the rotating lobe pushes up on the tappet again.

Valve Design. All four-stroke engines have two poppet valves per cylinder. The hole through which the air-fuel mixture enters the cylinder is called the **intake port**. The hole through which the burned gases leave is called the **exhaust port**. These ports must be sealed tightly by the valves during the compression and power strokes. Even the smallest leak will cause a loss of power.

Figure 19-19 shows a closeup view of a valve and seat assembly. This assembly controls the passage of gases into and out of the combustion chamber. The **face** is the bottom edge of the valve. It is tapered, usually at a 45-degree angle. The circular area around the port is called the **seat**. It is tapered to match the valve face. Both the valve and its seat are carefully machined. This produces the most leakproof fit possible.

The valves have different locations in different engines. Most small engines use a **flathead** design. In this design, the valves are located side-by-side, next to the cylinder. See Fig. 19-20. Another common arrangement is the **overhead** design. This design is also used widely on automotive engines.

Fig. 19-18 **The lobe on the camshaft causes the tappet and valve to move up. The valve spring moves the valve and tappet back down.**

300

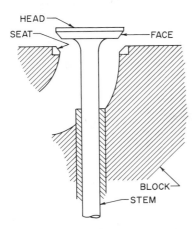

Fig. 19-19 The face of a poppet valve must fit snugly into the seat in the block. Damaged or worn valves must be serviced so they do not leak.

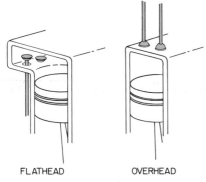

Fig. 19-20 Many small engines use a flathead, or *L-head*, arrangement for the valves. Some small engines use an overhead, or *I-head*, arrangement.

Tecumseh Products Co.

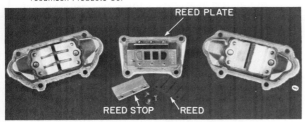

Fig. 19-21 This photo shows three reed valves. The center one has been disassembled to show the reed, reed stop, and reed plate.

Two-Stroke Cycle Operation

In most small two-stroke cycle engines, intake and exhaust are done without a valve train. Instead, the piston itself acts like a valve to open and close the intake and exhaust ports. Different methods are used for the intake of fuel. The most common method is the use of a reed valve.

Reed Valve. In a reed valve, the **reed** is a thin, flexible strip of steel. One end of it attaches to a small metal plate called a **reed plate**. The other end of the reed can move freely over a port in the plate. Many reed plates have several ports and several reeds. Many plates also have a thick metal piece called a **reed stop**. This stop keeps the reed valve from opening too far. The valve could be damaged otherwise. See Fig. 19-21. The reed valve is usually attached to the inside wall of the crankcase. Its job is to control the flow of the air-fuel mixture into the crankcase. The following paragraphs describe how the reed valve does this.

Engine Operation. The basic two-stroke cycle was described in Chapter 17. Therefore, some of the following material is going to be review. However, this time you'll see how the reed valve works in the cycle.

Figure 19-22 shows the basic two-stroke cycle. Remember, there are four piston actions — intake, compression, power, and exhaust — just like in a four-stroke engine. However, all of these actions take place in only two strokes. Intake and compression take place on the first stroke. Power and exhaust take place on the second stroke.

In *A* of Fig. 19-22, the piston is moving upward. It is passing the intake and exhaust ports. This seals the air-fuel mixture in the combustion chamber. The piston's upward movement also creates a partial vacuum in the crankcase. Atmospheric pressure then pushes air and fuel from the carburetor through the reed valve. This is how fuel enters the crankcase.

In *B* the piston has continued upward. It has compressed the fuel trapped in the combustion chamber. In the crankcase, the new air-fuel mixture has filled the vacuum. The reed valve has just closed. Crankcase pressure seals the new air-fuel mixture in the crankcase.

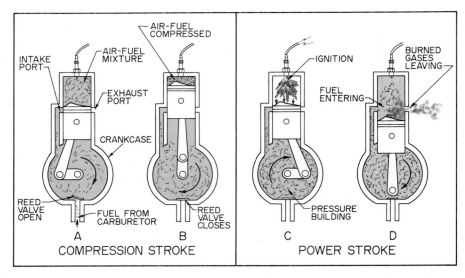

Fig. 19-22 **There are only two strokes per cycle in a two-stroke engine. The four drawings help you see that all four piston actions — intake, compression, power, and exhaust — take place in just two strokes.**

Ignition takes place in *C*. The burning gases are pushing the piston down on its power stroke. This movement compresses the air-fuel mixture trapped in the crankcase. In this way, pressure builds up in the crankcase.

In *D* the piston has passed the exhaust port. The burned gases are starting to escape. The piston has also passed the intake port. Crankcase pressure is pushing the new air-fuel mixture into the combustion chamber. The raised head of the piston is deflecting the air-fuel mixture up. The new mixture is pushing out the burnt gases. This completes the two-stroke cycle. The cycle begins again when the piston starts back up the cylinder.

Two-stroke intake and exhaust has advantages and disadvantages. It doesn't need as many moving parts as a four-stroke valve train. This is an advantage in cost and maintenance. The disadvantage is in its less efficient use of fuel. Some air-fuel mixture is lost as it drives out the exhaust gases. Another disadvantage is rougher and louder engine operation.

LUBRICATION SYSTEM

An engine's moving parts must be lubricated with oil. Without lubrication, an engine will soon fail. Lubrication serves the following purposes:

- **Reduces wear** — Oil produces a thin, low-friction surface. In this way, it reduces the amount of friction between parts that rub against each other. See Fig. 19-23.
- **Increases efficiency** — The low friction provided by lubrication allows more power to be used for work.
- **Absorbs shock** — As the piston moves up and down, the engine parts receive many sudden shocks. The oil acts to cushion these shocks.
- **Seals piston rings** — Oil fills the small gaps between the rings and the cylinder wall. This keeps combustion gases out of the crankcase.

In four-stroke engines, the oil also helps cool the engine and clean the engine. Read on to find out how two-stroke lubrication is different from four-stroke lubrication.

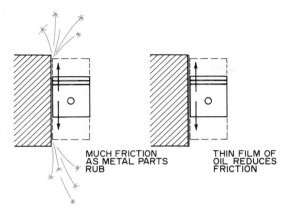

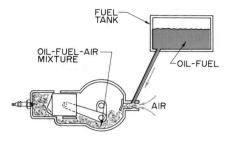

Fig. 19-24 Two-stroke engines are lubricated by an oil-fuel mixture.

Fig. 19-23 When metal parts rub against each other, the resulting friction causes wear. Oil greatly reduces the friction, allowing easy movement. This in turn reduces wear.

Two-Stroke Cycle Lubrication

The lubrication system for a two-stroke engine is very simple. You mix the oil with the fuel before filling the fuel tank. Oil and fuel then flow through the engine together. See Fig. 19-24.

The oil is much heavier and stickier than the fuel. As the oil-fuel mixture circulates, much of the oil sticks to engine parts. It coats the crankshaft, connecting rod, bearings, and cylinder wall.

Four-Stroke Cycle Lubrication

Four-stroke engines have a separate lubrication system. You don't add oil to the fuel. This is because the air-fuel mixture doesn't circulate through the crankcase.

Most small four-stroke engines have a **splash lubrication** system. This type of system lubricates the engine parts by splashing them with oil.

Figure 19-25 shows a typical splash system. Notice the **dipper** that is attached to the bottom of the connecting rod. The dipper dips into the oil stored in the crankcase. As the connecting rod rotates, so does the dipper. As it passes through the oil, the dipper splashes oil onto the moving engine parts.

Some small engines use a **pressure lubrication** system. Just as in an automotive engine, an oil

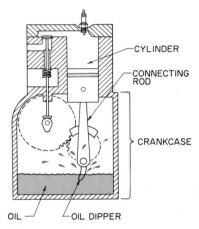

Fig. 19-25 In the simplest type of four-stroke lubrication, a dipper splashes oil on the engine parts. The dipper is either cast as part of the connecting rod or is bolted to it.

pump supplies the pressure to move the oil. (For more information on pressure lubrication, see *LUBRICATION SYSTEM* in Chapter 18.)

COOLING SYSTEM

One important thing to remember about small gas engines is that they are *heat engines*. If you recall, heat engines are not very efficient. Less than one-third of the combustion heat is actually converted into mechanical energy. The rest of the heat is just wasted. It is absorbed by the metal parts of the engine.

Too much heat can damage the engine. Therefore, there has to be a way to remove excess heat. Some combustion heat leaves the engine with the exhaust gases. But this isn't enough. The engine must also use a separate cooling system.

Most small engines use an **air cooling** system. Here's how the system works:

Combustion heat transfers from the combustion chamber to the cylinder head and cylinder block. Both the head and the block have **cooling fins** cast into them. See Fig. 19-26. The cooling fins increase the surface area of the engine. This provides extra contact between the outside air and the engine. The outside air passes over the fins, picks up the combustion heat, and carries it away.

Some small engines are used where there isn't any circulating air. These engines require a **cooling fan**. See Fig. 19-27. The cooling fan is part of the flywheel. As the flywheel spins, the fan blows air over the cooling fins. A **shroud** directs the air over the engine.

A few small engines use a **water cooling** system. Just as in an automotive engine, a **water pump** supplies pressure to move water through the engine. (For more information on water cooling, see *COOLING SYSTEM* in Chapter 18.)

Kawasaki Motors Corp., U.S.A.

Fig. 19-26 Fins on the cylinder head and the block allow the surrounding air to remove a great deal of heat. The air passing over this motorcycle engine cools the engine as the motorcycle moves.

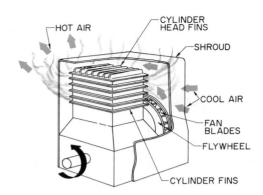

Fig. 19-27 The cooling fan is cast as part of the flywheel. The faster the crankshaft turns, the more air the fan blows across the fins.

FUEL SYSTEM

The fuel system brings air and fuel together and mixes them for combustion. A typical fuel system includes an air cleaner, a fuel supply, a carburetor, and a governor. See Fig. 19-28.

Air Cleaner

All engines need a large amount of air for combustion. Air enters an engine through the air cleaner. See Fig. 19-29. The air cleaner removes dirt and dust particles from the air. This keeps them from entering the engine. These particles can scratch and rub the cylinder wall and the piston rings. This can cause much wear and damage.

Air cleaners help quiet the rush of air into the carburetor. This makes the engine run quieter. If an engine backfires, the air cleaner also helps prevent fires. This is especially important in dry grassy areas and forests.

Most air cleaners are attached to the carburetor. The air passes directly from the cleaner to the carburetor. There are two common types of small engine air cleaners: dry-element and oil-foam.

Dry-Element Cleaner. The main part of a dry-element air cleaner is the **dry filter**. Dry filters are made of materials such as paper and felt. See Fig. 19-30. Tiny holes in the filter allow

Briggs & Stratton

Fig. 19-28   This photo shows the main parts of a small engine fuel system.

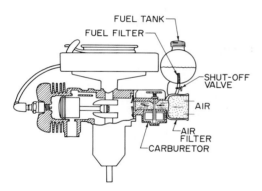

Fig. 19-29   Engines need a great deal of *clean* air. The air entering the engine must pass through an air cleaner before mixing with the fuel.

air to flow through the cleaner into the carburetor. The filter traps and holds particles of dust and dirt that can't get through the holes.

Oil-Foam Cleaner. Oil-foam cleaners use a filter made of a plastic sponge-like material. It is soaked in oil. See Fig. 19-31. Air enters the cleaner and passes through the filter. Dirt and dust particles stick to the oily filter. Only clean air enters the carburetor.

Fig. 19-30   Small particles of dust and dirt stick to paper filters. The cleaned air then passes through tiny holes and into the carburetor.

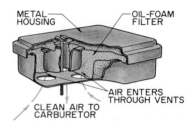

Fig. 19-31   The foam filter used in an oil-foam cleaner contains oil. As air passes through the foam, dirt sticks to the oil.

Fuel Supply

Some small engines use a **pressure-feed** fuel delivery system. This system uses a **fuel pump** and is much like the common automotive fuel system. (See *Fuel Pump* in Chapter 18.) However, most small engines use either a gravity-feed system or a vacuum-feed system. Both of these systems use a **fuel tank**, a **shut-off valve**, and a **fuel filter**.

Gravity Feed. The gravity-feed system is a simple way of supplying fuel. In this system, the fuel tank is located above the carburetor. See Fig. 19-32. The force of gravity pulls the fuel into the carburetor.

Vacuum Feed. The vacuum-feed system is sometimes called a **suction-feed system**. In this system, the fuel tank is located just below the carburetor. See Fig. 19-33.

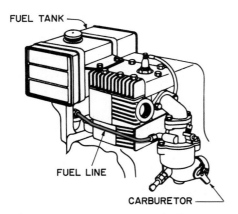

FUEL TANK

FUEL LINE

CARBURETOR

Fig. 19-32 **In a gravity-feed fuel system, the fuel tank is mounted above the carburetor. Gravity feeds the fuel from the tank to the carburetor.**

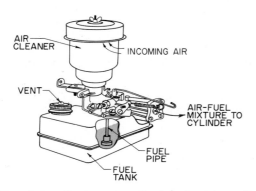

AIR CLEANER

INCOMING AIR

VENT

AIR-FUEL MIXTURE TO CYLINDER

FUEL PIPE

FUEL TANK

Fig. 19-33 **In a vacuum-feed fuel system, the fuel tank must be very close to the carburetor. A vacuum in the carburetor causes fuel to flow from the tank.**

Air enters through the air cleaner and produces a vacuum in the carburetor. Atmospheric pressure then pushes on the fuel in the fuel tank. Fuel passes through the **fuel pipe** and into the carburetor.

Carburetor

The carburetor is the most important part of the fuel system. Its main job is to mix air and fuel properly for combustion. The carburetor also controls the amount of air-fuel mixture entering the combustion chamber. See Fig. 19-34.

Small engines use many different kinds of carburetors. However, almost all carburetors use the same basic parts and operating principles. This section discusses these parts and principles. Study this basic information first. Then you'll be able to understand the different carburetor types that are described later.

Venturi. The main body of a carburetor is a tunnel-like passageway, or chamber. The chamber is wide at both ends. One end opens onto the air cleaner. The other end opens onto either the intake port or the intake manifold. (The **intake manifold** is the passageway to the engine.) The carburetor chamber is narrow in the middle. The narrow part is called the *venturi*. See Fig. 19-35.

To get to the combustion chamber, air must pass through the carburetor. There are two partial vacuums involved in the flow of air. One is **engine vacuum**. On the intake stroke, the piston moves down. This creates a low-pressure area above the piston. This vacuum allows atmospheric pressure to push air through the carburetor.

The second partial vacuum is produced in the venturi. As air passes through the venturi, it speeds up. This speed increase creates a vacuum. The fuel supply is vented to the atmosphere. Therefore, the venturi vacuum allows atmospheric pressure to push fuel from the fuel supply to the venturi. See Fig. 19-36.

The fuel enters the venturi through a small opening. The fast-moving air atomizes the fuel. (*Atomize* means to change from a liquid into a fine mist, or spray.)

The low pressure in the venturi also lowers the fuel's boiling point. This helps vaporize the fuel. *Vaporization* changes the atomized fuel into a gas. The fuel can then mix easily with the air.

There are two basic types of air-fuel mixtures: rich and lean. A **rich mixture** contains more fuel than the amount of air needed to burn it completely. A **lean mixture** contains more air than is needed to burn the fuel.

Fuel Nozzle. Fuel is pushed into the venturi through a fuel nozzle. See Fig. 19-36 again. The

Fig. 19-34 This drawing shows the basic functions of a carburetor. The carburetor mixes air with fuel. It also directs the mixture to the combustion chamber.

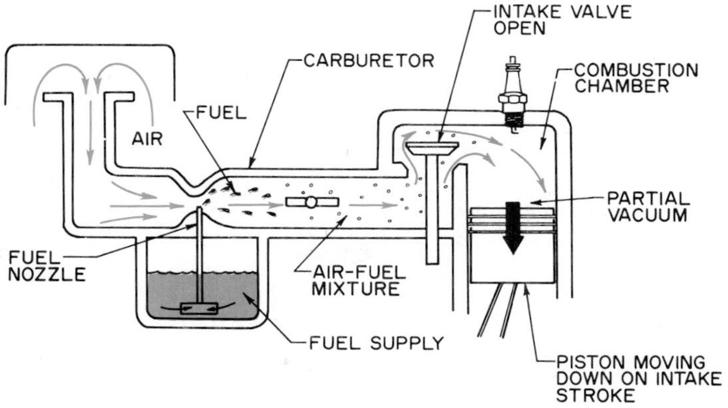

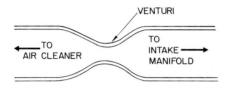

Fig. 19-35 The carburetor is basically an open chamber with a narrow middle called a *venturi*.

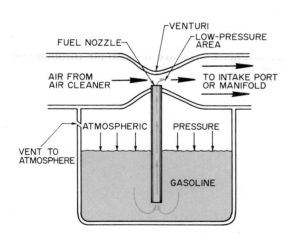

fuel nozzle is simply a tube. It leads from the fuel supply to the venturi. There is low pressure at this point. Atmospheric pressure, which is higher, pushes fuel through the nozzle. The fuel then mixes with the air in the venturi.

Throttle Valve. The throttle valve is a round, thin plate. It fits on a shaft at the cylinder end of the carburetor. See Fig. 19-37. Rotating the shaft turns the throttle to different positions. The throttle position controls the amount of air-fuel mixture entering the cylinder. In this way, the throttle controls the engine's speed.

Figure 19-37 shows the throttle valve in three different positions. In the *closed* position, the valve almost completely blocks off the car-

Fig. 19-36 Air passing through the carburetor speeds up as it passes the venturi. This creates a partial vacuum. Atmospheric pressure can then push fuel from the fuel supply into the venturi.

Fig. 19-37 The throttle valve controls the amount of air and fuel entering the engine. In this way, the throttle controls engine speed.

A. **Engine at idle**   *B.* **Engine at medium speed**   *C.* **Engine at full speed**

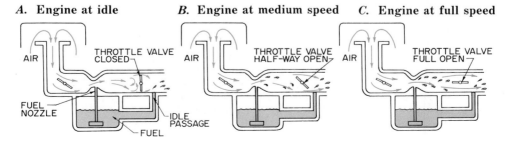

buretor. It lets only a small amount of air pass through. This small amount is not enough to pull fuel from the nozzle. However, a small amount of fuel *can* enter the carburetor through an **idle passage.** (Idling is explained later.)

In the *half-way open* position, the throttle valve is tilted at a 45-degree angle. See Fig. 19-37B. This time, more air can pass through the carburetor. The air creates a venturi vacuum. Therefore, fuel flows out of the fuel nozzle and into the engine. The engine speeds up to a medium speed.

In the *full open* position, the valve is at a right angle to its closed position. See Fig. 19-37C. This allows the greatest amount of air-fuel mixture into the engine. The engine speeds up to its maximum speed.

There are two basic ways to control the position of the throttle:

• The engine operator can move a **throttle lever.**
• A **governor** can control the throttle position automatically. (Governors are described later in this chapter.)

Choke Valve. The choke looks much like the throttle. It is a thin, round butterfly valve. It fits on a shaft at the air cleaner end of the carburetor. See Fig. 19-38. Closing the choke reduces the amount of air entering the carburetor. The operator can usually adjust the choke with a **choke lever.**

The choke's main job is to help start cold engines. Cold engines need a richer air-fuel mixture than warm engines do. That is, they need more fuel and less air. The choke in Fig. 19-38

is closed. It lets the smallest amount of air into the carburetor. This increases the engine vacuum at the fuel nozzle. The higher vacuum allows atmospheric pressure to push more fuel through the fuel nozzle. Once the engine is warm, the choke is opened.

Many small engines have a **primer** instead of a choke. The operator presses the primer to pump extra fuel to the carburetor. This produces the richer mixture needed for starting. Figure 19-39 shows a typical small engine primer.

Idle Fuel Adjustment. At times, an engine runs with the throttle closed. The rotating crankshaft does not perform any useful work. We say that the engine is **idling.** A special screw controls the proportion of air to fuel in the air-fuel mixture during idling. This screw is called the **idle fuel adjustment screw.** It is also called the **idle mixture screw.** See Fig. 19-40.

The idle fuel adjustment screw is a kind of "needle" valve. The operator turns it in or out. This reduces or increases the amount of fuel that enters the air stream.

The usual air-fuel ratio for idling is 7-10 to 1, by weight. This means that there are between 7 and 10 pounds of air for every pound of gasoline.

Deere & Co.

Fig. 19-39 Primers supply extra fuel to the carburetor to help start a cold engine. The parts labeled here are (A) the governor linkage, (B) the main fuel adjustment, (C) the idle fuel adjustment, and (D) the idle speed adjustment.

Fig. 19-38 The choke valve limits the amount of air entering the carburetor. This creates a vacuum at the fuel nozzle. Atmospheric pressure then pushes more fuel into the air stream for a richer mixture.

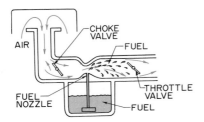

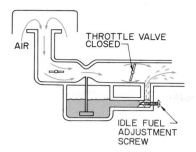

Fig. 19-40 The idle fuel adjustment screw controls the amount of fuel in the air-fuel mixture when the throttle is closed.

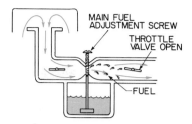

Fig. 19-41 During engine operation above idling speed, the main fuel adjustment screw controls the amount of fuel that passes through the fuel nozzle.

Main Fuel Adjustment. When the engine runs at speeds above an idle, two things happen. First, atmospheric pressure pushes fuel up through the fuel nozzle. Second, the fuel mixes with the incoming air. The amount of fuel that passes through the nozzle can be controlled. This is done with the **main fuel adjustment screw**. See Fig. 19-41.

The main fuel adjustment screw is a needle valve. The operator turns it in or out. This controls the proportion of air to fuel at medium speeds. The usual ratio is 15 to 1 (15 pounds of air to every pound of gasoline).

Idle Speed Adjustment. A special screw controls the amount that the throttle valve is open during idling. This screw is the **idle speed adjusting screw**.

The idle speed adjusting screw is located in an arm attached to the throttle shaft. As you turn the screw in, the end of it moves against the carburetor body. This causes the throttle valve to open. See Fig. 19-42. As the throttle opens, the idle speed (rpm) increases. If you turn the screw out, the idle speed decreases.

Fig. 19-42 The engine's rpm at idle is controlled by the idle speed adjusting screw. Turning this screw *in* increases the idle speed. Turning it *out* decreases the speed.

Float Carburetor

Many small engines use float-type carburetors. These carburetors have a **float assembly**. This assembly maintains a constant supply of fuel for the carburetor. Its main parts are a float, a float bowl, and a needle valve. See Fig. 19-43.

There is an air inlet at the top of the float bowl. This **vent** maintains atmospheric pressure in the bowl. The atmospheric pressure pushes the fuel from the bowl into the low pressure of the venturi.

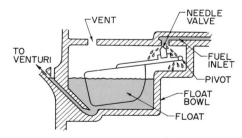

Fig. 19-43 The parts of the float assembly work together to provide a constant supply of fuel to the carburetor.

The **float bowl** holds the fuel. Fuel travels from the bowl to the venturi. As this happens, the level of fuel in the bowl goes down. As the fuel level drops, the hollow metal **float** drops. This movement slowly pulls the **needle valve** away from the **fuel inlet seat**. More fuel then enters the bowl.

As fuel enters the bowl, the float rises. It moves the needle valve back into the fuel inlet seat. Just as much fuel enters the bowl as leaves it. The result is a constant supply of fuel to the venturi.

For proper float operation, the engine must be upright. Some engines operate in different positions. Chain saw engines and weed trimmer engines are examples. These engines can't use float carburetors. Instead, they use diaphragm or vacuum-feed carburetors.

Vacuum-Feed Carburetor

Most vacuum-feed, or **suction**, carburetors have a simpler design than float carburetors. Vacuum-feed carburetors don't have a float assembly. The fuel goes directly from the fuel tank to the carburetor. Another difference is that only one needle valve is used for both the main fuel adjustment and the idle fuel adjustment. Figure 19-44 shows a typical vacuum-feed carburetor.

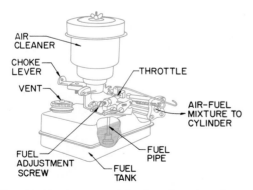

Fig. 19-44 A vacuum-feed carburetor does not have a float assembly. However, it works on the same principle as the float carburetor. Atmospheric pressure pushes fuel through the fuel pipe into the carburetor.

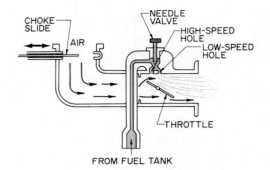

Fig. 19-45 This vacuum-feed carburetor uses only one needle valve for both high-speed and idle adjustments. This simple design works well on engines that run at fairly constant speeds.

Vacuum-feed carburetors also do not have a venturi. A specially placed throttle valve takes the place of the venturi. The throttle is in the carburetor at the end of the fuel pipe. See Fig. 19-45.

When the throttle valve is turned, it restricts the flow of air and creates a low-pressure area. This allows atmospheric pressure to push fuel up the fuel pipe and into the carburetor. Together, the throttle and the needle valve control the air-fuel ratio at all speeds.

Many vacuum-feed carburetors have a **sliding choke**. See Fig. 19-45 again. The operator slides the choke in or out. This kind of choke works the same way as a butterfly-valve choke.

Diaphragm Carburetor

Float carburetors and vacuum-feed carburetors must be upright to work properly. This is usually not a problem. Most engines run vehicles and machines that stay in an upright position.

However, some engines must operate in other positions. For example, chain saws are often turned to the side or even almost upside down. Their engines need carburetors that can work in any position. See Fig. 19-46. The most common carburetor used in these engines is called a *diaphragm carburetor*.

Diaphragm carburetors have a lot in common with other types of carburetors. They control engine speed with a throttle. They use a choke

Homelite Textron

Fig. 19-46 Chain saw engines must be able to work in positions other than upright. They use a diaphragm carburetor.

or a primer for starting. And they redirect fuel to an idle passage to produce a rich idling mixture.

Like float carburetors, diaphragm carburetors also maintain a constant fuel level. But this is where the big difference comes in. Instead of using a float, diaphragm carburetors use a **diaphragm** to control the fuel intake. This diaphragm is a flexible elastic disc. It fits inside the bottom of the carburetor. See Fig. 19-47.

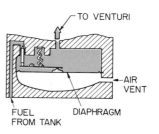

Fig. 19-47 This diagram shows the bottom part of a diaphragm carburetor.

Figure 19-48A shows how the carburetor looks before the engine is started. The fuel chamber is filled and the diaphragm supports the weight of the fuel. Notice the different parts of the carburetor. A **needle valve** fits into the **fuel inlet.** This valve controls the entry of fuel into the **fuel chamber.** The free end of the **inlet valve control lever** holds the inlet valve in place. A **spring** pushes down on the control lever and helps keep the diaphragm down. The pressure of the fuel above the diaphragm balances the pressure of the air below the diaphragm.

Now look at Fig. 19-48B. The engine has been started and there is a vacuum produced in the venturi. This vacuum reduces the pressure on the fuel in the fuel chamber. The force produced by the atmospheric pressure below the diaphragm is now greater than the force above the diaphragm. The greater force pushes the diaphragm up as the fuel enters the venturi.

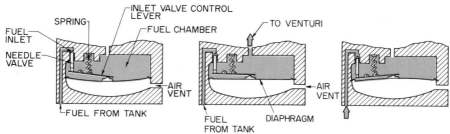

A. Diaphragm down —
fuel chamber full

B. Diaphragm rises —
fuel chamber empties

C. Diaphragm rises further —
fuel enters fuel chamber

Fig. 19-48 These drawings show how a diaphragm carburetor maintains a constant fuel supply. It does this regardless of its position.

In Fig. 19-48C the force of the atmospheric pressure below the diaphragm is enough to overcome the spring tension. The control arm pivots and allows fuel to enter the fuel chamber. Fuel comes from the fuel tank located nearby. The venturi vacuum is great enough to allow the atmospheric pressure in the fuel tank to force fuel into the fuel chamber.

During operation, the diaphragm stays at a balanced position. It lets fuel enter past the inlet valve at the rate the fuel is used by the engine.

Governors

The amount of load on an engine affects its speed. When the load increases, the engine must work harder. Working harder slows down the engine. When the load lightens, the engine doesn't have to work as hard. This causes the engine to speed up.

A device called a **governor** is used to even out changes in engine speed. The governor is connected to the carburetor's throttle. The governor automatically adjusts the throttle to maintain a constant speed. It also provides the safety feature of limiting the engine's maximum speed.

Two types of governors are commonly used on small engines: centrifugal and air vane.

Centrifugal Governors. Many governors use the principle of **centrifugal force**. This is the force that tends to push a rotating object out

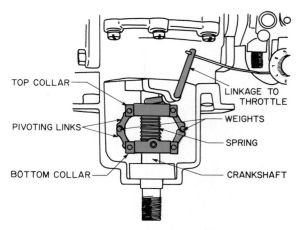

Fig. 19-49 **The bottom of a centrifugal governor is fastened to the crankshaft and can't move. However, the top is free to move up and down.**

from its center. Governors that use this force are called centrifugal, or **mechanical**, governors.

Figure 19-49 shows a common type of centrifugal governor. This type of governor fastens directly to the crankshaft. The top and bottom collars are connected by metal links that can pivot. The top collar is connected to the throttle by linkage. (**Linkage** is a system of levers.)

As the crankshaft turns, it produces centrifugal force. This force pulls out on the weights held by the pivoting links. As the links move out, the upper collar drops down the shaft. See Fig. 19-50A. The collar's downward movement pulls the throttle closed. This slows the engine down.

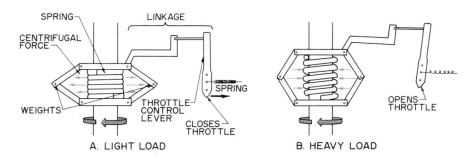

Fig. 19-50 **The operation of a centrifugal governor: In *A*, the engine is under a light load. This causes the engine to speed up, forcing the weights outward. The weights in turn move the throttle linkage to reduce speed. In *B*, the engine is under a heavy load, which slows down the engine. This reduces the centrifugal force. In turn, the spring tension moves the linkage to open the throttle.**

Suppose that a load is put on the engine. The engine has to work harder, so it slows down. The crankshaft doesn't turn as fast. Therefore, the centrifugal force on the governor decreases. This causes the links to come back against the shaft. The spring pushes the collar upward. See Fig. 19-50B. The upward movement of the collar opens the throttle. The engine then speeds up to its original speed. This helps the engine handle the heavy load.

A centrifugal governor allows the engine to maintain a fairly constant speed as the load changes. What you have seen is just one type of centrifugal governor. Other types use the same principle and do the same job.

Air Vane Governors. Like centrifugal governors, air vane governors change the throttle opening to fit the load on the engine. Air vane governors operate on air flow from the flywheel fan. The faster the flywheel turns, the more air flow it creates.

A small plastic or steel piece catches the flywheel air flow. This piece is called an **air vane**. The air vane attaches to the throttle with a linkage. A spring holds the throttle open. See Fig. 19-51.

As the engine speeds up, air pushes on the vane. The vane pulls the throttle closed. In this way, the governor limits the engine's speed. The operator can adjust the engine's top speed by turning the **adjusting nut**. This changes the spring tension and makes it easier or harder for the air vane to pull the throttle closed.

When the engine is under a heavy load, it slows down. The crankshaft doesn't turn as fast. Therefore, the flywheel slows down and throws off less air. With less air pushing against it, the air vane moves back to its original position. This allows the spring to pull the throttle open. The engine then speeds up to carry the load.

Exhaust

As burned gases leave an engine, they first pass through an **exhaust passageway**. (Sometimes this is called the **exhaust manifold**.) Then the gases pass through a **muffler**. The muffler is a device that reduces the noise of the escaping gases. Figure 19-52 shows the exhaust passageway and muffler for a four-stroke engine.

Here's how a muffler works: The gases that leave the engine are under high pressure. Pushing these gases straight into the air would make them expand suddenly. This expansion would produce a very loud noise. The muffler lets the gases expand *before* they reach the outside. This reduces the pressure, so there is less noise.

Fig. 19-52 **The exhaust passageway provides a way for exhaust gases to leave the engine. The muffler makes sure their exit is a fairly quiet one.**

Kohler Co.

Briggs & Stratton

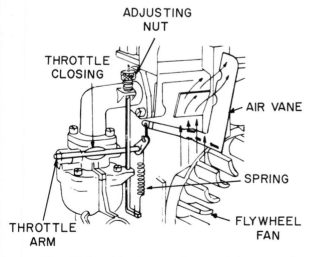

Fig. 19-51 **Air vane governors are operated by air coming off the flywheel fan. The air forces the vane back, and the linkage pulls the throttle closed.**

IGNITION SYSTEM

A small engine ignition system has many things to do. To begin with, it must either have a source of electricity or must generate its own electricity. Next, the ignition system has to move the electricity to the combustion chamber. There the electricity must produce a high-voltage spark to ignite the air-fuel mixture.

What's more, the ignition system has to produce the spark at exactly the right time. *And* it has to repeat the whole ignition process at least 1000 times each minute!

Small engines use three common types of ignition systems: magneto ignition, capacitive discharge ignition, and battery ignition. You'll learn about each of these systems as you read on. However, to really understand how these systems work, you should have some background in electricity and magnetism. Now might be a good time to review the basic principles described in Chapter 7, "Electrical Power."

Magneto Ignition

The magneto ignition is the system most widely used in small engines. Therefore, it's the system we'll spend the most time with. First you'll learn about the different parts of the system. Then you'll see how the parts all fit together in the **magneto cycle**.

The magneto system is really a pretty simple kind of ignition system. It produces its own electricity and doesn't need a battery. There are five basic parts in the system. These are the armature, the magnets, the breaker points, the condenser, and the spark plug. See Fig. 19-53.

Like an automotive breaker-point ignition, the magneto system has two separate circuits: the **primary circuit** and the **secondary circuit**. As the system operates, the breaker points open and close the primary circuit. This develops, or **induces**, a current in the secondary circuit.

Armature. The armature (see Fig. 19-53) consists of a wire coil wrapped around an iron core. The armature is mounted next to the flywheel. As the flywheel turns, magnets in the flywheel rotate past the armature. This induces a low-voltage current in the coil. The armature then

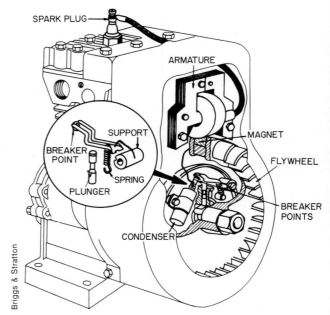

Briggs & Stratton

Fig. 19-53 **Most small engines don't have batteries. They produce their own electric current using just the parts labeled in this drawing. These parts make up the** *magneto ignition system.*

raises the voltage of the current. This higher voltage is needed for ignition.

The armature coil has two separate windings of wire. The **primary winding** is a low-voltage winding. It is heavy wire. The wire is wrapped around the core between 150 and 200 times. One end of this winding attaches to the breaker points. The other end attaches to the armature's core. During engine operation, current is first induced in the primary winding.

The **secondary winding** is a thinner wire than the primary winding. One end of the secondary winding connects with the spark plug. The other end attaches to the core.

The secondary winding wraps around the primary winding. It is usually wrapped about 20,000 times. This is about 100 times the number of windings in the primary winding. With this ratio, the armature can act as a step-up transformer. (A **step-up transformer** is a device for increasing voltage.) See Fig. 19-54.

It increases the low voltage in the primary winding by 100 times. The resulting voltage gives the current a tremendous push. The current needs this push to jump a spark plug gap and ignite the air-fuel mixture.

The **armature core** is made of several plates of soft iron. The iron concentrates the magnetic field around the coil. You will see why this is important when you read about the magneto cycle.

Magnets. The magneto system needs a magnetic field to operate. Magnets are used to generate a magnetic field around the armature. The magnets are either cast into the flywheel or attached to the flywheel. See Fig. 19-55. As the flywheel turns, the magnets repeatedly induce a magnetic field in the armature. The magnets also *reverse* this magnetic field. This results in an electrical current in the armature coil.

Breaker Points. The breaker points are two metal contact points located near the crankshaft. They are usually made of a heat-resistant metal. One of the points doesn't move. The other breaker point is attached to a moveable arm. See Fig. 19-56.

The breaker points act as an electric switch in the **primary circuit**. This is the path that current takes through the breaker points and the primary coil winding. When the points close (touch), current flows through the primary circuit. When the points open, the circuit is broken. This action sends high-voltage current through the **secondary circuit**. The secondary circuit consists of the secondary armature winding, the spark plug wire, and the spark plug.

The breaker points must open at just the right time to fire the spark plug. Two-stroke engines need a spark on every second piston stroke. Therefore, the points must open once for every complete rotation of the crankshaft. Some small engines do this by way of a **cam lobe** on the crankshaft. (This is the method shown in Fig. 19-56.) As the lobe turns, it opens and closes the points.

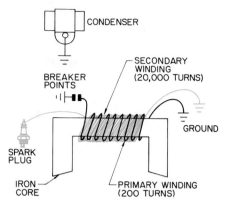

Fig. 19-54 **The armature has two windings. They work together to produce high-voltage electricity for ignition.**

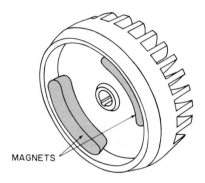

Fig. 19-55 **A typical small engine flywheel. The magnets are needed to operate the magneto system.**

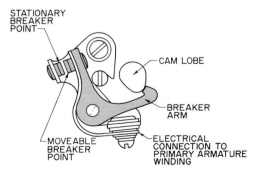

Fig. 19-56 **When the two breaker points touch, electric current flows through the completed circuit. When the points separate, the circuit breaks and current cannot flow.**

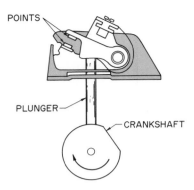

Fig. 19-57 In many small engines, the moveable breaker point is operated by a plunger. The plunger rides on the crankshaft.

Other small engines have a flat spot ground on the crankshaft. The flat spot serves the same purpose as a cam. A **plunger** rides on the crankshaft. It closes the points when it goes over the flat spot. See Fig. 19-57.

Some four-stroke engines use a **camshaft** to open the points. A camshaft is a lobed shaft separate from the crankshaft. The crankshaft turns the camshaft, and the cam lobe pushes the moveable point open. The camshaft opens the point once for every *two* rotations of the crankshaft.

Condenser. When the breaker points open, current in the primary circuit tries to jump the gap between them. If this happens, the magneto may not generate the high ignition voltage. Current jumping between the points also burns up the points. A condenser solves both of these problems.

The condenser is usually wired across the points. It acts as a temporary storage area for electric current. The condenser has a very low resistance to electrical current. As the points open, the current goes into the condenser. This keeps the current from jumping the gap. When the points are completely open, the condenser discharges its stored current back through the primary circuit to the electrical ground.

Small engines have different kinds of condensers. Most condensers consist of two long sheets of foil. A sheet of insulation material separates the two pieces of foil. The sheets are rolled up together. Then they are placed in a small metal cylinder. See Fig. 19-58.

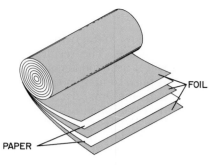

Fig. 19-58 The layers of foil in a condenser act as a storage area for current. This keeps the current from jumping across the breaker points.

Spark Plug. The spark plug conducts electricity into the combustion chamber. It then produces the spark that ignites the air-fuel mixture. There are hundreds of different sizes and kinds of spark plugs. However, the parts and operation of all plugs are much the same. Figure 19-59 shows the parts of a typical spark plug.

The top part of a spark plug is called the **terminal**. The wire that carries current from

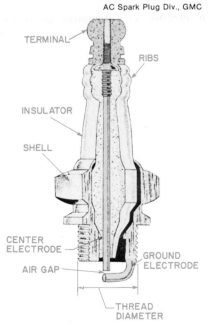

Fig. 19-59 The spark plug receives electricity from the secondary coil winding. Then it produces a high-voltage spark.

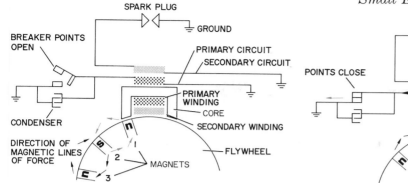

A. Points open — magnets approach armature

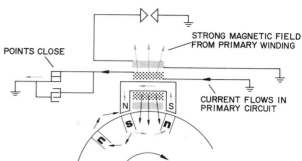

B. Points close — magnetic field develops in armature

Fig. 19-60 **The operation of a magneto ignition circuit. The sequence of actions shown here must be repeated thousands of times per minute to keep the engine running efficiently.**

Fig. 19-60 (Continued on p. 318)

the secondary coil winding attaches to this terminal.

The **center electrode** is a heavy metal wire connected to the terminal. It carries the current to the bottom of the spark plug. A porcelain **insulator** encloses all but the bottom tip of the center electrode. The insulator protects the electrode. It also keeps the high-voltage current from escaping.

The bottom part of the insulator is inserted inside a metal **shell**. The shell is much like a bolt. It has threads and flat places for a wrench. This is so the spark plug can be screwed into the cylinder head.

The **ground electrode**, or **side electrode**, is a stiff wire that extends from the bottom of the shell. This electrode makes a sharp turn to come close to the center electrode. The space between the electrodes is only about 1/32 of an inch (0.8 mm). High-voltage current jumps across this **air gap** to produce the ignition spark.

The Magneto Cycle

Have you read about step-up transformers in Chapter 7? How about the automotive ignition system described in Chapter 18? Both of these topics are good background for understanding the magneto ignition cycle. You may have to read through the following description a couple

of times to "get the drift." Just take your time and refer to Fig. 19-60 as you go along.

The flywheel in Fig. 19-60 has several magnets cast into it. However, we will look at just three of them. They are labeled 1, 2, and 3.

O.K., let's look at Fig. 19-60A. Notice that the points are open. Remember, the points are part of the primary circuit. When they are open, no current can flow through the circuit. Also notice the arrows between the three magnets. These arrows show the direction of the magnets' lines of force.

The flywheel is turning in a clockwise direction. As magnets 1 and 2 line up with the armature, they induce a magnetic field in the primary winding. See Fig. 19-60B. The arrows extending outward from the armature core show that the magnetic field cuts through both sets of armature windings.

Now, what happens when a magnetic field cuts through a conductor? Of course! It produces an electrical current in the conductor. In Fig. 19-60B there is a current produced in the primary circuit. At this time, the points are closed. This allows the current to flow through the primary circuit.

At the same time, the lines of force around the armature and magnets 1 and 2 are traveling in the same direction. The current in the primary circuit helps keep the lines of force going in the direction.

(*Fig. 19-60* continued)

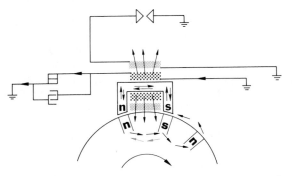

C. Magnetic poles repel each other — lines of force oppose

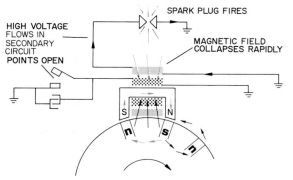

D. Armature fields collapses — spark plug fires

Next step: In Fig. 19-60C, magnets 1 and 2 have moved on. Now magnets 2 and 3 are lined up with the armature. The points are still closed.

Now, notice the lines of force again. This time there are lines going in opposite directions. They *oppose* each other. This is because the north and south poles of magnets 2 and 3 are repelled by the north and south poles of the armature. (Remember, like poles repel each other.) The result is that the magnets try to reverse the armature's magnetic field. However, the armature's field is too strong to be reversed.

The situation quickly changes when the points open. See Fig. 19-60C. As soon as the points open, the primary circuit is broken. This cuts off the current to the primary coil.

Now, remember, all this is happening very fast. As soon as the primary current is cut off, the armature's magnetic field starts to collapse.

The flywheel magnets are *already* trying to reverse the armature's field. They are exerting a force *against* the armature's field. Therefore, the magnets speed up the collapse of the armature field.

As the field collapses, it cuts through the secondary winding. This produces a high-voltage charge in the secondary circuit. The voltage is high enough to push the current to the spark plug. The current then sparks across the air gap to ignite the air-fuel mixture.

Simple, isn't it? To sum it all up: Rotating the flywheel magnets past the armature causes a repeated, almost constant build-up and collapse of a magnetic field. The opening and closing breaker points are an important part of this build-up and collapse. Opening and closing the primary circuit causes a surge of high-voltage current in the secondary circuit. This current sparks at the spark plug to ignite the air-fuel mixture.

Now that you understand all that, here's one more fact: In normal operation, the magneto cycle can produce over 1000 sparks every minute!

Capacitive Discharge Ignition

Many of the newer small engines have capacitive discharge (**CD**) ignition systems. These systems are also called **solid-state** ignitions. They have all the parts of the magneto system, except for breaker points. Instead of points, capacitive discharge systems have complex electronic parts. These parts perform the same switching that points usually do.

Capacitive discharge systems have several advantages over magneto systems. For one thing, there is no need to service or replace points. CD systems also produce higher and more constant voltage. Therefore, there is more dependable ignition. There is also longer spark plug life.

Like magneto systems, CD systems use flywheel magnets to produce electricity. CD systems have a capacitor (condenser), primary and secondary windings, and a spark plug. They also have three coils. These coils produce the high-voltage spark needed to ignite the air-fuel mixture. Like the magneto system, the CD system produces over 1000 sparks per minute.

Battery Ignition

Most small engines are used to power moving or portable devices, such as lawnmowers and chain saws. These devices need to be as lightweight as possible. For this reason, most small engines don't have batteries. A battery would add to the weight of the device.

However, some small stationary engines do use a battery ignition system. The advantage of a battery ignition is that it doesn't need a mechanical input to deliver electricity. When the engine is started, the battery sends current to the primary circuit. The breaker points break the circuit and induce current in the secondary circuit. This is the same thing a CD ignition or magneto ignition does. The difference is that you don't have to get the flywheel moving to produce a spark.

Once the engine is running, a generator or alternator supplies current. Some of this current is used for ignition. Some of it is stored by the battery.

For more information on the battery ignition system, see *IGNITION SYSTEM* in Chapter 18.

STARTING SYSTEM

An engine in good working condition needs two things to operate. First, it needs fuel. Second, it needs momentum (continuing motion). The operator supplies the fuel by filling the gas tank. Once the engine is running, it supplies its own momentum. An engine will keep itself running for as long as the fuel supply lasts.

An engine can *keep* itself running. However, it can't *start* itself. The engine must get its starting momentum from an outside source. This is the job of the starting system.

There are many different kinds of starting systems. The two basic types are manual (handpowered) and electric. Most small engines have a manual starting system.

Manual Starters

Manual starters are also called **hand starters**. The two basic types of manual starters are rope starters and crank starters.

The simplest of all starters is the **wind-up rope starter**. See Fig. 19-61. It has only two parts: a rope and a pulley. The pulley attaches directly to the crankshaft. The rope fits in a notch in the pulley. It winds around the pulley.

A hard pull on the rope turns the crankshaft through about two complete cycles. This is usually enough to start the engine. If a second pull is needed, the operator must rewind the rope by hand. An engine that doesn't start after two tries probably has something wrong with it.

A more complex version of the wind-up rope starter is the **recoil starter**. See Fig. 19-62.

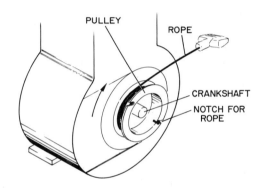

Fig. 19-61 **With a simple rope starter, the operator supplies momentum by pulling on the rope. This turns the crankshaft until the engine can provide its own momentum.**

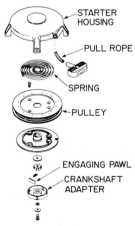

Fig. 19-62 **A recoil starter is also called a *rope-rewind* or *retractable* starter. The operator doesn't have to rewind the rope by hand for the second pull. However, there are a lot of parts involved!**

Tecumseh Products Co.

This starter uses a spring to rewind the rope. Pulling the rope winds a spiral **recoil spring**. Releasing the rope allows spring tension to wind the rope back onto the pulley.

Many motorcycles use a manual starter called a **kick starter**. The kick starter has a crank that is attached to the crankshaft. When the rider kicks the crank, it turns the crankshaft. Once the engine starts, the starter disconnects from the crankshaft. See Fig. 19-63.

Another type of crank starter is used on many lawnmowers. The operator uses a crank to wind a spring tight. A button or lever releases the spring tension. The tension then works like a rope starter to turn the crankshaft. See Fig. 19-64.

Electric Starters

When compared to manual starters, electric starters have both advantages and disadvantages. Electric starters make it much easier for an operator to start an engine. This is especially true for an engine that is not in perfect running condition. However, electric starters add weight to the output device. In addition, the parts are more complex. And electric starters cost more.

All electric starters have an electric motor called the **starting motor**. A battery is needed to operate this motor. Small engine starting motors are very much like the starters used on cars. When the operator turns the ignition key, the motor turns the crankshaft. Once the engine starts, the motor disengages from the crankshaft. (For more information on electric starters, see *STARTING SYSTEM* in Chapter 18.)

Fig. 19-63 **This type of starter is used to start motorcycles. Kicking the crank turns a gear. The gear turns the crankshaft to start the engine.**

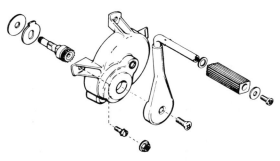

Tecumseh Products Co.

Briggs & Stratton

Fig. 19-64 **Turning the crank on this type of starter pulls a spring tighter and tighter. Releasing the spring provides the force to turn the crankshaft.**

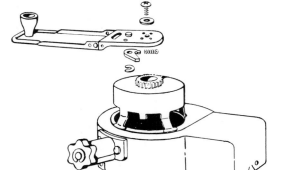

Stop Switch

There has to be a way to stop an engine, as well as start it. Most small engines are stopped by keeping the spark plug from firing. One way to do this is to ground the secondary circuit wire at the spark plug. (In this case, *grounding* means to connect the wire to a metallic part of the engine.) Another way is to ground the primary circuit.

Figure 19-65 shows a common mechanism used to ground the secondary circuit. The **stop switch** (or "kill" switch) is a small strip of metal. It is attached to the cylinder head, close to the spark plug. To stop the engine, the operator pushes the strip against the spark plug wire. This keeps the current from going to the spark plug. Instead, it goes into the metal of the cylinder head. The spark plug can't fire, so the engine stops.

In some cases, the stop switch is connected to the breaker points. The operator uses the switch to ground the primary circuit. This keeps the magnetic field in the armature from collapsing. As a result, no spark occurs and the engine stops.

Briggs & Stratton

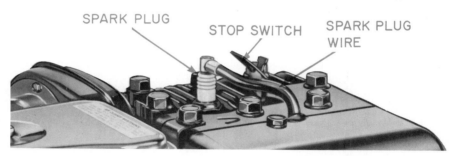

SPARK PLUG

STOP SWITCH

SPARK PLUG WIRE

Fig. 19-65 **You can stop an engine by grounding the secondary circuit with a stop switch.**

STUDY QUESTIONS

1. List the six major small engine systems.
2. What is the main job of the mechanical system?
3. The cylinder and the crankcase are usually manufactured as a single part. What is this part called?
4. Describe the piston.
5. What is the main job of the compression rings?
6. What is the main job of the oil control rings?
7. What engine part connects the piston to the crankshaft?
8. What are bearings used for in a small engine?
9. What does *rpm* mean?
10. True or false: Most lawn mowers use a horizontal crankshaft.
11. Describe the flywheel.
12. Four-stroke engines use a group of parts to open and close the valves. What are these parts called?
13. What engine parts does the crankshaft use to drive the valve train and control its speed?
14. The air-fuel mixture enters the cylinder through one hole. Exhaust gases leave through another. What are these holes called?
15. What is the most common method of fuel intake used on two-stroke engines?
16. Name the piston actions in the two-stroke cycle.
17. List four purposes of the lubrication system.
18. True or false: Four-stroke engines use a mixture of oil and fuel to lubricate the engine.
19. What engine parts increase the surface area of the engine?
20. What are the two common types of small engine air cleaners?
21. What are the two common types of small engine fuel supply systems?
22. What is the narrow middle part of the carburetor called?
23. What two partial vacuums are used to draw air through the carburetor?
24. What is the purpose of the throttle valve?
25. What is the purpose of the choke valve?
26. Name the three basic carburetor air and fuel adjustments.
27. Describe how a float assembly maintains a constant supply of fuel to the carburetor.
28. What is another name for the vacuum-feed carburetor?
29. What do diaphragm carburetors use to control the fuel intake?
30. What is the purpose of the governor on a small engine?
31. Name the two common types of governors.
32. What engine part reduces the noise of the escaping exhaust gases?
33. Name the three common types of ignition systems used on small engines.
34. Name the five basic parts of the magneto system.
35. What two circuits are used in the magneto system?
36. What two electrodes does a spark plug have?
37. What is the spark plug air gap?
38. What two things does an engine need to operate?
39. What engine part is turned by the starter?
40. What happens when the operator pushes the stop switch against the spark plug wire?

ACTIVITIES

1. Prepare a report that shows as many different uses of small engines as you can find. Clip out pictures of small engines in use from magazines and newspapers. You can also get information by:
   - Visiting stores and asking for brochures on small engines and the equipment they operate.
   - Writing to small engine manufacturers.
   - Observing residential, commercial, agricultural, and recreational uses of small engines.

2. Conduct a survey of the small engines used in your home or neighborhood. Write up your findings in a report that includes (1) the condition of the engines and (2) the condition of the equipment they operate.

3. Choose one of the small engines you investigated in Activity 2. With the approval and supervision of your teacher, demonstrate how the engine powers its output device.

20

Small Engine Safety

Small engines are not especially dangerous-looking objects. However, they can cause *all* kinds of accidents and be *extremely* dangerous. Small engines can start fires that could result in severe burns. They can also produce harmful gases. What's more, small engines can cause electrical shocks, bad cuts, and broken bones!

Yes, small engine work *can* involve some dangerous situations. But don't let the potential dangers scare you away. There is a way to avoid accidents. And this way is to develop a "safety-first" attitude.

Be serious about safety and the possibility of accidents. The safe way is usually the right way. Sure, you can work faster if you forget about safety. But if you hurt yourself, the time you save is worthless.

This chapter will tell you about two types of safety rules. First, there are rules to follow all the time in the school shop. Second, there are special rules for working with small engines. Follow these rules and be alert to the possibility of accidents. This is the way to a good experience with small engines. See Fig. 20-1.

GENERAL SHOP SAFETY

Most school shops have a list of general safety rules posted somewhere. The following rules are some of the more common ones. They may not be very exciting, but they *are* very important. Following these rules will help you develop your safety-first attitude.

- **Dress for the job** — Don't wear a tie or loose clothing. If you have long hair, tie it back or cover it. This is especially important when operating equipment with rotating parts. Loose clothing and long hair can catch in the moving parts. They can pull you into the parts. Jewelry, such as rings and watches, can also catch in moving parts. You should remove all jewelry.
- **Protect your eyes** — Wear eye protection at all times when working in the shop. When working with power machines such as grinders, wear both safety glasses and a face shield or goggles. This will give your eyes double protection. Remember that your eyes can't be replaced. They deserve the best protection you can give them. See Fig. 20-2.
- **Be neat and orderly** — Don't leave parts or tools on the floor. They can cause someone to trip. Always store sharp-pointed or sharp-edged tools as soon as you finish with them. Watch for heavy objects that can fall and injure feet or legs. For example, be sure an engine is tightly mounted before you begin working on it. See Fig. 20-3.

Fig. 20-1 **This shop scene demonstrates a *safety-first* attitude. By paying attention to safety rules, you are more likely to have a good shop experience.**

324

325

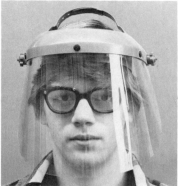

Fig. 20-2 If you work without eye protection, your eyes are always in danger. Goggles, glasses, and face shields keep your eyes safe from harm.

Fig. 20-3 Keep your work area neat and organized. Mount the engine securely before starting work. This will keep you from having to "wrestle" with the engine as you work on it.

Fig. 20-4 Always use the right tool for the job. Using the wrong tool can result in damage to the tool, damage to the equipment, or injury to yourself.

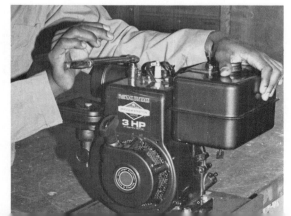

- **Always use the proper tools** — For example, don't use an adjustable wrench or pliers in place of a socket or box-end wrench. See Fig. 20-4. Adjustable wrenches and pliers tend to slip more easily than box-end or socket wrenches. And slipping tools usually cause hand injuries. They can also ruin the nut or bolt. Using the right type and size of tool will prevent slipping.
- **Use tools properly** — Tool safety means more than just choosing the right tool. First, your hands shouldn't be so greasy that you can't hold the tool firmly. Second, you should always maintain a good balanced position as you use the tool. But before you use *any* tool, think about any problems that might arise. For example, when loosening a stuck bolt or nut, push the wrench with an open hand. If the bolt loosens suddenly, you won't bang your knuckles. See Fig. 20-5.
- **Work at a steady pace** — Never hurry as you work. Hurrying leads to awkward and dangerous situations. It also gives you less time to think ahead and avoid accidents.
- **Ask for help when you need it** — If you aren't sure how to do something, check the operator's manual or ask your instructor. Use only the tools and machines that you have been trained to use safely.

Fig. 20-5 **You can avoid an accident by thinking ahead. Here the boy is using his palm to push the wrench. By doing this, he avoids hitting his hand on the engine when the bolt loosens.**

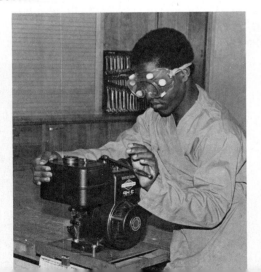

SMALL ENGINE SAFETY

There are safety rules for small engine work, just as there are rules for general shop work. The following pages give the most important rules. Be sure to follow these rules at all times when doing small engine work.

Preparing to Use an Engine

Before starting and running an engine, read the operator's manual carefully. The manual will explain hazards associated with running the engine. It will also point out special safety precautions that will help you.

Before you start the engine, check all safety guards and devices. Make sure they are in good condition. See that the guards are in their proper places. The operator's manual usually provides information on the proper use and maintenance of all safety devices. See Fig. 20-6.

Handling Gasoline

Most small engines run on gasoline or a mixture of gasoline and oil. Gasoline is highly flammable (easily set on fire), and it is explosive. You must be *extremely* careful when working with or around gasoline.

One very important rule is to always keep gasoline in a closed container. Gasoline left in

Fig. 20-6 **Learn all you can about the small engine you are assigned to. This way, you'll be able to work with the engine safely.**

Fig. 20-7 Always keep gasoline in a sturdy, tightly closed container. A spark can ignite gasoline fumes and cause an explosion.

an open container will give off fumes, or vapors. Under certain conditions, a spark can ignite gasoline vapors. This could cause an explosion. You can easily prevent this by storing gasoline in the proper type of container. See Fig. 20-7. Do *not* store gasoline inside your house or in a crowded area. Also, be sure to keep the cap on the engine's fuel tank.

Here are some additional rules to follow when working with gasoline:

- **Keep all flames and heat sources away from gasoline** — Know where the proper fire extinguisher is. And know how to *use* it. A fire may start even with the best of precautions. Be prepared!
- **Avoid breathing gasoline vapors** — Gasoline vapors are toxic (poisonous). Make sure your work-place is ventilated (has fresh air passing through it).
- **Avoid contact with gasoline** — Many gasolines contain added chemicals, such as tetraethyl lead. These are poisonous when absorbed through the skin. Wash your hands thoroughly if gasoline splashes on them.
- **Never use gasoline to clean parts** — As you read ahead, instructions will call for cleaning engine parts with a **solvent**. For our uses, a solvent is a petroleum-based fluid that dissolves dirt, oil, and grease. Gasoline is too dangerous to use as a solvent. On the other hand, kerosene is a good and fairly safe solvent.

- **Clean up gasoline spills** — Wipe up all gasoline spills *immediately*. Then place the wiping cloth in a closed, metal container. If you spill gasoline on your clothes, remove the clothing immediately.

Filling Fuel Tanks

Pouring fuel into a small engine fuel tank can be a dangerous situation. Remember, gasoline is highly flammable. It doesn't take a lot of heat to ignite it.

Engines get very hot when they are running. Obviously, you should not pour fuel into the tank of a running engine. This would be extremely dangerous. Even a stopped engine that is still hot could ignite the gasoline. Wait until the engine is cool to refill the tank.

Take care as you add fuel to the tank. Touch the spout of the gasoline can to the edge of the tank. This will prevent a spark of static electricity. The spark could ignite the fuel.

Do not fill a fuel tank completely. Leave at least an inch of space at the top to allow for expansion. Try not to spill the gasoline as you pour. See Fig. 20-8. If you *do* spill or overfill, wipe up all the gasoline immediately.

Fig. 20-8 Using a fuel container with a special pour spout makes tank-filling easier and neater.

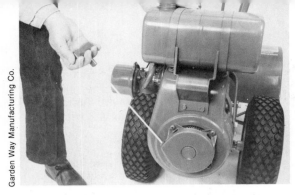

Fig. 20-9 **This is the proper way to hold and pull a starting rope. Pull the rope out rapidly. Let it rewind slowly. Be ready for kickbacks and misfires — they could pull on your hand.**

Starting an Engine

Some small engines start with a starting rope. You should take special precautions when starting this type of engine. Never wrap the rope around your hand. Hold the rope so that you can release it easily by opening your hand. See Fig. 20-9. "Kickbacks" and engine misfires can cause a sudden pull on the rope. They can pull your hand into the engine if the rope is wrapped around your hand. You must be able to let go immediately.

Some small engines start with a crank. When you turn a crank, wrap your fingers around the handle. But keep your thumb next to your index finger (first finger). *Don't* wrap your thumb around the handle. You could be injured by a kickback or misfire. See Fig. 20-10.

Exhaust Gases

Small engines produce several different poisonous gases. The most dangerous is carbon monoxide. Even a small dose of this gas can make you very ill. Large doses cause death. Each year many people die from carbon monoxide poisoning.

The key to avoiding injuries from exhaust gases is **proper ventilation**. Exhaust gases are harmless as long as you have plenty of fresh air. Never operate engines inside closed buildings. In fact, don't run them in *any* place where exhaust gases can build up. The only exception to this rule is a closed building that is equipped with a system for carrying the gases outside.

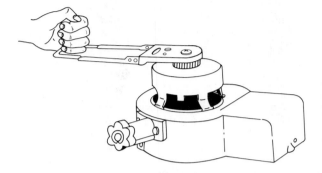

Fig. 20-10 **The correct way to hold the handle of a crank starter.**

See Fig. 20-11. In all cases, however, avoid breathing exhaust fumes as much as possible.

Running Engines

You should follow precautions when you work with a running engine. Remember the general shop rule of not wearing loose clothing. This is especially important when operating an engine. You must also keep your hair, hands, and legs away from the moving parts of the engine.

There are several other precautions to take with running engines:

- **Heat** — The cylinder and the exhaust system become very hot during operation. You can burn yourself *severely* by touching these parts. Don't get close to hot engines unless it is absolutely necessary. Allow the engine to cool first. *Never* cover a hot engine with flammable materials such as rags and plastic sheets.
- **Electrical shock** — There is always the possibility of high-voltage shocks from the ignition system. These shocks are especially dangerous for persons with heart trouble. Avoid touching the parts of the ignition system while the engine is running.
- **Batteries** — Be careful when handling a small engine battery. Batteries contain acid.

Fig. 20-11   **When working on a running engine indoors, use an exhaust system to conduct dangerous carbon monoxide gas to the air outside.**

The acid can ruin your clothing and burn your skin. It can *blind* you if it gets into your eyes. Batteries also give off hydrogen while they are being charged. Like gasoline, hydrogen is highly flammable. It ignites easily and will explode. Keep batteries away from sparks, heat, and flames.

- **Noise** — Small engines can be very noisy. Never operate an engine that has a faulty exhaust system or doesn't have a muffler. Protect your ears from excessive noise by wearing ear protection. Special ear muffs and ear plugs will help you avoid injury.

- **Unattended engines** — Never leave a running engine unattended. Be especially careful if children are present. Always turn off the engine before you leave the area.

STUDY QUESTIONS

1. What's the best way to avoid accidents in the school shop?
2. List seven general shop safety rules.
3. Why should you not store gasoline in an open container?
4. List five safety rules for working with gasoline.
5. True or false: The best time to fill a small engine's fuel tank is when the engine is hot and running.
6. Why shouldn't you wrap a starting rope around your hand when starting an engine?
7. How can you avoid the dangers of engine exhaust gases?
8. True or false: It's best to wear loose clothing when working with small engines.
9. Which parts of the engine get hot enough to cause severe burns?
10. What is potentially dangerous about the small engine ignition system?
11. Why do you have to be careful when handling small engine batteries?
12. True or false: Small engines don't make enough noise to require any special ear protection.

ACTIVITIES

1. Select a lawn mower or other power device operated by a small gas engine. Make up a list of the possible hazards involved in starting the engine. Include any special problems, such as the danger of a rotating lawn mower blade.

2. Prepare a list of small engine safety rules and procedures to follow in and around your home. Think of the different types of power devices as you make your list.

Maintaining and Troubleshooting Small Engines

21

As small engines operate, many of the moving parts rub against each other. The parts gradually wear down. Dust and dirt work their way into the engine and speed up the wearing-down process. There is also the high temperature of combustion. This heat puts great stress on engine materials.

Sooner or later, all engines develop problems from friction, heat, dust, and dirt. However, you can do a lot to slow down the wear and tear. This chapter covers four types of engine procedures:

- **Routine maintenance** — This type of maintenance prevents engine problems from happening.
- **Engine storage** — This procedure keeps problems from developing while the engine is in storage.
- **Troubleshooting** — This is a step-by-step process for finding the source of engine problems.
- **Repair procedures** — These are general guidelines for conducting small engine repairs.

ROUTINE MAINTENANCE

Most small engines provide trouble-free operation for long periods of time. However, you *do* have to care for them properly. Most small engines require routine maintenance on the following systems and parts:

- Oil supply
- Cooling system
- Air cleaner
- Fuel filter
- Crankcase breather
- Spark plug
- Carburetor
- Battery

Every engine manufacturer provides instructions for routine maintenance of its engines. When you buy a small engine or a device using one, you should receive an **owner's manual**. This manual will help you maintain the engine

Fig. 21-1 **Operating and service manuals are designed to help owners maintain and service their small engines.**

properly. Most manufacturers also print service and repair manuals for their engines. These manuals are usually available by request only.

Service manuals usually include more detailed information about service and repairs than owner's manuals. See Fig. 21-1. They often contain a parts list with descriptions to help you identify and order parts. You can usually get additional information at retail stores dealing in small engines, repairs, and parts.

Many small engines have the same basic service and maintenance procedures. This chapter presents the most common ones.

Checking and Adding Oil

Two-stroke engines are usually lubricated by the oil in the oil-fuel mixture. No other engine lubrication is needed. The proper oil-fuel ratio is usually printed on the engine or in the owner's manual. Some two-stroke engines can use a pre-mixed oil-fuel solution.

Four-stroke engines are lubricated with oil from the crankcase. Some oil can be lost while the engine is operating. Sometimes the oil slips past the piston rings into the combustion chamber, where it is burned. Another problem is worn or damaged gaskets and seals. These allow oil to leak from the engine. Both problems result in a lower oil level in the crankcase. For this reason, you should check the oil level before you start the engine. To get an accurate reading, be sure the engine is on a level surface.

The procedures for checking the oil level vary from engine to engine. Some engines have **dipsticks**, which have marks to show the oil level. See Fig. 21-2. To check the oil in these engines, first shut off the engine. Then pull out the dipstick. Wipe it dry. Slide the dipstick back into the crankcase, then remove it again. You'll see whether oil needs to be added.

Many small engines don't have dipsticks. On these engines, you usually check the oil by removing the **oil filler plug**. See Fig. 21-3. The oil level should reach the bottom of this plug. If the oil level looks low, add oil until it reaches the plug level. Then replace the plug.

When you add oil to the crankcase, add it slowly. This will keep air from being trapped

Garden Way Manufacturing Co.

Fig. 21-2   **On engines with dipsticks, the oil level must be kept between the ADD and FULL marks.**

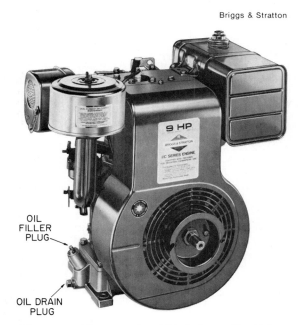

Briggs & Stratton

OIL FILLER PLUG

OIL DRAIN PLUG

Fig. 21-3   **Many engines don't have dipsticks. Check the oil level on these engines by removing the oil filler plug. The oil level should be up to the bottom of the plug.**

in the oil. Don't add too much oil. Overfilling can cause carbon to build up in the combustion chamber and on the spark plug. Overfilling can also cause oil to be thrown from the crankcase breather. (The crankcase breather is described further on in this chapter.) Each of these results can damage the engine. This will interfere with its performance.

Be careful to keep out dirt when you are adding the oil. One purpose of lubrication is to keep the engine free of "foreign" particles. Dirt does *not* belong in the engine. It can damage parts and reduce power. Adding dirty oil defeats the cleaning function of lubrication. Therefore, you must keep all funnels, oil spouts, and storage containers clean. Clean away any dirt before you open the oil container.

Changing the Oil

The oil picks up carbon and dirt as it lubricates the engine. Over a period of time, the oil collects more and more impurities. The oil gradually loses its ability to lubricate properly. This is why you must change the oil regularly. You must remove the old, dirty oil and replace it with new, clean oil. The manufacturer's instructions will tell you how often to change the oil.

Drain the oil from the engine while the oil is warm. Warm oil will flow more freely. This does two things. It allows more of the oil to be drained. It also allows the oil to carry more impurities out of the engine.

The oil drains from an **oil drain plug** on the bottom of the crankcase. See Fig. 21-4. Place a container under the drain plug to catch the oil. Then remove the plug. After all the oil has drained, replace and tighten the plug.

To add new oil, first remove the oil filler plug. Put a funnel in the hole. Pour the new oil in slowly. Fill the crankcase and replace the plug. If the engine has a dipstick, use it to check the oil level. Otherwise, fill the crankcase up to the plug. Be careful to keep dirt out of the engine. See Fig. 21-5.

Which Oil To Use

You can't use just *any* oil to lubricate your engine. You have to use the *right* kind of oil. This will provide the best engine performance. There are two things to consider when selecting oil for an engine. One is the oil's viscosity. The other is the oil's service rating.

Viscosity. Viscosity is a measure of the thickness of an oil. It is determined by the rate at which the oil flows. For example, oils that are thick and heavy do not flow easily. These oils have a **high viscosity**. Thin, light oils flow more easily. They have a **low viscosity**.

Garden Way Manufacturing Co.

Briggs & Stratton

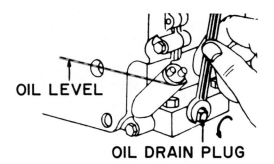

Fig. 21-4 **Drain oil from the engine by removing the oil drain plug. The plug location depends on the type of engine.**

Fig. 21-5 **Use a clean funnel to fill the crankcase with oil.**

The Society of Automotive Engineers (SAE) has set up a viscosity numbering system. The numbers used in this system are called **grades**. Thin oils have low grades, such as *SAE 10*. Heavy oils have higher grades, such as *SAE 40* or *SAE 50*. Some numbers include a *W*. This means that the oil can be used in freezing *winter* conditions. The SAE number is usually printed or stamped on the oil can. Engine manufacturers recommend certain viscosities of oil for use in their engines at different temperatures. Manufacturers generally recommend a high-viscosity oil for summer. They recommend a low-viscosity oil for winter.

Multi-viscosity oils fulfill both the low- and high-viscosity requirements. For example, in cold weather, SAE 10-40 provides the same lubrication as SAE 10. In warm weather, it provides the lubrication of SAE 40. With multi-viscosity oil, you don't have to change the oil as the temperature changes.

Service Ratings. The American Petroleum Institute classifies oils according to their quality. The Institute uses six letter classifications to indicate how well an oil performs in lubrication tests. These ratings are *SA, SB, SC, SD, SE,* and *SF*. SA is the lowest (poorest) rating. SF is the highest (best) rating.

Engines that must perform in difficult conditions need a higher class of oil. Engines exposed to less wearing conditions can use lower classes. Manufacturers usually recommend certain classes of oil for their engines. You are always safe, of course, when you use the better oils.

Cooling System

People often use small engines in dusty, dirty conditions. For example, lawn mower engines must operate with grass and dirt in the air. These materials can stick to the cooling fins. The cooling system can get clogged. And if the engine doesn't cool properly, it can overheat. This can damage the engine seriously. Therefore, you should clean the cooling system regularly. How often a system needs cleaning depends on the operating conditions.

Do *not* clean the cooling system while the engine is hot. Let the engine cool before you clean any part of it.

Many small engines use a **blower housing** to keep out dirt and other matter. The first step in cleaning the cooling system is to remove the housing. See Fig. 21-6. After removing the housing, clean all dirt from the flywheel vanes. See Fig. 21-7. Also clean the cooling fins on the engine cylinder and head. Wipe away all the dust. Then replace the housing.

Fig. 21-6 **Remove the blower housing to service the cooling system. On the engine above, three bolts hold the housing on the engine.**

Fig. 21-7 **After removing the blower housing, clean dirt from the cooling fins. This will allow cooler engine operation.**

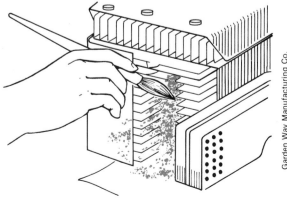

Garden Way Manufacturing Co.

Air Cleaner

The air that enters the engine first passes through the air cleaner. The air cleaner removes dirt from the air. This keeps the dirt from damaging the engine. You must clean the air cleaner regularly. There are two common types of air cleaners: dry-element and oil-foam.

Dry-Element Cleaners. Figure 21-8 shows how to check a dry-element air filter. You can knock the dirt free by tapping the filter sharply on a hard surface. You can also clean these filters with compressed air. Blow the air through the filter opposite the direction of normal air flow. This will remove much of the dirt collected in the filter. See Fig. 21-9.

> **SAFETY NOTE:** Be careful not to point the compressed air hose toward other people when cleaning the filter.

Fig. 21-8 **To check a dry-element air cleaner, remove the filter cover.** Garden Way Manufacturing Co.

Fig. 21-9 **You can clean a dry air filter by tapping it on a hard surface or by blowing compressed air through it.**

The filter may still be dirty. Sometimes you can clean it with a non-sudsy detergent and warm water. Afterwards, rinse the filter inside and out with clean water. See Fig. 21-10. Allow the filter to dry completely before using it again. (It may take a day or two to dry out completely.) If the filter is still clogged, replace it with a new filter.

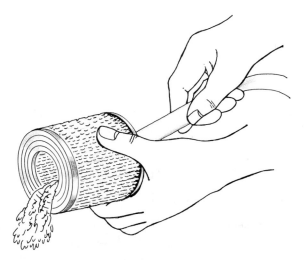

Fig. 21-10 **If compressed air doesn't clean the filter, wash it with mild detergent and water. Then rinse with clean water.**

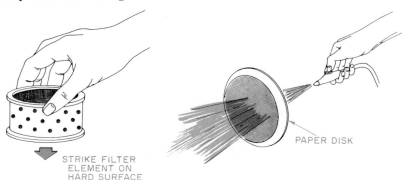

STRIKE FILTER
ELEMENT ON
HARD SURFACE

PAPER DISK

Oil-Foam Cleaners. Oil-foam cleaners strain air through an oil-soaked foam filter. To service this type of cleaner, first lift off the cover. Then remove the foam filter. See Fig. 21-11. Wash the filter thoroughly in clean kerosene or liquid detergent and water. After washing the filter, dry it with a cloth. Then work some clean oil into the foam. Squeeze out any excess oil. Then put the filter and cap back in place. See Fig. 21-12.

Fuel Filter

Many small engines use a fuel filter to clean the gasoline before it reaches the carburetor.

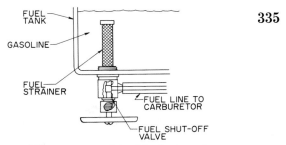

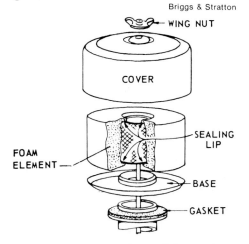

Briggs & Stratton

Fig. 21-11 The parts of an oil-foam air cleaner.

Fig. 21-12 There are four steps in maintaining an oil-foam cleaner.

Briggs & Stratton

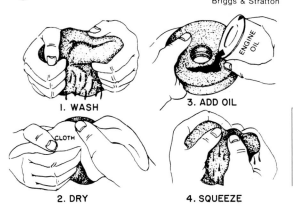

1. WASH
2. DRY
3. ADD OIL
4. SQUEEZE

A. **Filter mounted inside fuel tank**

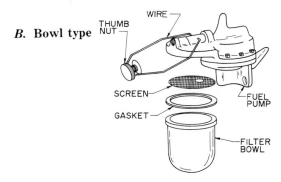

B. **Bowl type**

Fig. 21-13 Two common types of fuel filters.

The filter usually contains a fine wire screen. This screen traps particles of dirt that might clog small openings in the carburetor.

On many engines, the fuel filter is in the fuel tank. It is usually placed on the bottom, where the fuel line attaches. Older engines have a filter screen placed between the fuel pump and a glass bowl. A wire loop and thumb screw hold the bowl in place. See Fig. 21-13. Some new engines have plastic filters. These are placed "in-line" between the fuel supply and the carburetor.

The in-line type of filter can't be cleaned. However, you should check it regularly to make sure it isn't blocked. First disconnect the fuel line at the carburetor. Then run fuel from the line into a small container. (If the engine has a fuel pump, you must crank the engine to make the fuel flow.) If the flow is slow, replace the filter with a new filter.

SAFETY NOTE: Before testing the fuel flow, be sure to ground the spark plug (connect it to a metal part of the engine). This will keep the engine from starting.

Fuel tank filters and glass bowl filters can be cleaned. If the fuel tank is above the filter, shut off the fuel supply with the shut-off valve. Then drain the fuel from the filter in a safe area. (With some in-tank filters, you may have to drain the tank first.) Next, take the filter apart. Clean all the parts with solvent. Make sure the parts are dry before putting the filter back together.

Crankcase Breather

In four-stroke engines, combustion pressure constantly pushes burned gases past the piston rings. The gases pass into the crankcase. Exhaust gases must not be allowed to build up in the crankcase. Otherwise, the pressure of the gases will increase. This pressure build-up can force crankcase oil out of the engine, past gaskets and seals.

Crankcase breathers prevent pressure build-up in the crankcase. They vent (release) the exhaust gases before the pressure can get too high.

There are several different types of crankcase breathers. Refer to the service manual to find out what type is used on your engine. All types need to be cleaned regularly. First, remove the breather assembly from the engine. See Fig. 21-14. Clean the parts in solvent and dry them thoroughly. Then re-install the breather. (It's a good idea to replace the breather gaskets at this time, too.)

Spark Plug

Burning fuel leaves carbon deposits on the spark plug. You should remove these deposits regularly. Use a **spark plug socket wrench** to remove the plug from the engine. See Fig. 21-15. The socket will help keep the plug from breaking during removal. If the plug breaks, you must replace it.

Scrape the carbon deposits from the plug with a pocket knife or a wire brush. Then clean the plug with solvent. Never use an abrasive-type cleaner on small engine spark plugs.

Look at the plug carefully. The electrodes may be burned away. The porcelain insulation may be cracked. In either case, you should replace the plug. See Fig. 21-16. Adjust the gap

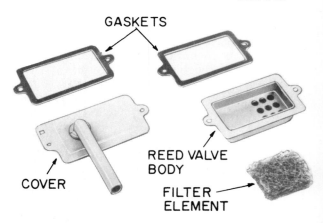

Fig. 21-14 These are the parts of a crankcase breather. To maintain the breather, simply clean the parts.

Fig. 21-15 Use a spark plug socket wrench to remove a spark plug.

between the electrodes to the manufacturer's specifications. To do this, first check the gap with a **wire gage**. See Fig. 21-17. Do *not* use a leaf-type feeler gage. Leaf gages can be inaccurate. See Fig. 21-18. Make the adjustment by bending the outer electrode. You can do this with the "gapping" part of the wire gage (see Fig. 21-17 again).

There may be a gasket between the spark plug and the cylinder. Inspect the gasket before replacing the plug. If the gasket is cracked or damaged, replace it.

Fig. 21-16 **Damaged spark plugs should be replaced.**

Prestolite

A. Spark plug in good condition

B. Spark plug with burned electrode

C. Spark plug with cracked porcelain

Also inspect the wire that leads to the spark plug. Wipe it clean and look for cracks or other damage. When you remove this wire, hold it by the connector that fits over the plug. Then pull. Never hold just the wire when pulling. This could separate the wire from the connector.

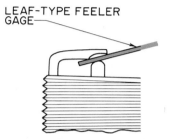

LEAF-TYPE FEELER GAGE

A. **Inaccurate measurement**

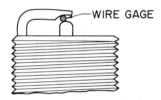

WIRE GAGE

B. **Accurate measurement**

Fig. 21-18 **Leaf-type feeler gages can't measure the electrode gap properly. To get an accurate measurement, use a wire gage.**

Fig. 21-17 Use a wire gage to check spark plug gap.

AC Spark Plug Div., GMC

Carburetor

There are three types of carburetor adjustments: (1) the main fuel adjustment, (2) the idle fuel adjustment, and (3) the idle speed adjustment. Different carburetors have the adjusting screws in different places. Figure 21-19 shows one arrangement.

The carburetor fuel adjusting screws are usually needle valves. The point of a needle valve is small and fragile. It can easily be bent or distorted. Therefore, do *not* use excessive force when adjusting these valves. You can't make a proper adjustment with a damaged needle valve.

Not all small engine carburetors have the adjustments just described. In fact, some carburetors are not adjustable. With these carburetors, the settings are made at the factory. They can't be changed. Adjust only carburetors that have adjusting screws for your use. Always follow the directions provided by the manufacturer.

Main Fuel Adjustment. Adjust the main fuel screw when the engine is warm. First, open the throttle half-way. Run the engine without a load.

Turn the main fuel screw in or out to get the smoothest possible operation. Usually, only partial turns of the adjusting screw are necessary.

Idle Fuel Adjustment. The idle fuel adjustment should also be made while the engine is warm. Be sure the choke is open. Close the throttle until the engine idles. Then turn the idle fuel screw in or out until the engine runs at its smoothest.

Idle Speed Adjustment. Make this final adjustment after setting the main and idle fuel screws. Let the engine idle while you turn the idle speed screw. Use a **tachometer** for this adjustment if one is available. (A tachometer measures the engine's speed in rpms — revolutions per minute.) Set the engine speed to the manufacturer's specification.

If you don't have a tachometer, you can still make a fairly good adjustment. Just set the screw to the lowest speed at which the engine runs smoothly without stalling (shutting off).

NOTE: Different manufacturers may give different directions for adjusting the idle speed. To be sure, always check the service manual.

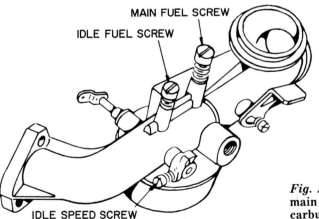

Fig. 21-19 **This diagram shows the three main adjusting points on one type of carburetor.**

Letting the "Pros" Do It

Cars and trucks are routinely serviced, or "tuned-up." Applied to cars and trucks, the word *tune-up* has special meanings. It refers to routine maintenance of the electrical system and fuel system. It also includes the study of engine performance and the identification of possible troubles. A tune-up usually includes changing the points, the spark plugs, and the condenser. The ignition system is also checked and adjusted. It is repaired if necessary.

Small engines are not treated the same way. Instead, the manufacturer recommends **periodic** (regular) maintenance. Owners may take their engines to a **service center**. A service center tune-up usually includes the following:

- Servicing the cooling system
- Cleaning the air cleaner, fuel filter, and crankcase breather
- Servicing (or replacing) the spark plug

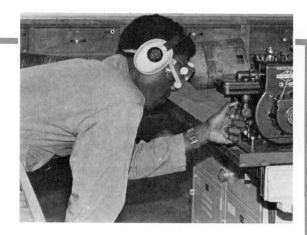

- Adjusting the carburetor
- Changing the oil
- Servicing the battery

The service center mechanic will also check the engine for proper operation. From this checking, he or she can find any existing or potential problems.

Battery

An engine with an electric starter usually has a battery. The battery powers the starting motor. Battery service usually includes the following:

- Inspecting and cleaning
- Adding water
- Checking and charging

If your engine has a battery, you should check and service it regularly. Manufacturers usually provide an **equipment manual** for a device powered by a small engine. Refer to this manual for battery servicing procedures.

ENGINE STORAGE

If you won't be using an engine for some time, you should store it. How you store an engine can affect its future condition. The following tips on storage will help keep your engine in good working condition.

It is important that the engine be kept dry during storage. If you store the engine outside, protect it with a waterproof cover. Don't make the cover so tight that air can't circulate over the engine. Proper ventilation is important to prevent excessive condensation of moisture.

You should also drain the carburetor to prevent varnish build-up. First close the shut-off valve, then open the carburetor drain plug. Some carburetors don't have a drain plug. In this case, run the engine at a fast idle until it slows down noticeably. When this happens, shut off the ignition immediately.

For short-term storage, fill the fuel tank with fresh fuel. This will keep moisture from condensing in the tank.

For storage of a month or more, you must drain the fuel tank, fuel filter bowl, and fuel lines. Let these parts dry. Then replace the tank cap and the fuel filter bowl. Draining the system prevents gum and varnish from forming in the system.

Next, remove the spark plug. Pour one or two tablespoons of engine oil into the cylinder. Crank the engine to distribute the oil. Replace the plug. Then turn the flywheel slowly by hand. As you reach the top of the compression stroke, you should feel some resistance. The flywheel may even bounce back slightly. Leave the engine in this position. The cylinder wall and piston rings are now protected from corrosion.

Engines with electric starting systems require further preparation. Disconnect the cables and be sure the battery is charged. Keep the battery in a warm place to prevent freezing.

Usually you should also prepare the piece of equipment powered by the engine. It may have belt-driven units or belts in a clutching system. Loosen all the belts. This will keep them from stretching out. If the equipment has tires, raise it off the ground. This will keep the engine weight off the tires. There will be no flat spots on the tires after storage.

TROUBLESHOOTING

Troubleshooting means checking an engine to find out what part or condition is causing poor performance. The purpose of troubleshooting is to find the trouble so that it can be corrected.

A small engine requires three conditions for good operation:

- **Carburetion** — The proper air-fuel mixture must be delivered to the combustion chamber.
- **Compression** — There must be proper compression of the air-fuel mixture.
- **Ignition** — A strong spark must be delivered to ignite the air-fuel mixture.

An engine may have a problem in any one of the three areas. If something is wrong, the engine will run poorly. Or, it may not run at all! Nearly all engine trouble can be traced to carburetion, compression, or ignition. As you read on, you'll see that there are many possible causes for problems in these three areas. Some of the causes are easy to correct. For example, if the engine doesn't run, it might simply be out of gas. Other causes are harder to correct.

You may decide to repair engine problems yourself, with the help of a service manual. Or, you may choose to have the more difficult work done at a small engine service center.

Fig. 21-20 **You can check ignition by trying to generate a spark. Hold the spark plug wire 1/8″ from the spark plug while turning over the engine.**

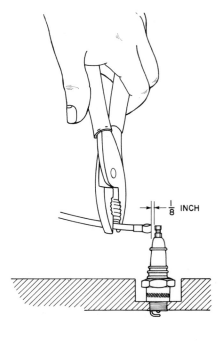

Ignition

Check the ignition by trying to generate a spark. To do this, you must first remove the spark plug wire. (Before removing the wire, make sure it is clean and dry. A dirty or damp wire can give you a shock.) Use insulated pliers to grip the wire an inch or two from the end. Hold the wire about 1/8 inch from the plug. Then crank the engine with the starting mechanism. See Fig. 21-20.

You should see a spark between the plug and the wire as you crank the engine. If there isn't a spark, something is wrong with the ignition system. Try holding the plug wire near the cylinder head while cranking the engine. See Fig. 21-21. If you get a spark this time, the spark plug is bad. If you don't get a spark, the problem could be any of the following:

1. Incorrect breaker point gap
2. Dirty or burned breaker points
3. Stuck or worn breaker point plunger
4. Shorted or open secondary circuit wire
5. Faulty magneto system
6. Condenser failure
7. Sheared flywheel key
8. Worn flywheel bearings

Carburetion

When checking carburetion, first see if there is fuel in the tank. Most fuel tank caps have small vent holes. See if the vent holes are clear. A clogged vent hole keeps atmospheric pressure from pushing fuel into the carburetor. See Fig. 21-22. If the fuel tank is above the engine, make sure the shut-off valve is open. See if the choke valve operates freely.

> **SAFETY NOTE:** Remember that you are working with gasoline, which can burn or explode. Always work on a cool engine, away from open flames or heat. Work in a well-ventilated area.

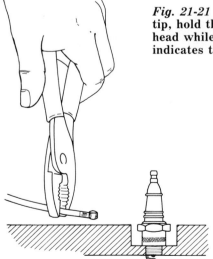

Fig. 21-21 **If there is no spark at the plug tip, hold the plug wire close to the cylinder head while cranking the engine. A spark now indicates that the plug is bad.**

Fig. 21-22 **If an engine isn't getting fuel, check the fuel supply. Also clean the gas cap vent hole with a fine wire or pin.**

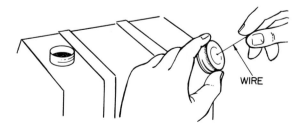

Garden Way Manufacturing Co.

If the engine won't start, remove the spark plug, and inspect it. The spark plug can provide important information about the engine. A good plug will look like the one in Fig. 21-23. If the plug is *wet* or *black* (Fig. 21-24), check the following:

1. Overchoking or sticking choke plate
2. Excessively rich fuel mixture
3. Water in fuel
4. Carburetor fuel valves stuck open

If the plug is *white and dry* (Fig. 21-25), check the following:

1. Leaking carburetor mounting gasket
2. Dirty fuel filter
3. Carburetor fuel valves stuck shut
4. Faulty fuel pump (if used)
5. Excessively lean fuel mixture
6. Fuel line kinked or smashed closed

You can do a simple check to determine whether fuel is getting to the cylinder. First, remove the spark plug. Then pour a small amount of gasoline into the cylinder. If the engine fires a few times after you replace the plug, check the possible causes of a white and dry plug.

Compression

To check compression, first disconnect the high-voltage wire to the spark plug. Leave the spark plug in place. Then try to turn the flywheel by hand. The flywheel should bounce back as you reach the top of the compression stroke. If it doesn't, or if it offers only small resistance to turning, the compression is low.

Champion Spark Plug Co.

Champion Spark Plug Co.

Champion Spark Plug Co.

Fig. 21-23 This spark plug is fairly clean. Its insulator is brown or tan-colored. This plug indicates a proper air-fuel mixture.

Fig. 21-24 A wet or black spark plug indicates that there is too much fuel going to the engine.

Fig. 21-25 The insulator on this plug is white and dry. This means that the engine isn't getting enough fuel.

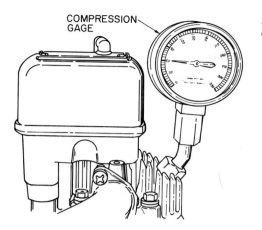

COMPRESSION
GAGE

Fig. 21-26 Use a compression gage to check cylinder compression.

You can make a more accurate compression test with a **compression gage**. See Fig. 21-26. First remove the spark plug. Then press the rubber part of the compression gage into the spark plug hole. While holding the gage in this position, crank the engine over. Continue cranking the engine until the gage needle stops moving. The gage reading will show the amount of compression in the cylinder. Compare the reading to the manufacturer's specifications.

The following problems can cause low cylinder compression:

1. Loose spark plug
2. Loose cylinder head bolts
3. Blown (damaged) head gasket
4. Poor valve clearance adjustment
5. Valve(s) sticking open
6. Cracked or warped cylinder head

It's possible that the engine trouble may not be caused by any of the above problems. In this case, you will probably have to **overhaul** the engine. This means "tearing down" the engine and checking all the parts. Major problems that can cause compression loss include the following:

- Burned valves or valve seats
- Badly worn or scored cylinder wall
- Worn or broken rings
- Broken connecting rod
- Warped valve stems
- Cracked cylinder

REPAIR PROCEDURES

During troubleshooting, you may discover that the engine requires a major repair. Some problems, such as burned breaker points, are more easily repaired than others. However, each repair procedure requires that you disassemble and re-assemble a part of the engine.

It's important to follow the engine manufacturer's procedures for any major repairs. Many procedures are the same for most small engines. However, there are some procedures that are used on only a few engines. The service manual for the engine you are repairing is the best source of information.

The following are general procedures to follow in repairing a small engine:

- Obtain the manufacturer's service manual.
- Find the repair you need to make in the manual.
- Study the procedures before you do anything. This way, you'll know whether you have all the tools you need to complete the repair.
- Assemble all the tools and cleaning supplies you'll need.
- Repair the engine as described in the manual.
- Clean and mark the engine parts as you make the repair. This allows you to re-assemble the engine in the reverse order of disassembly.
- After completing the repair, test the engine to make sure it runs properly.

STUDY QUESTIONS

1. What small engine systems and parts require routine maintenance?
2. True or false: An owner's manual generally contains more detailed information than a service manual.
3. Where can you find the proper oil-fuel ratio for a two-stroke engine?
4. Why is it important to check the oil with the engine on a level surface?
5. What's the reason for changing the engine oil regularly?
6. What two things are important in selecting oil for an engine?
7. Do multi-viscosity oils have a low viscosity or a high viscosity?
8. What's the main reason for servicing the cooling system?
9. Describe two ways of cleaning a dry-element air cleaner.
10. Which of the following types of fuel filters can be cleaned: in-line filter, fuel tank filter, glass bowl filter?
11. What is the function of the crankcase breather?
12. What tool should you use to remove a spark plug?
13. True or false: Leaf-type feeler gages give a more accurate reading of the spark plug gap than wire gages do.
14. Name the three types of carburetor adjustments.
15. What three things does battery service usually include?
16. Why should you drain the fuel system for long-term storage?
17. What three conditions does a small engine need for good operation?
18. What can the spark plug appearance tell about the fuel mixture?
19. Why is it important to consult the service manual when doing major engine repairs?

ACTIVITIES

1. Change the oil on a small four-stroke engine. If you don't have an engine at home, list the steps you would follow in changing oil. Use the procedures described in this chapter.
2. Check and clean the air filter and fuel filter on a small engine. Identify the type of filters used on the engine. If you don't have an engine, list the steps you would follow. Identify the type of filters used on your "pencil and paper" engine.
3. Select a small engine at your home or at a friend's home. List the steps needed to prepare the engine for winter storage (or, if it is a snow blower, for summer storage). If you don't have an engine available, visit a local store and prepare your report for one of the engines on display.

Advances in Energy, Power, and Transportation

What new energy sources will be available in the future? What new transportation devices will be developed? Will solar energy be a major source of power, or will something even more exciting emerge? The next three chapters will help you explore some of these questions *and* help you develop innovative (new and different) ideas of your own.

Solar Energy

Solar energy is one of our most promising alternative energy sources. Each day the sun provides us with 500 times more energy than we use. We could meet the United States' daily energy demand by capturing only one-tenth of this energy. Ideally, we could capture this energy with collectors on only 2 percent of our nation's surface. As you can see, solar energy has a very great potential.

The sun can provide energy for two common uses:

- **Heating** — We can use solar energy to heat both water and buildings. Over one-quarter of our energy usage is for these purposes. Therefore, this is an important area in which to develop solar energy.
- **Electricity** — Electricity has the most uses of any of our forms of energy. We can convert solar energy directly into electricity. We can also use it to generate steam, which can drive turbine generators. Scientists know *how* to use solar energy to generate electricity. But they must do more research and development to make solar-produced electricity economical.

HEATING WITH SOLAR ENERGY

People are already using solar energy for heating. This is happening all over the world. Fuels are becoming more scarce and expensive. Therefore, solar heating is becoming more popular.

Using the sun for home heating isn't new. The Greeks and Romans used the sun to heat homes nearly 2000 years ago. Greek builders positioned houses so that they would capture the sun's heat in the winter. The Romans used glass to let sunlight into their homes. The glass kept the sun's heat from escaping. The Romans also used the sun to heat private and public baths. Roman law even made it a crime to block the sunlight striking a person's home!

The twelfth-century Anasazi Indians also used solar energy. Figure 22-1 shows how they constructed their homes under cliffs. These cliff dwellings are still standing in the southwestern United States. The dwellings captured the maximum amount of low-angle winter sun. In the summer, the cliff shielded the homes from the intense high-angle summer sun.

In the United States, solar water heating dates back to the nineteenth century. Figure 22-2 shows an 1896 solar water heater on the

Fig. 22-1 **Early American Indian tribes used solar energy to heat their homes. These cliff dwellings were heated by the low winter sun.**

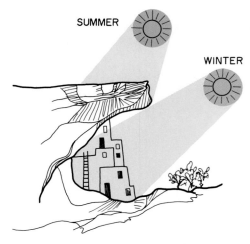

Pacific Gas & Electric

Fig. 22-2 **This photograph, taken in 1896, shows an early type of solar water heater. This system was called the "Climax" by its manufacturer.**

roof of a California home. Four water tanks collected heat from the sun. They supplied hot water for baths and other household uses. A wood stove provided additional hot water for laundry and heat during the winter.

During the early part of the twentieth century, more people used solar water heating systems. By 1941, at least 60,000 solar water heaters had been installed. However, after World War II there was a great deal of inex-

pensive electricity and natural gas. This caused a decrease in the use of solar water heaters. At one point, there were only about 5000 solar water heaters in use. However, gas and electricity prices have been increasing. Therefore, more people are using solar water heating systems again.

The Movement of Heat Energy

You'll learn about several different kinds of solar heating systems in this chapter. However, you should first know how heat energy moves from one place to another.

Heat moves naturally from warm areas to colder areas. It always seeks to balance its temperature. Suppose the air outside your home is colder than the air inside. In this case, the inside heat will seek every possible route to flow to the outside. When it is hotter outside than inside, heat will try to flow into your home.

Heat has three ways of moving from one place to another: conduction, radiation, and convection.

Conduction. The movement of heat through a substance is called *conduction*. Figure 22-3A shows heat moving through a solid wall by conduction. The denser the material, the more

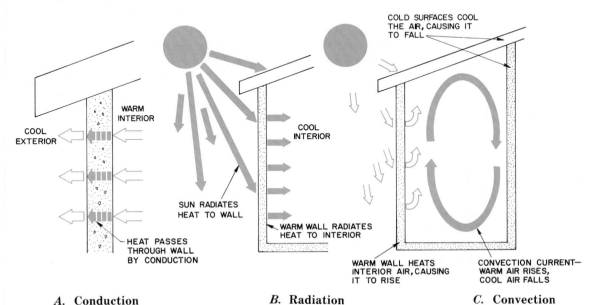

A. **Conduction**   *B.* **Radiation**   *C.* **Convection**

Fig. 22-3 **Heat moves by conduction, radiation, and convection.**

quickly the heat can move through it. (*Density* refers to how closely the molecules of a substance are packed together.) For example, concrete allows heat to pass more quickly than wood does.

Radiation. Heat can also move by electromagnetic waves. These heat waves can pass through space and air without being absorbed. This form of heat transfer is called *radiation*. Through radiation, heat moves from warm objects to cooler ones. It does this without warming the space in-between. Figure 22-3B shows how the sun heats a house by radiation.

Heat and light travel from the sun to the earth by radiation. An object gives off heat when solar radiation strikes it. We can trap this heat inside a building by taking advantage of the *greenhouse effect*.

You may remember the greenhouse effect from Chapter 4. Radiant energy passes freely through glass. However, when it hits the interior of a building, the radiant energy changes into heat energy. Heat energy has a longer wavelength than radiant energy does. It can't pass through the glass. In this way heat collects inside the greenhouse. See Fig. 22-4.

Convection. A third way that heat may be transferred is by *convection*. Convection is the moving of heat from warm areas to colder areas by air, water, or other fluids. Heated air or water will always move to a colder area. Once there, the fluid will lose its heat.

Figure 22-3A shows a house being heated by convection. First, solar heat passes through the wall by conduction. Then the wall radiates heat to warm the inside air. The warm air rises to the ceiling. When it touches the ceiling and opposite wall, the air transfers its heat to these cooler surfaces. The cooled air then sinks to the floor. Now, all this time, more air is being heated on the "sunny side" of the house. It rises, transfers heat, and sinks in a regular cycle. This movement of air is called a **convection current**.

Later on, you'll see how solar homes can use liquid convection currents to heat water.

Preventing Heat Loss

There are two basic requirements for the successful use of solar energy. First, the energy must be collected when it is available. Second, the energy must be held onto as long as possible. There is no place on earth where the sun shines *all* of the time. As one scientist put it, the sun is "a part-time performer in a full-time world." It's easy to heat a house with solar energy on sunny days. But a truly successful system captures and stores enough energy to supply heat at night and on overcast days.

Energy conservation is very important in solar heating systems. A drafty and poorly insulated house is hard to heat with solar energy. The house will be warm when the sun shines. However, the heat loss at night can be greater than the amount stored during the day. Therefore, the first step in creating a good solar heating system is to *minimize the heat loss*.

Figure 22-5 shows the many places in a house from which heat can escape. This heat loss can be reduced by insulating roofs, walls, and floors. Caulking cracks also helps. These conservation steps serve two purposes:

Fig. 22-4 **This glass-walled building illustrates the greenhouse effect. Short-wave energy from the sun passes freely through the glass. However, once it hits an object, it changes into longer heat waves. The long waves can't pass back through the glass. Therefore, the heat is trapped inside.**

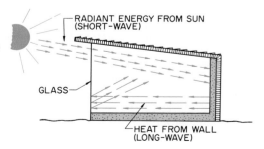

RADIANT ENERGY FROM SUN (SHORT-WAVE)

GLASS

HEAT FROM WALL (LONG-WAVE)

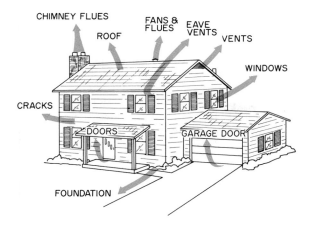

Fig. 22-5 This diagram shows heat losses from the home. Some heat loss is desirable during the summer. However, all heat loss should be minimized during the winter.

- They prevent heat loss during the winter.
- They keep heat from entering the house during the summer.

Passive Solar Heating

There are two types of systems used for solar heating: active and passive. **Active systems** collect solar energy. They also transfer the heat mechanically to where it is needed. **Passive systems** use building and landscaping design to collect, store, and transfer solar heat. In more technical terms, passive solar heating is *a system into which heat energy flows by natural means*. These natural means (ways) include conduction, radiation, and convection.

Passive Home Heating — Direct-Gain. The simplest kind of passive solar heating is heating living space with sunlight. See Fig. 22-6. This is called a **direct-gain** heating system. It collects heat directly from the sun.

To use a direct-gain system, a house must have large windows facing south. This allows the winter sun to heat the inside through the greenhouse effect. The glass traps the sun's radiant energy inside the house. The residents use curtains or insulated shades to control the amount of heat entering the home.

The house must also have some type of **thermal collector** for heat storage. The house

in Fig. 22-6 uses a heavy masonry (brick or stone) floor to collect and store heat. The thermal collector also radiates heat back into the living area. See Fig. 22-7. Heat storage is the key to successful passive solar heating. Other thermal collectors include water-filled containers, concrete, and adobe.

Pacific Gas & Electric

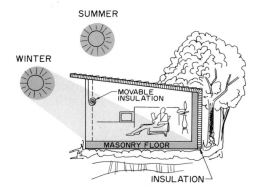

Fig. 22-6 A direct-gain solar heating system collects heat in the living area.

Mekler/Ansell Associates, Inc.

Fig. 22-7 In this home, sunlight shines through south-facing windows and heats a tiled floor. The tile then radiates heat into the living space.

In the northern hemisphere, the summer sun is much higher than the winter sun. To keep out the summer sun, houses with direct-gain systems use an **overhang**. (This is the same principle used by the Indian cliff-dwellers.)

Depending on the climate, direct-gain systems can provide part or all of the needed heat. Radiant energy will provide some heat even on cloudy days. However, the amount of heat will be much lower than on sunny days.

Most solar-heated homes have a secondary, or **standby**, heat source. This source is usually a gas or electric furnace or a wood stove. The standby source provides heat during extended cloudy conditions or extended cold spells.

Passive Home Heating — Indirect-Gain. Indirect-gain solar heating works differently than direct-gain heating. In an indirect-gain system, sunlight strikes a **thermal mass** located between the sun and the living area. For example, the thermal mass in Fig. 22-8 is a masonry wall. This wall collects the sun's energy during the day. A large glass window keeps the heat energy from radiating outside. The wall can then heat the living area by both radiation and convection. Here's how it works:

As the wall heats up, it radiates heat to the living area. The warmed air forces cooler air through the vents at the bottom of the wall. See Fig. 22-8A. In the space between the window and the wall, the cool air warms up and rises. It flows back into the living area through the top vents. A convection current heats the living area throughout the day.

The vents provide temperature control. The home-owner can close them when the house is warm enough. The vents can also be closed at night. At night, the warm wall radiates heat into the living area. See Fig. 22-8B.

Figure 22-9 shows how a house can be cooled in the summer with an indirect-gain system. Note that the bottom vent is open. Two other vents are open, too — one at the top of the window and one at the back of the house. The air between the window and the wall heats up and rises. It escapes to the outside through the top vent. This action pulls cooler air into the house through the back vent.

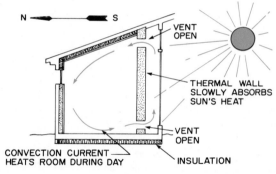

A. **Heating during the day**

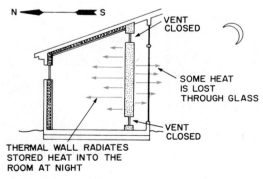

B. **Heating at night**

Fig. 22-8 **Indirect-gain home heating.**

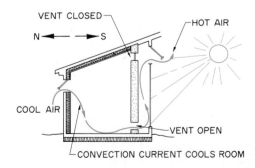

Fig. 22-9 **Indirect-gain heating systems can also be used to help cool the home. The sun creates a convection current that draws cool air through the house.**

Another form of indirect-gain system is a greenhouse attached to the home. See Fig. 22-10. The greenhouse provides both heat and a place to grow vegetables. Here's how it works:

Rocks and water-filled drums provide the thermal mass to collect heat energy. The greenhouse provides heat to the home by convection. Vents control the flow of heat and isolate the greenhouse when necessary. (For example, the home may be warm enough, or the greenhouse may be cool.)

Greenhouses can become overly hot during the summer. They should have vents that allow hot air to escape. Many greenhouses used for home heating also have shades that can close off the summer heat. This keeps the home cooler during the summer. A solar home can combine different types of indirect-gain designs. For example, the home in Fig. 22-11 uses both a greenhouse and a thermal wall.

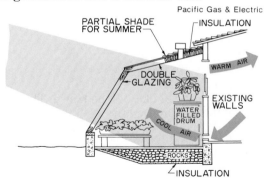

Pacific Gas & Electric

Fig. 22-10 **In this indirect-gain system, the attached greenhouse provides heat to the rest of the house.**

Fig. 22-11 **This solar home uses a greenhouse to help provide heat. The sun's heat is absorbed by a masonry wall inside the greenhouse. The wall then radiates heat into the home.**

Solar Design Associates

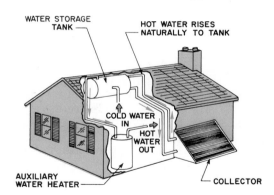

Fig. 22-12 **A thermosiphon water heating system.**

Passive Water Heating. A passive solar system can also be used to heat water. A passive solar water heating system consists of a water storage tank, a solar collector, and the necessary water pipes. See Fig. 22-12. Here's how it works:

First, local water pressure pushes cold water into the storage tank. The water flows down through a pipe to enter the collector. The collector absorbs the sun's heat and transfers it to the water. The warm water then rises to the storage tank through a different pipe. Meanwhile, more cold water enters the collector to replace the warm water.

When hot water is needed, it is taken from the top of the storage tank. As the water is used, it flows through a water heater. This heater operates only when the solar-heated water isn't hot enough. Otherwise, the solar-heated water just passes through the water heater and into the home's water supply.

The principle at work in this water heating system is called **thermosiphoning**. It is simply the circulation of water by a convection current.

Solar Collectors

Before we go on to active solar heating systems, let's take a look at a device that can be used in both passive systems and active systems. This device is the *solar collector*, or **solar panel**. A solar collector absorbs heat from the sun. It also transfers the heat to water or air inside of it.

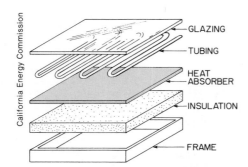

Fig. 22-13 An exploded view of a liquid-type solar collector.

Figure 22-13 shows an exploded view of one type of solar collector. Notice that it has five main parts. The **frame** supports the other parts of the collector. The **insulation** keeps the absorbed heat from radiating back to the surrounding air.

The **heat absorber** is an important part of the collector. It is the part that collects the sun's energy and converts it into heat. The heat absorber may be made of aluminum, copper, steel, or a combination of metals. Some absorbers are painted a flat black color. Others are coated with an absorbent material. This is done to increase the amount of solar energy absorbed.

Some heat absorbers, like the one in Fig. 22-13, have **tubing** fastened to them. Water circulates through this tubing. Other heat absorbers do not use tubing. Instead, they are made of two sheets of metal bonded together. There are narrow channels between the sheets. This allows the water to flow through the absorber.

Some heat absorbers are made to heat water, as just mentioned. Others are made to heat air. This second type simply has an air space above the metal plate. The air circulates through this space.

Solar collectors are sealed on top with **glazing**. Glazing is glass or fiberglass that is transparent or translucent. (A **transparent** substance is one that you can see through. A **translucent** material admits light, but it isn't as clear as a transparent material.) Through the greenhouse effect, the glazing traps the sun's heat in the collector.

Depending on the climate, the glazing may be single-pane, double-pane, or even triple-pane.

The more panes, the greater the reduction of heat loss.

Solar collectors should always face south when used in the northern hemisphere. They should also face directly toward the sun. This allows the collector to collect the most heat. The best angle for setting a collector depends on the location of the house. In the northern part of the country, the winter sun is very low in the sky. The collector must be set at a steep angle to collect the most sunlight. The further south the house, the lower the collector angle should be.

Active Solar Heating

Like a passive solar heating system, an active system collects solar energy. The difference is that an active system moves the energy mechanically to where it is needed. Active systems usually cost more than passive systems. However, they can provide better temperature control. Active systems are also more efficient. See Fig. 22-14.

Active solar heating systems usually have four main parts:

- A collector or collectors to trap the sun's heat
- A storage unit to hold onto the heat until it is needed
- A distribution system to deliver the heat to where it is needed
- Controls to regulate the system

Fig. 22-14 This home uses an active solar heating system. It has two banks of solar collectors on the side of the roof facing south.

See Fig. 22-15. Most active heating systems also have a **back-up** heating system. This may be an oil or gas furnace. The back-up system provides heat during cloudy or very cold weather.

There are two basic kinds of active solar heating systems. One kind uses hot water to heat the home. The other kind uses hot air. The two kinds of heating systems have much in common. We will look at each system separately.

Active Home Heating — Hot Water. Figure 22-16 shows a typical hot-water heating system. Water circulates through solar collectors on the roof. The sun heats the water. A pump then

California Energy Commission

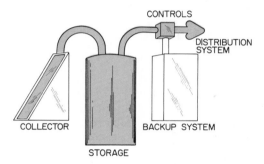

Fig. 22-15   The main elements of an active solar heating system are shown here in color.

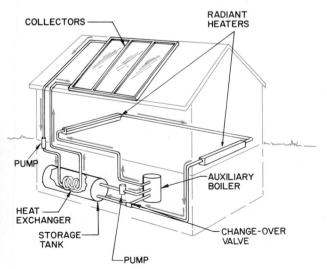

Fig. 22-16   An active solar hot-water heating system.

pumps the water to a **heat exchanger** in the storage tank. (A heat exchanger is a device that transfers heat from one fluid to another.) The hot water from the collector heats the water in the storage tank. The water is then pumped back to the collectors to be reheated. This cycle keeps the water in the storage tank hot.

When heat is needed in the home, a second pump sends heated water from the storage tank through baseboard **radiant heaters**. These heaters radiate heat from the water to the living area.

In cold or cloudy weather, the storage water may not get warm enough to heat the home. When this happens, a **change-over valve** shifts the system to an auxiliary boiler.

The design shown in Fig. 22-16 is just one of many possible active solar hot-water systems. Some systems use water pipes beneath the floor instead of baseboard heaters.

Active hot-water systems have another use besides home heating. They can also provide hot water for washing, bathing, and so on.

Active Home Heating — Hot Air. In a hot-air heating system, the sun heats air as the air flows through solar collectors. First a fan slowly blows air through the collectors. The air absorbs heat. Then it travels to a storage area in the basement. The storage area holds the heat until it is needed. See Fig. 22-17A.

Fig. 22-17 Solar hot air heating system. (Continued on p. 354)

A. **Heat moving from collector to storage**

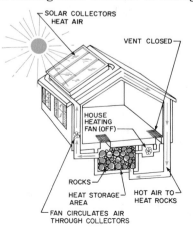

At night, when the house needs heat, another fan circulates air through the storage area and around the house. See Fig. 22-17B. An auxiliary heater provides extra heat if needed.

Hot-air heating systems are quite flexible. If the house needs heat during the day, air from the collectors can be circulated through the house without going to storage. See Fig. 22-17C.

Hot-air heating systems can also be used for summer cooling. During the night, outside air can be circulated through the heat storage area. This lowers the temperature of the rocks. During the day, the warmer air from inside the house can be cooled by circulating it over the rocks. The effectiveness of the system depends on how much cooling occurs at night.

(*Fig. 22-17* continued)

B. Heat moving from storage to heat the house

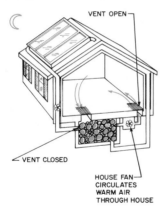

C. Heat moving from collector through the house

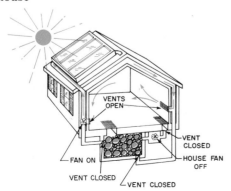

ELECTRICITY FROM SOLAR ENERGY

The generation of electricity from solar energy has been possible since the nineteenth century. However, all of the early generating devices produced only small amounts of electricity. They were also limited to special uses such as operating electrical controls.

The search for a way to convert large amounts of solar energy into electricity has intensified during recent years. Many electricity generating projects have been tried. The most successful ones are now producing electricity for our use. They are also being used to plan future solar energy generating plants.

The following sections describe both the conversion systems already in use and those that are under development.

Solar Cells and Batteries

Solar cells are devices that convert sunlight directly into electricity. Solar cells are also called **photovoltaic cells**. These cells use **photons** (light energy) to produce electricity. Solar batteries are simply groups of solar cells. Putting many cells together increases the electric output. See Fig. 22-18.

During the 1950s, Bell Telephone Laboratories developed the basic solar battery commonly used today. Most solar batteries have an efficiency of about 11 percent. This means that the battery can convert 11 percent of the solar energy striking its surface into electricity.

Fig. 22-18 **This solar battery contains 3600 individual solar cells. Together they can produce 13.1 watts of electricity.**
Bell Laboratories

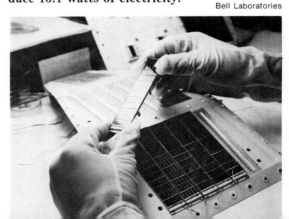

Solar Cell Construction. Solar cells are a type of semi-conductor. (Semi-conductors are covered in Chapter 7.) These cells have no moving parts. They are made mainly of silicon.

In the construction process, silicon is mixed with a small amount of arsenic. This mixture is then formed into a large crystal. The crystal is cut into thin wafers about 1/25 of an inch *(1 mm)* thick.

The wafer is then exposed to a vapor containing boron. The boron penetrates into the wafer about 1/10,000 of an inch *(1/400 mm)* deep. Next, two wires are attached to the wafer. One wire attaches to the silicon center. The other attaches to the boron-penetrated surface.

The silicon/arsenic center of the wafer has a different electrical characteristic than the boron part of the crystal. See Fig. 22-19. The silicon/arsenic layer is labeled **P** (positive). The boron layer is labeled **N** (negative). When light strikes the solar cell, it produces a voltage difference between the P and N layers. About 0.45 volt is developed. This electrical current, though small, can be used to do work. See Fig. 22-20. It takes over 100 solar cells to light one 100-watt light bulb. Also, solar cells are expensive to manufacture. As a result, electricity from solar cells is much more expensive than electricity generated at fossil fuel plants.

Scientists are now doing research in two areas to improve solar cells:

• Developing more efficient cells
• Developing less expensive cells

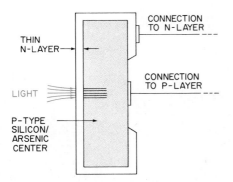

Fig. 22-19 **The construction of a solar cell.**

THIN
N-LAYER

CONNECTION
TO N-LAYER

LIGHT

CONNECTION
TO P-LAYER

P-TYPE
SILICON/
ARSENIC
CENTER

SOLARTS

Fig. 22-20 **When exposed to light, the solar cell on the left provides electricity to spin the model windmill.**

Solar cells cost less to manufacture now than they did a few years ago. Someday we may have solar cells that can produce electricity as cheaply as fossil fuel plants.

Uses of Solar Cells. Figure 22-21 shows a system that may be the home solar energy system of the future. This system combines solar electricity generation with solar hot-air heating. Using a single energy source for two useful purposes is called **cogeneration**. The panels on the roof contain solar cells. The cells convert sunlight to electricity. The electricity is either stored in storage batteries or is used in the home right away. Any excess electricity is sold to the power company for other uses.

Air circulates through the solar collectors as described earlier under *Active Home Heating — Hot Air*. The **heat pump** in the system provides both heat in the winter and air conditioning in the summer. The heat pump runs on electricity from either the solar cells or the power company.

Redrawn with permission of Shell Oil Co.

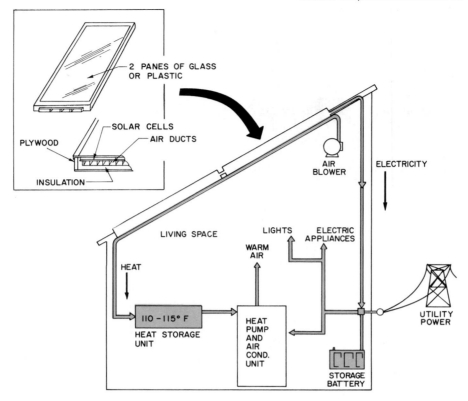

2 PANES OF GLASS OR PLASTIC

SOLAR CELLS

AIR DUCTS

PLYWOOD

INSULATION

AIR BLOWER

ELECTRICITY

LIVING SPACE

LIGHTS

ELECTRIC APPLIANCES

WARM AIR

HEAT

110 – 115° F

HEAT STORAGE UNIT

HEAT PUMP AND AIR COND. UNIT

STORAGE BATTERY

UTILITY POWER

Fig. 22-21 This home solar energy system supplies both electricity and hot air for space heating.

Figure 22-22 shows a solar power plant. This plant uses solar cells mounted on motor-driven trackers. Each tracker is 32 feet square and contains 256 solar batteries. The plant is fully automated. It starts itself up at sunrise and shuts down at sunset. The trackers slowly turn throughout the day to "track" the sun. This makes for the most efficient use of the solar cells.

Solar cells can also provide power in outer space. Figure 22-23 shows the solar cells used on a satellite orbiting the earth.

Solar Energy for Electrical Generators

The sun's heat can also be used to produce steam to power electrical generating plants. Special collectors are used to generate the high temperatures needed to produce steam.

Fig. 22-22 A solar (photovoltaic) plant that produces electricity.

NASA

Fig. 22-23 This satellite orbited the earth gathering basic information about the universe. Its electrical equipment was powered by panels of solar cells.

One generating system uses mirrors to reflect sunlight and concentrate it onto the top of a water tower. See Fig. 22-24. The heat of the concentrated sunlight converts water into high-temperature steam. The steam is then used to operate turbine generators.

The mirror plant is called Solar One. It has operated in California's Mojave Desert since 1982. It can produce up to 10 megawatts of electrical power. This is enough power to supply the needs of a town of 6000 people.

Fig. 22-24 Special mirrors called *heliostats* surround the central water tower at this solar power plant. They concentrate sunlight onto the water tower to produce steam.

Pacific Gas & Electric

Electrical Energy from Outer Space

In outer space, solar energy is constant. There are no clouds to block the sun's rays. If we could capture solar energy in space, then send it to the earth, we would have a constant supply of energy. Two systems have been proposed to accomplish this: orbiting mirrors and orbiting solar cells. Both systems would collect solar energy for the production of electricity.

Orbiting Mirrors. Collecting solar energy on earth is an inconsistent process. Collection is most efficient when the sun is directly above the collectors and shining straight down. This only happens around noon each day, and then only in the summer. Placing mirrors in orbit around the earth could solve this problem.

The proposed system for putting mirrors in orbit is called the **Space-Orbiting Light-Augmentation Reflector Energy System (SOLARES)**. For this system, astronauts would assemble giant lightweight mirrors in space. Then they would put the mirrors into orbit. The mirrors would constantly adjust to reflect sunlight onto a single location on earth. See Fig. 22-25. There is no night in outer space. Therefore, the mirrors would reflect sunlight constantly. The sun's energy would then be used to produce electricity in one of the ways described earlier.

NASA

Fig. 22-25 This artist's rendition shows how a ring of orbiting mirrors would reflect sunlight constantly to a collector site on earth.

Orbiting Solar Cells. Another proposed way to collect solar energy in space is the **Solar Power Satellite (SPS)**. This system is similar to the solar energy collection systems used in many of our manned and unmanned space flights. These systems use photo cells to convert sunlight into electricity.

Figure 22-26 shows an artist's view of the SPS system. First the satellite would convert sunlight into electricity. Then it would transmit this energy to earth. The energy would be in the form of **microwaves** (high-frequency radio waves).

Both SOLARES and SPS would capture solar energy on a constant basis. However, neither system has been developed to the testing stage.

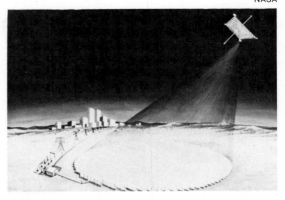

NASA

Fig. 22-26 Located 36,000 miles above the earth, the Solar Power Satellite would "beam" energy down to a collector site in the form of microwaves.

Solar-powered Transportation: Radiant Energy Leads the Way

As you learned in Chapter 1, solar energy is both heat energy and light energy. This **radiant energy** may hold some very interesting surprises for us in the future. We may be able to use the sun's energy for *transportation*, either directly or indirectly. Here's how:

- One indirect way to use radiant energy is to first use it to produce electricity. The electricity can then be used to produce hydrogen by the electrolysis of water. (Chapter 1 described this process.) Hydrogen is very flammable and can be used as a substitute for gasoline. Also, it is easy to convert an automobile engine to run on hydrogen. In fact, the U.S. Postal Service uses a vehicle that runs on hydrogen.

- What about direct ways to use solar energy? Well, vehicles equipped with photovoltaic cells have already been built and operated. For example, a solar-powered plane called the *Gossamer Penguin* was flown across the English Channel. And solar-powered cars have been driven across country. The *Quiet Achiever* shown in the photo is one example.

- Another direct solar transportation method is straight out of science fiction. However,

many practical inventions were once only science fiction. **Solar sailing** is a proposed space exploration system that would actually be powered by sunlight! Photons would fall onto a vast lightweight sail and provide a push. The force of the push would be very small — as you might imagine. But this force would build up and become greater than the thrust of the most powerful rocket engine.

Someday, long-distance space vehicles may have solar sails. The voyager spacecraft is racing for the outer reaches of our galaxy. With the constant sunlight of outer space, it could push on to the ends of the universe!

Yes, whether in our own transportation system or in the vastness of outer space, radiant energy may truly lead the way into the future.

STUDY QUESTIONS

1. What two common uses can the sun provide energy for?
2. Name three ways in which heat can move from one place to another.
3. What two things are needed for the successful use of solar energy?
4. What is the first step in creating a good solar heating system?
5. Name the two basic types of solar heating systems.
6. What is the key to successful passive solar heating?
7. Why do solar-heated homes have a stand-by heat source?
8. What methods of heat transfer are used in an indirect-gain heating system?
9. Define *thermosiphoning*.
10. Name the five main parts of a liquid-type solar collector.
11. Name the four main parts of an active solar heating system.
12. What are the two basic types of active heating systems?
13. Fill in the blanks: Solar cells convert ____ directly into ____.
14. Fill in the blanks: Groups of solar cells are called ____ ____.
15. How much voltage does a single solar cell produce?
16. Define *cogeneration*.
17. Name two proposed systems for collecting solar energy in outer space.
18. List three possible ways to use solar energy to power transportation devices (directly or indirectly).

ACTIVITIES

1. Identify the uses being made of solar energy in your community. You could do this by visiting your local power company office or by interviewing a builder or dealer who specializes in solar energy. (Check the yellow pages in your local telephone book.)
2. If you can identify one or more uses of solar energy, select one use for a brief report. In your report, explain how solar energy is being used in place of other energy sources. If you could not identify any present uses, prepare a report on the future of solar energy in your community. (The local power company may have some information.)
3. See if a local hobby store or department store sells solar kits. These "do-it-yourself" kits operate motors, cars, and other devices. Obtain and assemble a kit. Then demonstrate your project in class. (If enough students build solar cars, you could plan a "solar race.")

Nuclear Power

Nuclear energy is the energy released when matter is changed into energy. The amount of energy we can obtain from nuclear reactions is almost unbelievable. Albert Einstein showed the relationship between matter and energy with the equation $E = mc^2$. This equation reads *Energy equals mass times the square of the speed of light.* In simple terms, this means that tiny amounts of matter have the potential to produce giant amounts of energy.

Notice the phrase *nuclear* energy. The conversion of matter into energy takes place in the *nucleus* of the atom.

There are two types of nuclear reactions: fission and fusion. In **fission**, energy is produced when atoms are split apart. In **fusion**, energy is produced when atoms are combined. In both reactions, a small amount of matter is changed into a large amount of heat and light energy.

In this chapter, we will look at both fission and fusion. Fission is an important source of power at the present time. Fusion is a possible energy source for the future.

NUCLEAR FISSION

Figure 23-1 shows the fission of uranium. The nucleus of a uranium atom is being split by a neutron. The nucleus is *fissioned*. That is, it is split in two. This process produces energy. It also produces two or three free neutrons. To continue the reaction, at least one of the neutrons must split another uranium nucleus.

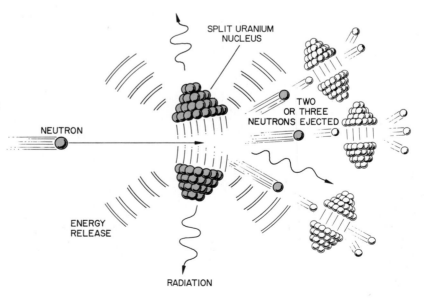

Fig. 23-1 **Nuclear fission is the splitting of the nucleus of a uranium atom with a neutron. Energy is released in the form of heat. Freed neutrons then cause a *chain reaction* to split other nuclei.**

SPLIT URANIUM NUCLEUS

TWO OR THREE NEUTRONS EJECTED

NEUTRON

ENERGY RELEASE

RADIATION

When this process continues, a "chain reaction" is produced.

In a nuclear reactor, millions of uranium nuclei are split into lighter materials each second. A single fission reaction can produce more than 20 million times as much energy as a single chemical reaction. This energy is mainly heat. The heat is used to boil water to produce steam. The steam is then used to operate a turbine generator.

Nuclear Fuel

Uranium is a natural radioactive substance found in the earth. It has two major forms, called **isotopes**. One of these isotopes, uranium-235 (U-235), is rare. It makes up less than 1 percent of the uranium found. The other isotope is uranium-238 (U-238). This is the most common type of uranium. However, by itself U-238 is not suitable for use in present-day reactors. First it has to be enriched with the rare isotope, U-235.

The uranium enrichment process is done in large, complex plants. This process is also the first step in creating nuclear weapons. Therefore, the plants are closely guarded. Enriched uranium is 97.5 percent U-238 and 2.5 percent U-235. This is the fuel of nuclear reactors.

Look back at Fig. 23-1. Notice that each fission reaction releases two or three neutrons. Only one neutron is needed to sustain a chain reaction. The remaining neutrons may do one of two things:

- They may combine with U-238 to form a substance called **plutonium**. See Fig. 23-2. Plutonium is *fissionable*. That is, neutrons can split its nuclei to release energy. In this way, plutonium can serve as reactor fuel.
- The neutrons may be absorbed by control rods to regulate the reactor's energy output. This process will be described later.

Power from Nuclear Energy

At the present time, the fission of uranium is the only developed source of nuclear power. This source supplies just under 3 percent of all our power. This amounts to 13 percent of the electricity produced in the U.S. each year. Nuclear power is produced in the form of electricity at **nuclear plants**. See Fig. 23-3.

Fig. 23-2 **When a neutron strikes the nucleus of U-238, a new substance — plutonium — is formed.**

Once a nuclear plant is constructed, its power output is less expensive than power from other sources. One pellet of uranium costs about $7 at 1985 prices. It has the energy potential of three barrels of oil ($84) or one ton of coal ($29). All fuel costs continually rise. However, oil and coal prices are expected to increase at greater rates than uranium. Therefore, in comparison, nuclear fuel is inexpensive. However, nuclear power plants cost much more to build than coal- or oil-burning plants. Nuclear plants also cost more to operate.

Light-Water Reactors

As you will learn, nuclear reactors use a liquid to transfer heat. Reactors that use ordinary water to transfer heat are called *light-water reactors*. (There are some heavy-water reactors, but they are not common. To transfer heat, heavy-water reactors use a rare type of water called **deuterium oxide**.)

Fig. 23-3 **This nuclear power plant in Plymouth, Massachusetts, generates 1100 megawatts of electricity. (One megawatt meets all the power needs for 1,000 people.)**

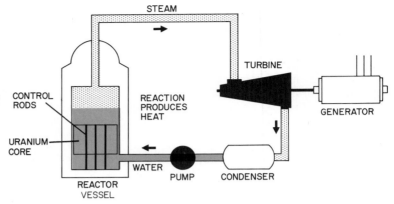

Fig. 23-4 A boiling-water reactor produces steam to operate a turbine and produce electricity.

There are two types of light-water reactors: boiling-water reactors and pressurized-water reactors.

Boiling-Water Reactors. The **core** of a nuclear reactor is the part that contains the fissionable material. In a boiling-water reactor, the core is placed inside a water-filled container called a **vessel**. See Fig. 23-4.

The heat produced by the fission reaction boils the water in the vessel. Steam collects at the top of the reactor. As the steam continues to be produced, the pressure increases. This pushes the steam into the turbine. The rotating turbine drives an electrical generator.

After the steam leaves the turbine, a condenser cools it. This changes the steam back into water. A pump returns the water to the reactor vessel. The water-to-steam-to-water cycle operates constantly to produce electricity.

Pressurized-Water Reactors. You may remember from Chapter 1 that water boils at

212° F *(100° C)* at sea level. However, it can reach much higher temperatures if it is kept under pressure. This is what happens in a pressurized-water reactor.

Figure 23-5 shows a pressurized-water reactor. Water surrounds the uranium core in the reactor vessel. The water is kept under high pressure. This permits the water to heat up well beyond its normal boiling point. A pump then moves the hot water to a heat exchanger.

A **heat exchanger** is a device that transfers heat from one fluid to another. See Fig. 23-6. The reactor produces high-temperature, high-pressure water. This water passes through the coiled pipe inside the heat exchanger. The water surrounding the pipe is at a lower pressure and temperature. Heat from the reactor water passes through the walls of the pipe and heats the low-pressure water.

The heat of the reactor water changes the low-pressure water into steam. The steam then

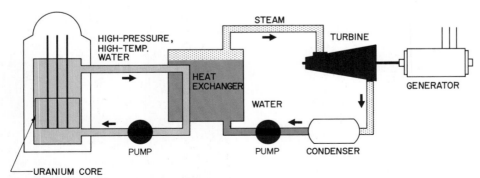

Fig. 23-5 Pressurized-water reactors use a heat exchanger and two separate water systems.

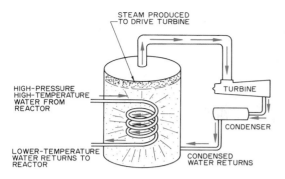

STEAM PRODUCED
TO DRIVE TURBINE

HIGH-PRESSURE
HIGH-TEMPERATURE
WATER FROM
REACTOR

TURBINE

CONDENSER

LOWER-TEMPERATURE
WATER RETURNS TO
REACTOR

CONDENSED
WATER RETURNS

RADIOACTIVE SIDE   NON-RADIOACTIVE SIDE

Fig. 23-6 In a pressurized-water reactor, the heat exchanger connects two separate water systems. Heat from the radioactive side produces steam in the non-radioactive side. This steam is used to drive a turbine generator.

drives a turbine generator. A condenser condenses the steam. A pump then sends the water back to the heat exchanger to repeat the process.

What happens to the high-pressure water? After passing through the heat exchanger, it returns to the reactor to be reheated.

As you will learn, one of the harmful by-products of fission is **radiation**. In a pressurized-water reactor, the heat exchanger acts as a barrier against radiation. It keeps radiation from passing to the turbine side of the system.

Pressurized-water reactors are used in both power plants and ships. See Fig. 23-7.

Controlling Reactor Output. The reaction in nuclear reactors is usually controlled with **control rods**. Control rods are made of boron, silver, and other materials. Their job is to absorb free neutrons. This limits the chain reaction.

Power plant workers can slow down or stop the reaction by inserting the control rods into the uranium core. They can speed up the reaction by pulling the rods out. (Of course, this is all done by remote control.)

Reactor Charges and By-Products. The fuel for a nuclear reactor is called the **charge**. The charge consists of packs of 1/2-inch diameter **fuel rods**. Together, many fuel rods make up a **fuel unit**.

Figure 23-8 shows a fuel unit being lowered into a reactor. A typical reactor core consists

Fig. 23-7 This nuclear-powered aircraft carrier uses a small pressurized-water reactor.

of many fuel units. Thousands of fuel rods are used in the core. A typical charge costs more than 10 million dollars ($10,000,000)!

One harmful by-product of a fission reaction is **radiation**. See Fig. 23-1 again. Radiation consists of sub-atomic particles and high-energy rays. Radiation is dangerous. It makes reactor water radioactive. Special shields protect workers from the reactor's radioactivity. These shields are made of lead and concrete.

Another by-product of a fission reaction is plutonium-239. The formation of plutonium was described earlier. Plutonium is a fissionable material. Therefore, it takes part in the chain reaction. In this way, it helps provide energy for the production of electricity.

A typical reactor requires a fuel change once every three years. During the three-year period, the U-235 is slowly used up. However, the plutonium slowly builds up. As it builds up, the

Fig. 23-8 Lowering a fuel unit into a nuclear reactor. The photograph was taken looking down. To get an idea of the size of the reactor, notice the worker in the upper right portion of the picture.

plutonium is also used as fuel. By the end of the three-year period, there is actually more plutonium taking part in the chain reaction than U-235.

After the fuel is removed, it is separated into three materials. They are uranium-238, plutonium, and water products.

The uranium-238 can be enriched with U-235 and re-used in the reactor. At the present time, there is only limited use for plutonium and waste products. However, one special type of nuclear reactor uses both U-238 and plutonium as fuel. This reactor is called the **breeder reactor**.

Using breeder reactors is one way to deal with some of the by-products of light-water reactors. However, breeder reactors cost much more to build. They also produce, or "breed," more plutonium than is used for fuel. Plutonium is an extremely dangerous substance. Therefore, many people are concerned about the production of more of it.

Safe Operation — The Controversy

The nuclear age began when two atomic bombs were exploded in Japan during World War II. Nuclear power has been *controversial* since its beginning. That is, people have strong feelings for and against nuclear power. The concerns now center on three major topics, or **issues**:

- The danger of radioactivity
- The disposal of nuclear wastes
- The creation of large supplies of plutonium easily used in making nuclear weapons

Radioactivity. Great efforts have gone into nuclear reactor safety. However, there is still the danger of accidental release of radiation. In routine plant operation, there is little or no release. A well-operated nuclear power plant is safe and very quiet. However, accidents can happen. The 1979 accident at the Three Mile Island plant in Pennsylvania released some radiation. The effect of this release will not be known for years.

Scientists generally agree that all radiation is harmful. However, they also point out that we are constantly exposed to natural radioactivity. The sun and radioactive materials in the earth are two natural sources of radiation. When operating correctly, a nuclear power plant does not add much to the normal radioactivity of an area.

Nuclear Wastes. Nuclear wastes consist of new substances created during fission. Nuclear fuel is processed after it is used. During this processing, the wastes are separated out for disposal. Many people are opposed to nuclear power because of the danger involved in the disposal of nuclear wastes. Even supporters of nuclear power agree that we have not given enough attention to the proper disposal of nuclear wastes.

Nuclear wastes are very radioactive. They decay (release radiation) for hundreds of years before they become harmless. Therefore, nuclear wastes disposal must be a long-range solution. The problem is to dispose of the wastes safely. We must protect not only ourselves but our children and grandchildren.

Misuses of Plutonium. Plutonium can easily be made into nuclear bombs — a serious threat to the world. Unfortunately, there is already a large supply of plutonium. (You will remember that it is a by-product of light-water reactors.) As nuclear fuel is reprocessed, the plutonium supply will grow. Plutonium stockpiles must be guarded against theft by terrorists.

Decision Making. Our nation is faced with the need to constantly make decisions about nuclear power. We already have vast stockpiles of nuclear wastes and plutonium. Soon we may have nearly 150 nuclear plants in operation. We must all clearly understand the benefits and problems of nuclear power. Nuclear power will continue to be controversial. As citizens, we will have to make responsible decisions about its use.

At the present, there are no new nuclear power plants being planned in the U.S. Of the ones now under construction, two things are being done. Some plants are being completed and put into operation. Others are being delayed or cancelled.

Other countries are expanding their nuclear power capacity. For example, France is building 10 large new plants each year. By 1990, 50 percent of the electricity produced in France will be from nuclear plants. Other countries are planning similar increases in nuclear power. In the U.S., the only increase will come from the plants now being built. By 1990, we can expect about 15 percent of our electricity to come from nuclear plants.

NUCLEAR FUSION

Fusion is often called the "ultimate energy source." Fusion is a nuclear reaction in which two lightweight nuclei join to form one heavier nucleus. See Fig. 23-9. When this happens, a tremendous amount of heat and light energy is released. This is the same reaction that happens in the sun.

Scientists understand the theory of fusion reactor operation. They have also successfully conducted fusion experiments. However, the construction of an experimental reactor is still far away. The research problem is to develop a *continuous* and *controlled* fusion reaction.

The Fusion Reaction

The fuel for a fusion reaction is hydrogen. Hydrogen is very abundant. It can be extracted from ordinary water. For hydrogen atoms to join, their nuclei must be forced together at extremely high temperatures and pressures. These temperatures and pressures are like those inside the sun. The problem is to create sun-like conditions on earth. These conditions are extremely difficult to produce.

Scientists produce a fusion reaction by combining two isotopes of hydrogen: deuterium and tritium. The common hydrogen atom has a nucleus with one proton and no neutron. When a neutron is added to the proton, the hydrogen becomes **deuterium**. When a second neutron is added, **tritium** forms. The deuterium occurs naturally. It makes up a small part of the hydrogen in water. Tritium, however, is produced by bombarding the element **lithium** with high-energy neutrons.

In the fusion process, deuterium and tritium are heated millions of degrees. They become a hot gaseous mixture called **plasma**. At a high-enough temperature, the isotopes combine to

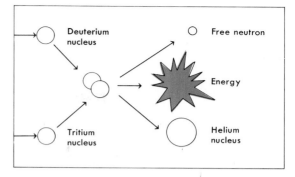

Fig. 23-9 In a fusion reaction, two isotopes of hydrogen — deuterium and tritium — combine to form *helium*. This reaction releases a huge amount of energy.

form helium nuclei. Each fusion reaction releases a fantastic amount of energy. See Fig. 23-9 again.

Containing the Fusion Reaction

The main problem in sustaining a fusion reaction is to *contain* the tremendously hot plasma. The plasma really cannot be contained in a solid container. Even if the solid were heated, it would be cold compared to the plasma. Therefore, the solid would cool the plasma, and the reaction would not continue.

Scientists are experimenting with two ways of containing the hot plasma: magnetic confinement and inertial confinement.

Magnetic Confinement. Russian scientists were the first to develop the magnetic method of confinement. Figure 23-10 shows a magnetic confinement test reactor. This type of reactor is also called a **tokamak**. *Tokamak* is the Russian word for *strong current*.

Fig. 23-10 A tokamak test reactor. The electromagnets create a powerful magnetic field. The magnetic field contains the hot plasma inside the donut shape.

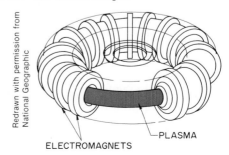

ELECTROMAGNETS   PLASMA

365

A tokamak uses huge electromagnets to generate a powerful magnetic field. This field is needed to contain the plasma. The reactor also requires a large transformer. The transformer induces an electrical current in the plasma. This current does two things:

- It creates another magnetic field that helps hold the plasma in the reactor.
- It heats up the plasma to the temperature at which fusion can take place.

Inertial Confinement. The inertial confinement process involves using tiny glass spheres filled with deuterium and tritium. Each of the spheres is a small fuel pellet. The idea is to bombard the fuel pellets with pulses of high-energy light. The light quickly raises the temperature of the fuel. The heat produces tiny fusion explosions as the two isotopes combine. Of course, energy is released during the explosions. Using many fuel pellets can provide a tremendous amount of energy.

The high-energy light used in inertial confinement is produced by a **laser**. Lasers are described in detail in the next chapter. You will also see a picture of lasers being used in the inertial confinement process.

Scientists have not yet produced the kind of sustained and controlled reaction they are looking for. Most researchers believe that a successful fusion reactor will not be built until after the year 2000. A commercial reactor may not exist until 2050.

Reactors in Orbit

The need for large amounts of electrical power for space projects is growing. Most satellites now need only about one kilowatt of power. This amount is easily produced by solar cells.

The space shuttle now allows us to place more complex and larger satellites into orbit. These satellites may need 100 or more kilowatts of power. This much power is beyond the capability of existing space power stations.

Scientists are studying the possibility of placing small nuclear reactors into orbit around the earth. These reactors would be about the size of a small car. They would be put into orbit by astronauts from the space shuttle. The reactors would then be connected to satellites to provide power.

The possibility of nuclear reactors in space causes concern among many people. They fear that a faulty reactor might spread nuclear waste over part of the earth. In fact, this very thing happened in 1978 and 1983. The Soviet Union had nuclear-powered satellites in orbit. They accidentally re-entered the earth's atmosphere and spread some contamination. This is why safety is such a big concern. As with any new technology, there are both advantages and disadvantages to consider.

NUCLEAR AUXILIARY POWER

The nuclear wastes produced by nuclear reactors can supply small amounts of power for special situations. Only a very small amount of waste products can be used in this way. There are two devices that can use nuclear wastes to develop power: the nuclear battery and the thermoelectric coupling.

Nuclear Battery

The nuclear battery is a simple device. It uses a beta-emitting radioactive material. (A **beta particle** is a high-speed electron. To *emit* means to give off.) If the material emits enough electrons, electrical current can be produced. See Fig. 23-11.

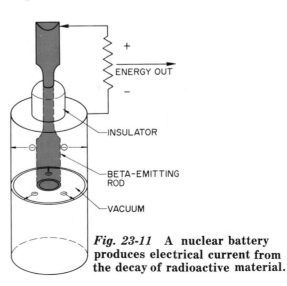

Fig. 23-11 **A nuclear battery produces electrical current from the decay of radioactive material.**

The central part of the battery is a rod coated with the beta-emitting material. The battery container is vacuum-sealed. Electrons flow from the rod to the container wall. This produces current. Voltages from nuclear batteries can be very high. However, the actual power output of electricity is very small.

Nuclear batteries are presently used to provide power for ocean buoys and other aids to navigation. These devices require only small amounts of electricity. With nuclear batteries, the devices can be placed in remote locations.

Thermoelectric Coupling

A thermoelectric coupling can be used to generate electricity from any heat source. "Any heat source" includes the decay of radioactive materials. A thermoelectric coupling, or **thermocouple**, uses two different materials joined together. When one end is heated, an electrical current is generated. See Fig. 23-12.

A thermocouple system is very inefficient. Only 1 to 5 percent of the heat is converted into electricity. However, thermocouples have advantages that make them very useful sources of electricity. They have no moving parts, and they are silent.

Nuclear thermocouples are called **radioisotopic thermoelectric generators,** or **RTGs.** These generators are widely used to provide power for satellites.

Fig. 23-12 **A thermocouple converts heat into electricity.**

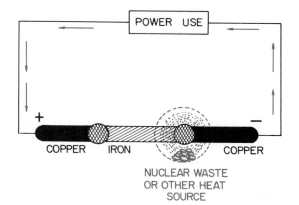

STUDY QUESTIONS

1. What is nuclear energy?
2. What are the two types of nuclear reactions?
3. What does it mean when something is *fissioned*?
4. What is done with the heat from fission in a nuclear reactor?
5. Name the two major isotopes of uranium.
6. Name two types of light-water reactors.
7. What is a heat exchanger?
8. How is reactor output controlled?
9. Name a harmful by-product of the fission reaction.
10. What happens in a fusion reaction?
11. What is the main research problem in nuclear fusion research?
12. Name two devices that can produce small amounts of power from nuclear wastes.

ACTIVITIES

1. Determine the source of the electricity used in your community. Is it a coal-burning power plant, a nuclear power plant, a hydro-electric power plant, or some other source? Find out the location of the power plant. What is the normal kilowatt output of the plant? (Hint: You can obtain this information from your local power company.)

2. Determine the location of the nearest nuclear plant. Are there nuclear plants under construction near your community? Prepare a report comparing the advantages and disadvantages of nuclear power plants.

Emerging Transmission and Transportation Systems

Using light to transmit power and perform work is already a proven technology. Concentrated light beams **(laser beams)** have many uses in industry, medicine, and research. We can expect this list to grow rapidly.

When materials become very cold, their properties often change. The study of materials at very low temperatures is called *cryogenics*. As some conductors become very cold, they become "superconductors." They conduct electricity with almost no power loss due to friction. This discovery has opened many possibilities in transmission and transportation. The possibilities include superconducting generators and magnetic-powered trains.

Cars, planes, and trains are all being improved in design and fuel efficiency. As fossil fuels become scarce, new types of engines and fuels

will be developed. In the future, hydrogen may replace gasoline. Electric motors may replace internal-combustion engines.

Each of the above topics — lasers, cryogenics, and new transportation devices — is covered in this chapter.

LASERS

A laser beam is a concentrated beam of light that travels in a very narrow, straight path. See Fig. 24-1. The word *laser* comes from **L**ight **A**mplification by **S**timulated **E**mission of **R**adiation.

Up until recently, power transmission has involved only mechanical devices, fluids, or electrical wire. Lasers are different. They can transmit power through the atmosphere without using any of these things. Also, the heat of laser beams can be controlled. It can range from very low temperatures to temperatures as hot as the sun. Lasers were first developed in 1960. They have already become an important device in power transmission.

Principles of Laser Operation

Laser beams consist of electromagnetic waves. These waves are at or near the wavelength of visible light. Normal light consists of waves of a variety of lengths. This produces mixed, or **multiple-direction**, light. The waves in a laser beam have almost identical lengths. Therefore, a laser beam can travel in a straight line for miles with very little spreading. This identical wavelength characteristic is called **coherence**.

Radio wavelengths are measured in yards (or meters). Television wavelengths are measured in feet (or tenths of a meter). The laser wave-

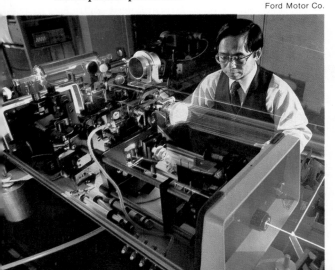

Fig. 24-1 Laser light beams (concentrated beams of light) are being used here to study atmospheric pollution.

Ford Motor Co.

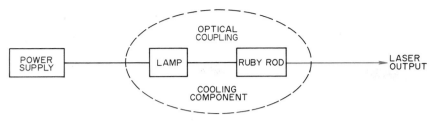

Fig. 24-2 **The components of a typical ruby laser.**

length is very short. It is measured in millionths of an inch. With their short wavelength, lasers can carry more messages than radio or television signals can. This makes them excellent carriers of telephone conversations.

The first laser beam was produced by a device that used synthetic rubies. The **ruby laser** is still in common use. Laser beams can also be produced by devices that use liquids, gases, and semiconductors. The result is the same: These lasers all produce an intense beam of light that travels in a straight line with almost no spread. Figure 24-2 shows the parts of a typical ruby laser.

In operation, the laser's power supply provides an electric current. This current produces a flash in the lamp. The flash develops electromagnetic waves along the ruby rod. One end of the rod is coated with a reflecting film. The other end is partially coated. The light reflects back and forth between the reflecting surfaces. The light finally escapes past the partially coated end as a high-intensity, coherent beam of light. The beam can be focused with a lens and put to work.

Lasers in Communication

Up until recently, laser beams have always been transmitted through the atmosphere. The use of lasers in communication systems has introduced a new form of transmission. This new form is called **fiber optics**. Lasers in fiber-optic systems use a semiconductor crystal. The laser is about the size of a pocket calculator. It produces one milliwatt of power. This is about 60,000 times less than the power used by a 60-watt light bulb.

In operation, a fiber-optic laser flashes bursts of light along hair-thin strands of flexible glass. Over 1000 telephone conversations can be carried over a single strand.

Fiber-optic systems have several important advantages over regular telephone systems. Cables made up of glass strands take up much less space than copper cables. Also, light signals need to be regenerated (strengthened) only about every 6 miles *(10 km)*. This is much farther than the regeneration distance in regular systems. There are plans to join the United States and Europe with a fiber-optic system.

Fiber-optic systems can also work well alongside electrical conductors. Fiber-optic strands and electrical wires can share space without either system affecting the other. Therefore, fiber-optic lines can be run anywhere that electrical lines are run. See Fig. 24-3.

Fig. 24-3 **Technicians here are preparing to install fiber-optic cable beneath the Chicago business district. The cable slides into a protective case already installed underground.**

Courtesy of Bell Laboratories

Fiber-optic telephone lines are being tested in a number of places throughout the country. The increased message-carrying capacity of fiber optics will have tremendous impact. It will permit transmission of motion pictures, television, and complex computer information. (All of these types of transmissions are now possible. However, they require many more telephone circuits.) The greater capacity of fiber optics will also make video telephone calls possible. In fact, we may see classes taught and conferences held using fiber-optic telephone lines.

Lasers in Industry

Lasers are being used more and more in industry. They can drill precision holes without distorting surrounding materials. In drilling, the laser first melts the material. It then vaporizes the melted substance. This method is used to drill holes in materials such as rubies and diamonds. These materials are too hard to be drilled with regular tools. Lasers can also drill very soft materials. For example, holes in nipples for babies' bottles are drilled with lasers.

Lasers can also be used to join metals by brazing or welding. Very small, thin metal pieces can be joined without any distortion from heat.

Lasers can also be used to *cut* materials. A laser can cut extremely complex designs in very hard materials. See Fig. 24-4. In a production situation, the laser can be programmed to repeat the cuts.

Heat treating of metals is also possible with lasers. Wherever heat is needed, lasers can provide it with a precision impossible with other methods.

Future Uses of Lasers

In Chapter 23, you learned about the inertial confinement method of producing nuclear fusion. This method requires the use of the world's most powerful lasers. One laser used in research is called NOVA. When it is fully developed, NOVA will produce a beam of light with 150 *trillion* watts of power. The beam will last for just a fraction of a second. However, in this mini-second, it will duplicate the temperature of the sun. The result, of course, will be a fusion reaction. Several different inertial confinement projects are now underway. See Fig. 24-5*.

There are many other uses for lasers. In eye surgery, they already permit operations never before possible. In the future, lasers will be used to open blocked arteries, and even to slow or stop cancer!

Lasers may also have a future role in national defense. Lasers stationed in space or mounted

*This work (shown in Fig. 24-5) was supported by the U.S. Department of Energy Office of Inertial Fusion under contract DE-FC08-85DP40200 and the Sponsors of the Laser Fusion Feasibility Project at the Laboratory for Laser Energetics.

Fig. 24-4 **Here a production laser is being used to cut saw blades out of steel.**

Fig. 24-5 **In this experimental fusion device, 24 lasers are used to produce a fusion reaction. They concentrate a 12-trillion-watt beam of light onto tiny glass spheres filled with deuterium and tritium.**
Laboratory of Laser Energetics

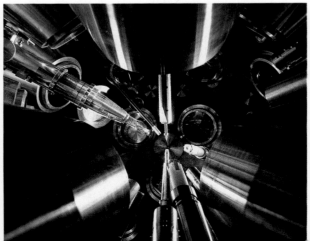

National Geographic Association

on mountain tops could be used to intercept enemy missiles.

The list of uses for lasers grows each year. Keep in mind that lasers harness *light*, a basic source of energy. At one time in history, people harnessed steam. The result of this was the Industrial Revolution. Someday, lasers may be seen as one of the greatest developments of the twentieth century.

CRYOGENICS

Cryogenics is a science that deals with the properties of materials at extremely low temperatures. Cryogenic temperatures range from −150° F *(−101° C)* to absolute zero (−460° F or *−273° C)*. As the temperature drops into this range, air turns into a liquid and then freezes solid. Most metals become brittle enough to shatter like glass. Other metals, such as lead and mercury, react differently. At cryogenic temperatures, lead can be made into springs and clear-tone bells. Mercury can be made into a hammer.

There are many industrial and commercial uses for cryogenic refrigeration. Super-coldness is used to prepare freeze-dried foods and freeze living cells. It is used in the manufacture of computer circuits and in space research. Cryogenic refrigeration also has an important future in the generation and transmission of power, and even in transportation.

In general, metal conducts electricity better as it becomes colder. This is because the electrons have less interference from molecular motion. In some metals, a strange thing happens as the temperature drops below −418° F *(−250° C)*: All detectable electrical resistance suddenly vanishes. The metal becomes a **superconductor**. In one experiment, a superconductor conducted a current for two years without additional current input. The current flow was then stopped by warming the metal.

Scientists are mainly interested in possible uses for superconductors:

- The production of super-efficient electrical generators and magnets
- The super-efficient transmission of electricity

Cryogenic Generators and Magnets

Figure 24-6 shows an experimental superconducting generator. It can produce 18 million watts of alternating current. This is enough power for a community of 20,000 people. The superconducting generator is only half the size and weight of a regular generator. However, it is much more efficient.

Superconducting magnets can develop tremendously powerful magnetic fields using very small amounts of current. In the section titled *ADVANCES IN TRANSPORTATION*, you will learn about the use of superconducting magnets in a new type of rail transportation.

Cryogenic Power Transmission

Savings in transmitting electricity with superconductors can be very great. Superconductors can transmit electricity for long distances without power loss. Superconducting wires are also much smaller than ordinary wires.

At the present time, power plants must be located fairly close to the area of use. Fuel must be transported to the plants, sometimes over great distances. With a superconducting transmission system, electricity could be generated at power plants near the fuel production site. Then the current could be transmitted through superconductors to the areas of use.

Fig. 24-6 **Preparing a superconducting generator for use.**

High Technology Magazine

ADVANCES IN TRANSPORTATION

In the future, we will see many changes in transportation. In the next century, fossil fuels may become more scarce. As they become scarce, they will also become more expensive. We are already seeing the development of more efficient automobiles and airplanes. Fuels that can be manufactured using alternative forms of energy will become more common. (Hydrogen is one example.)

New technology is being applied to several transportation modes. These modes include rail, highway, air, and even space transport!

Rail Transport

You may recall from Chapter 12 that most trains use diesel-electric engines. Compared to what rail transport could be, these trains are slow and noisy. They also produce a rough ride. These disadvantages can be overcome by using a new transportation idea: levitation. (To *levitate* means to *rise in the air*.) Levitation is accomplished by applying cryogenics to traditional rail transport.

Fig. 24-7 **A Maglev train rides on "rails" of electromagnetic waves. A magnetic force lifts the train above the track and then pushes it forward.**

In one **Maglev** (**Mag**netically **lev**itated) system, superconducting electromagnetic coils are mounted on the bottom of the train. Other coils are mounted on the road bed. When the train moves over the road coils, it induces a current in them. The road coils become electromagnets with the same polarity as the train coils. The two types of coils then repel each other. The repulsion force is strong enough to raise the train. See Fig. 24-7.

Trains using a Maglev system can reach speeds of over 300 miles an hour. They also provide a quiet and smooth ride. Several countries are at work on Maglev systems, including Japan and Germany. The British have already established the first commercial Maglev railway route. In the U.S., studies are being done on the **feasibility** of setting up Maglev routes. (An idea or plan is *feasible* when it is practical and economical.)

Highway Transport

Great advances have been made in producing more efficient automobiles. Today's cars are smaller, and their engines are more fuel-efficient. Cars can now travel twice as far on a gallon of fuel than they could in the 1960s.

Magnetbahn Transrapid

Future changes in highway transport will involve developing engines that use renewable sources of energy. Engines that burn methanol and synfuels are already in use. (These fuels were described in Chapter 1.) Engines that use hydrogen as a fuel have also been built.

Another major change would be the replacement of internal-combustion engines with electric motors. These motors would be powered by fuel cells or batteries.

Fuel Cells. Fuel cells are much like regular batteries. Both fuel cells and batteries convert chemical energy into electrical energy. Both devices also use an **electrolyte** (a solution or current-carrying paste). However, the overall chemical reaction in a fuel cell is very simple: Hydrogen and oxygen combine to produce water and electricity.

Figure 24-8 shows the basic parts of a fuel cell. The cell shown is the most common type. Hydrogen and oxygen are both fed into the cell from high-pressure containers. The hydrogen is the fuel of the fuel cell. The oxygen acts on the hydrogen and is called the **oxidizer.**

The fuel cell has two electrodes: the **fuel** (negative) and the **oxident** (positive). There is an electrolyte between the electrodes.

In operation, hydrogen is fed constantly into the cell. At the fuel side, the hydrogen atoms release electrons. The hydrogen, less its electrons, then travels through the electrolyte. When the hydrogen particle reaches the positive electrode (oxident), it combines with the oxygen to produce water.

Meanwhile, the electrons released by the hydrogen atoms must be replaced. The replacement electrons come from the positive electrode (oxident). To maintain an electrical balance, the electrons released at the negative electrode (fuel) travel through the external circuit. This produces a current to power the electrical device. See Fig. 24-8 again.

Notice that the only by-product of fuel cell operation is water. In addition, fuel cells are very efficient. They convert about 70 percent of the available chemical energy into electricity. (This compares to an efficiency rate of about 20 percent for automobile engines.) Fuel cells are also very dependable. However, more research

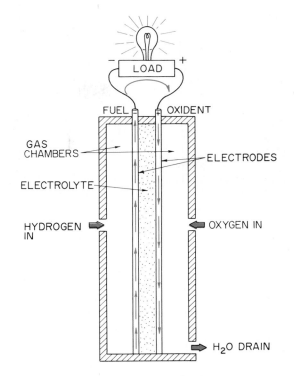

Fig. 24-8   **The basic parts of a fuel cell.**

and development is needed to make fuel cells a practical substitute for heat engines.

Electric (Battery) Vehicles. Electric cars are not new. In the early days of the automobile, electric vehicles outnumbered and outperformed cars with internal-combustion engines. However, by 1920 electric cars had disappeared.

Now, because of a growing shortage of fuel, electric cars are beginning to reappear. The United States Postal Service uses hundreds of EVs (electric vehicles) on its delivery routes. For many years, EVs have been used within large companies, on golf courses, and for a variety of industrial uses. See Fig. 24-9.

Now, electrical power is being considered for passenger cars. Electric vehicles don't have problems with performance. The electric motors used in them are efficient and powerful. The

Hyster Co.

Fig. 24-9 **This electrically powered fork lift is quiet and nonpolluting.**

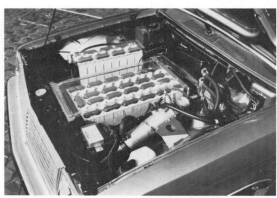

Delco-Remy Div., GMC

problem is the same one that caused the electric car to disappear in the 1920s: The driving range between battery charges is too short.

In an electric car, batteries store the energy for driving. The range of the car is limited by battery capacity. It is also limited by the number of batteries a car can carry. However, an electric car with a range of 80 miles *(128 km)* could be used for 90 percent of all daily trips. This range will probably be accomplished with the use of new lighter-weight batteries. See Fig. 24-10. In addition, some electric cars will be equipped with a **regenerative braking system**. With this system, the car itself recharges the batteries. The power for recharging is produced by a generator that connects with the wheels during braking.

With further research and development, electric cars will offer two major advantages over present automobiles. The benefits would be in cost-effectiveness and pollution control.

Air Transport

In the past, changes in aircraft design have often increased speed or size. In the future, the design emphasis will be different. Designers will strive for aircraft that are more efficient, comfortable, and safe.

Fuel economy especially is a major concern of airline companies. Commercial planes in the U.S. use more than 10 billion gallons of jet fuel each year. Fuel is one of the largest expenses of air travel. Therefore, an aircraft that is five to ten percent more efficient can mean a great increase in profits.

Two designs that would provide better fuel economy are the propfan and the canard.

Propfan Design. Most present-day jet planes are powered by turbofan engines. These engines propel the aircraft at speeds of 500-600 miles per hour. In contrast, regular propeller-driven ("prop") planes have a top speed of 450 miles per hour.

An engine with a propfan design combines the advantages of the prop plane with those of the turbofan aircraft. A propfan engine is actually the same as a turboprop engine (see Chapter 17). The difference is in the propeller blades. They have a different design, and there are more of them.

A propfan engine has eight to ten blades. The blades are shorter than regular propeller blades. Also, they are bent back. See Fig. 24-11. The shorter blades can turn faster than regular blades. The additional number of blades means more power for the aircraft.

The result of the propfan design is an aircraft that can fly as fast as a present-day jet airliner. The big advantage is that the new craft will use 20 percent less fuel.

Canard Design. The word *canard* is French for *duck*. Canard planes look somewhat like a duck in flight. The wings of these planes are placed at or near the rear of the aircraft. See Fig. 24-12. Canard planes don't have a regular tail. The rear of the plane may have the vertical part of the tail. But the horizontal part of the tail is placed in front of the wings.

The canard design results in planes that are light, fast, fuel-efficient, and very safe. The first canard planes were small and were designed for sports pilots. Now, small business planes are being planned. Another possible use of the canard design is in military aircraft.

Other Aircraft Designs. In the future, planes may have wings that can be pivoted. This will provide the best flight for any particular flying conditions. Aircraft may also be able to convert from a plane to a helicopter in just a few seconds. See Fig. 24-13.

Space Travel

The space shuttle will continue to improve our ability to explore space. Shuttle crews are now able to place satellites in very accurate orbits. Other crews can return later to service or repair the satellites.

With the shuttles, we will also be able to construct a large space station in orbit. Technicians at this station will be able to do many things. They will plan and conduct scientific experiments, manufacture products, and even assemble the equipment for a space flight to Mars.

Langley Research

Fig. 24-11 A propfan airliner will be able to fly as fast as today's airliners, but will use 20% less fuel.

Rutan Aircraft Factory

Fig. 24-12 A canard plane has the wings toward the rear of the craft. A small "tail" is usually placed near the front.

Bell Helicopter Textron

Fig. 24-13 This airplane combines the advantages of a helicopter and a regular airplane.

The Father of the MMU

When astronauts take a space walk, they wear a special kind of backpack. It is a personal rocket device called a **Manned Maneuvering Unit**, or MMU. This device lets the astronaut maneuver (move around) in outer space.

The rocket pack started as the idea of a young engineer. This creative person was Charles E. Whitsett, Jr., an Air Force officer. In 1961 Charles began dreaming of a way to "walk" in space without being attached to a spacecraft.

By 1971 Whitsett had developed the **prototype**, or first model, of the MMU. The **M509** used nitrogen gas jets to move the astronaut from place to place. Of course, this rocket pack could only work in the weightlessness of space. The nitrogen jets produced only a small amount of force on earth.

Whitsett further developed the M509 into the present MMU. Operating the MMU is simple. An astronaut can learn to fly one in just five minutes. An MMU has levers on each of its two arms. The astronaut controls 24 nitrogen jets by moving the levers. A microcomputer in the MMU helps the astronaut control the jets.

An astronaut can use an MMU to stay in one position. He or she doesn't have to hold on to something stable. Instead, the astronaut can work in space with both hands free.

Whitsett worked hard on the MMU. To test it, he spent more than 100 hours flying an MMU simulator. He turned his dream of the '60s into a successful tool for the '80s and beyond.

STUDY QUESTIONS

1. What is a laser beam?
2. How do the electromagnetic waves in a laser beam differ from those in normal light?
3. What new form of transmission has been introduced through the use of lasers in communication?
4. True or false: Fiber-optic strands and electrical wires cannot share space without affecting each other.
5. What two things happen to material that is drilled with a laser?
6. How will lasers be used in fusion research?
7. What is cryogenics?
8. What happens to some metals as the temperature drops below −418° F?
9. List two possible uses for superconductors.
10. What does Maglev stand for?
11. What two types of electrical devices would power motors used in electric vehicles?
12. What two designs would provide better fuel economy in air transport?

ACTIVITIES

1. Check with local industries in your area to see whether any companies are using lasers. Prepare a report describing the uses you find.
2. Repeat activity 1 for cryogenics.
3. Visit a local car dealer. Find out the changes made to vehicles in recent years to improve fuel economy. See if the dealer knows about any future plans to conserve even more fuel. Prepare a report describing your findings.

SECTION VIII

Career Planning and Development

Millions of people earn their living working in energy, power, and transportation. Some of these careers are very glamorous. The glamorous careers include airline pilot, astronaut, solar research engineer, airline flight attendant, and cruise ship officer. Other careers are less glamorous, but they provide enjoyable and profitable employment. These careers include sales, management, design, and manufacturing occupations.

Choosing a career is a serious matter. You'll spend a major part of your life working. If you enjoy your work, you will probably be more successful, and are more likely to have a happier life. Use this section of the book to consider your career options. It will provide information on some special skills, teach you the process of choosing a career, and get you acquainted with a few of the hundreds of careers in energy, power, and transportation.

Career Exploration

Your choice of a career will be one of the most important decisions you will ever make. As an adult, you can expect to spend one-quarter or more of your time at work. One-half of your time is used for sleeping, eating, and daily chores. Therefore, your career will use as much of your time as anything else you do.

Your career also affects your personal life. Many friendships are made at work. Also the amount of money you earn will help determine your leisure-time activities. Make your career choice carefully!

There are hundreds of careers in energy sources, transportation, power transmission and control, and the use of power. These careers cover a wide range of interests and abilities. One of them may be right for you.

Later in this chapter, you'll learn about the various careers. However, you should first learn about the work involved in *choosing* a career. To find the life-work that is right for you, you have to be willing to spend some time and effort.

CHOOSING A CAREER

At some stage in life, everyone must make a career decision. A few people decide on a career very early. They work toward it without changing. These people are the exceptions. Most people investigate many different careers before choosing one. Even then, they may enter a career, work in it for a while, and then decide to change careers. Many of these people will admit that they didn't plan ahead. They didn't choose their careers carefully. Or they may not have adequately explored the responsibilities or requirements involved.

It's true that you can't really know *all* the details about a career until you are involved in that career. However, planning ahead is a good idea. It will help you avoid choosing or "drifting into" a career that you won't enjoy at all. See Fig. 25-1.

As you plan your career, think of it as a long-term commitment. Your first career may not be your choice for life. However, many people stay in the first career area they enter for all or most of their lives. Therefore, it's important to choose a career area carefully.

This chapter offers some general pointers on how to go about making career decisions. You may not be ready to decide on a career right now. But it isn't too early to start exploring the different possibilities. See Fig. 25-2.

Fig. 25-1 **Entering a highly technical career such as engineering requires careful career planning. This engineer is inspecting a solar flat-plate collector system.**
Argonne National Laboratory

Fig. 25-2 It's never too early to start thinking about a career. As you do an activity, think about whether you would enjoy a career in that area.

Knowing Yourself

You have interests, skills, and values that are unique (special). Your personal characteristics are the best guides you have in choosing the right career.

Many people never take the time to understand themselves. They don't know what their strengths and weaknesses are. They may not know what kind of activities they enjoy most or do best. They even may not be aware that there are some activities that they *don't* enjoy! See Fig. 25-3.

Knowing what you like and dislike is a matter of experience. To avoid choosing an unsatisfactory career, you must first *know yourself*. Take a good look at your abilities, interests, and values. This "soul-searching" can tell you a lot about the type of career you would be most happy and productive in.

Your Abilities. Think about your abilities in school, in sports, and in recreational activities. Also consider your abilities at doing household chores or handling the responsibilities of a part-time job. You'll start to see what things you do well and what things you do poorly. You may find that you are better in math and science than you are in English. Or you may be better in industrial arts than in physical education. See Fig. 25-4.

Compare your abilities with those of other students your age. As you make these comparisons, recognize that you can strengthen your abilities. You can improve your skills — if you are interested. *Don't* underrate yourself. *Do* develop an understanding of your own strengths and weaknesses.

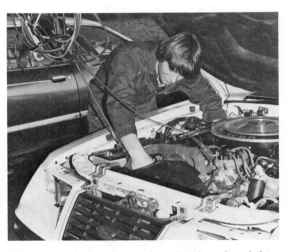

Fig. 25-3 A key factor in selecting the right career is knowing what you like and don't like to do. This person might do well in a career in auto mechanics.

Fig. 25-4 The courses you excel in at school can tell you a lot about your abilities. This person may enjoy a career in which she can use her hands to build things.

Your Interests. Make a list of the things you like to do and the things you don't like to do. Try to relate these likes and dislikes to possible careers.

You may enjoy being outdoors and working by yourself. Or you may enjoy working as part of a group on school assignments. You may be interested in how things work. Or you may like to read and write stories. See Fig. 25-5.

You should see that your interests are a major factor in choosing a career. For example, you may be interested in working by yourself. In this case, you may eventually eliminate careers that involve working with people, such as teaching. You may find being a mechanic, an engineer, or a drafter more satisfying.

Interests, like abilities, can be developed. There are always activities you haven't tried. Keep an open mind as you list your likes and dislikes. As your abilities in an area improve, your interest in the area may also grow. Abilities and interests are often closely related.

Your Values. Values are the things that you feel are important in life. Values include family, religion, freedom, and many others. Career values may include security, good working conditions, money, prestige, respect, the ability to set your own standards, the chance to help and serve others, or the opportunity to protect the environment. See Fig. 25-6.

Your values are an important guide in choosing a career. You should be aware that values can conflict. For example, you may be interested in protecting the environment and in making money. However, you may find that you can make money more easily in business than in environmental protection. As you consider different careers, think carefully about the things that are most important in your life.

Finding Out about Careers

To make a wise career choice, you must compare your personal characteristics and needs with the requirements of the career. But how do you find information about careers you're interested in? There are many answers to this question. The following are some of the best ways to find career information:

Fig. 25-5 If you like to read and study, you may want to consider a career that requires a college education.

Fig. 25-6 Your values are important no matter what you do. Avoid careers that conflict with the things you consider important.

Start at School. School provides a great opportunity to explore various career areas. By taking different courses, you will have many experiences related to many different careers. For example, think about the things you learned while reading this book. Did you enjoy learning about electrical systems? If you did, you may enjoy a career in the electronics field. See Fig. 25-7.

School may also help you learn which career areas you *don't* like. You can learn even from bad experiences. These experiences will help

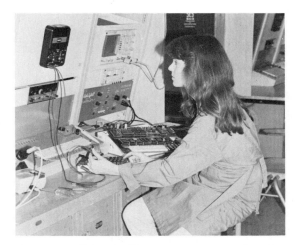

Fig. 25-7 **School offers many career-related experiences. In an electronics course, a student does many of the same operations performed in electronics-related careers.**

Fig. 25-8 **A part-time job is a good way to learn about a career. You can see first-hand the responsibilities involved.**

you focus your attention on the things you enjoy most or do best. However, don't ignore the basics, such as reading, writing, math, science, and social studies. Knowledge in these areas is important no matter what career you go into.

Learn at Work. Summer and part-time jobs are excellent sources of career information. Each job you have will expose you in some way to a new career area. Of course, you'll be on the

bottom, looking up. But you'll still get a first-hand look at the requirements for a career. Many part-time jobs also lead to full-time employment after graduation. See Fig. 25-8.

Do Research. Many careers are hard or impossible to experience through school or part-time work. Engineering and management are two examples. Research is a good way to learn about these careers. You can learn a great deal by sending for employment information, government publications, and career manuals. Your school library and the public library probably have a reference area for career information. Your school counselors are also good sources of career information. Using these different resources is a fast way to learn about many different careers. See Fig. 25-9.

Go on Field Trips. Your school may have a science club, an industrial arts club, or a Junior Engineering Technical Society. Often these groups take field trips to local industries. You can learn a lot about careers by taking field trips. You can actually see where people work and what they do. Try to imagine yourself doing the activities you see. Would you like that type of work? Don't be afraid to ask questions to learn more about the duties of the workers.

Talk with Workers. Most people are more than willing to talk about their careers to interested students. They will be especially helpful if you have questions to ask. Write your questions down before-hand so that you'll be

Fig. 25-9 **Your school library and community library have many publications that will help you find out about careers.**

prepared. You'll usually find that the more interested you are, the more information and help you'll receive. Don't try to hide the fact that you're considering a similar career. Most workers are flattered that someone is interested in doing the same work they do. See Fig. 25-10.

Making Career Decisions

Choosing a career is not a "one-time" activity. Instead, it is an ongoing process. As you consider possible careers, try to learn as much as you can about each one. Find out the working conditions, entry requirements, and major duties. Match these with your own abilities, interests, and values. Also identify the requirements that don't match your personal characteristics. For example, you may be an average or below-average student. However, the career you're interested in may require a college education. Keep in mind that you can improve your abilities. The desire for the career may help you improve your school performance. Think about the good and bad points of possible careers. You may not come up with a perfect match, but you may come close.

Making a career decision is a process of elimination. You eliminate career areas and narrow your choices until only a few careers remain. Don't be impatient. Some people choose a

career very early. Other people don't choose a career until adulthood. Also, your choice doesn't have to be final. You can always re-evaluate it and change it. In the end, choosing a career is an individual decision. Accept suggestions and information from parents and friends. But make the decision yourself.

CAREERS IN ENERGY, POWER, AND TRANSPORTATION

There are many careers in energy, power, and transportation. Careers exist in all parts of the country. The following divisions will give you an idea of some of the career possibilities. The careers mentioned are only a few of the hundreds possible.

Generation of Electricity

Electrical power is extremely important in our modern world. Not many people have ever been to a power plant. However, thousands of plants operate across the country to supply us with electricity. These plants need people in many different careers to operate them. **Engineers**

Pacific Gas & Electric

Fig. 25-10 Here a power company engineer is giving students hands-on experience of some of his responsibilities and duties.

Commonwealth Edison Co.

Fig. 25-11 This engineer is monitoring the production of electricity at a geothermal power plant.

plan and design more efficient power plants. They also direct the operation of existing plants. **Plant supervisors** oversee and direct plant operations and maintenance. **Technicians** and **tradespeople** construct, maintain, and repair electrical plants. They also work on the transmission lines that distribute power to homes and industry. See Figs. 25-11, 25-12, and 25-13.

Indiana & Michigan Power Co.

Fig. 25-12 **This equipment operator works at a nuclear power plant.**

Courtesy of Bell Laboratories

Fig. 25-13 **Maintaining and repairing power lines is a job that is vital to the well-being of the community.**

Energy Exploration and Research

At the present time, energy exploration is a very active career field. Oil companies employ many people in this area. These people search much of the world for sources of oil, natural gas, and other fossil fuels. See Fig. 25-14. There are career opportunities in energy exploration for **geologists, chemists, surveyors, cartographers** (map-makers), and **engineers**. See Fig. 25-15.

Sun Company, Inc.

Fig. 25-14 **These workers are using scientific methods to search for new sources of oil.**

Courtesy of Texaco, Inc.

Fig. 25-15 **This chemist is one of many people doing research on petroleum products.**

Fossil fuel exploration is going full speed. Great efforts are being made to ease our dependency on fossil fuels. As a result, there are many new careers in the research and development of alternative energy sources. Alternative energy sources include geothermal energy, solar energy, wind energy, synfuels, and energy from the oceans. Figures 25-16 and 25-17 show people working in careers in alternative energy.

Transportation

Our nationwide transportation system requires many people to put it into operation and to keep it going. Transportation workers are all trained to do specific jobs.

Many careers are available in transportation management. Railroads, airlines, and trucking companies all need people to manage their activities. Two careers in management are **air traffic controller** and **truck dispatcher**.

Of course, there would be no transportation without vehicle operators. These operators include **airplane pilots, railroad engineers, flight engineers, truck drivers,** and **ship captains.** See Fig. 25-18.

Other transportation workers prepare vehicles for movement. These people include **mechanics, baggage handlers,** and **dock workers.** See Fig. 25-19.

Still other people provide services to passengers. These people work as **flight attendants, ship stewards, sales agents, reservation clerks,** and **computer operators.**

U.S. Department of Energy

Fig. 25-17 **This laboratory researcher is working on a project to make solar cells more efficient and economical.**

Fig. 25-18 **Becoming a railroad engineer takes time and training.**

Fig. 25-16 **Here specially trained workers are assembling a large wind turbine.**

Pacific Gas & Electric

Fig. 25-19 Being a dock hand demands physical strength and a willingness to work hard.

Fig. 25-20 This environmental specialist is taking an air sample to determine the level of pollution.

Conservation and the Environment

In the past 10 years, conservation and environmental control have become more important. People have recognized the growing shortage of fossil fuels. They have also seen the effects of fossil fuel pollution. There are now more jobs than ever before in conservation and environmental control. The career positions include **environmental control specialist, conservation specialist, chemist,** and **scientist.** People in these positions look for ways to conserve energy and keep the air and water free from pollution. See Fig. 25-20.

Automated Control Systems

Automation and robotics are two of the fastest-growing career areas in the nation. The combination of instruments, computers, and automated manufacturing is changing how we build and assemble products. See Fig. 25-21.

Robotic systems are still quite new. Their future growth will depend on research and development. Many new and exciting careers will emerge.

Business

There are also many career opportunities for people interested in business. Careers include positions in sales, accounting, public relations, and management. See Fig. 25-22.

Fig. 25-21 People are needed to test and repair electronic equipment.

Fig. 25-22 Salespeople must be knowledgeable about the product they sell.

What's Your Career Type?

Terms such as *technician* and *engineer* are names for **career types**. Career types are broad categories of the type of work a person does. Career types also indicate the level and kind of education needed for a certain job. The following types are the most common ones:

- **Production worker** — Production workers usually build products in a factory. These products may include cars, trucks, airplanes, fluid power cylinders and valves, electric motors, switches, generators, and lawn mowers. Some production jobs can be very repetitive, such as assembly line work. However, many people enjoy this kind of work.

  Entry into many production positions doesn't require much prior training. You usually learn your duties on the job.

- **Mechanic** — Mechanics are most commonly employed in vehicle repair areas. An **auto mechanic** fixes cars when they need repair. A **diesel mechanic** repairs diesel engines. There are also **aircraft mechanics**, **boat mechanics**, **motorcycle mechanics**, and even **small-engine mechanics**.

  A mechanic is a *skilled* worker. He or she may learn the trade both in school and in an apprenticeship or a similar on-the-job training program. Many mechanics take **certification tests** after receiving their basic training. Being certified makes a mechanic more employable.

- **Technician** — *Technician* and *mechanic* are career types that have much in common. Both involve the service and repair of power system equipment. Mechanics usually work on transportation equipment. Technicians work with other types of equipment.

- **Technologist** — A technologist has more training than a technician. Technologists can get added training in one of two ways: They may complete a four-year college technology program. Or they may receive additional on-the-job training.

Technologist is a relatively new career type. The two main reasons for the development of this type are: (1) the continued growth of knowledge and (2) the need for engineers to become planners and designers. As engineers began to work at higher levels, technologists took over some of their duties. Technologists work at a level mid-way between that of the technician and the engineer.

- **Engineer** — Engineers are graduates of four-year colleges. Their education requires a great deal of math and science, plus specialized engineering courses. People interested in engineering should be strong in math. Mathematics is often called the "language of the engineer."

  In the areas of power and energy, engineers plan, design, and develop power systems and system parts. An engineer's tasks may vary from planning manufacturing operations to researching and developing new products. As you can see, *engineer* is a very broad career classification. There are many different types of engineers. For example, you could become a mechanical engineer, an electrical engineer, or a production engineer.

Career decisions are important. Read and think about the types of careers available.

An Energy Career Ladder

Public utility companies help their customers conserve energy. One method is to provide customers with energy audits. Patricia Holderman began her career as an **energy auditor.** Being employed in that position started her on her climb up an energy "career ladder." A *career ladder* is the framework of advancement or promotion within a company or industry. Getting a better job is taking a step up the career ladder.

Patricia's first job was making energy audits in private homes. Then she was promoted to commercial auditor, her first step up the career ladder. The photograph shows Patricia performing a commercial energy audit. In this photo she is noting the steam pressure used to heat a building. During the audit, she collected a variety of information.

Audits include information on heating and air conditioning, lighting, insulation, and losses of energy through such places as open doors. After being collected, the information is studied and recommendations made which will conserve energy. Patricia's work helped her company save money. And, more importantly, precious energy is being conserved.

Patricia's training for the work included a Bachelor of Science Degree in Environmental Studies.

She emphasized Energy Resources Management in her studies. Today's starting salary in the position of energy auditor is $20,000-$22,000 per year.

Patricia's next step up the career ladder was to her present position of Supervisor of Residential (home) Audits. She has earned two job promotions during the brief 2-1/2 years she has been employed by PG&E. That's two steps up the career ladder already. And her work in Energy Resources Management has contributed in a positive way to her company and to society in general.

STUDY QUESTIONS

1. How can you avoid choosing an unsatisfactory career?
2. What three personal areas can you look at to find out more about yourself?
3. Name five ways to find information about careers.
4. List six general career areas involving energy, power, and transportation.
5. List five career types.

ACTIVITIES

1. Prepare a list of your abilities, interests, and values. Also include your strengths, weaknesses, likes, and dislikes. Self-evaluation is very difficult. Some people evaluate themselves too low; others too high. Try to be as objective and honest as you can.
2. Identify several careers in which you are interested. Gather information about the requirements for these careers.
3. See if you can match your career information (Activity 2) with your knowledge of yourself (Activity 1). Identify the good and bad matches. A perfect match is unusual. However, this exercise will help you get started in your career planning.

387

Special Career Skills

There are many skills that will help you with your career. **Entrepreneurship** is one of them. An entrepreneur ("ahn-tray-pruh-noor") is a person who starts a new business. It is important to know *how* to start your own business. It is also important to know *when* and *if* you should start your own business.

Leadership is another special career skill that you can develop. If you develop your leadership abilities, you will be able to advance further in your chosen career.

ENTREPRENEURSHIP

Each year many new businesses are started. Some are successful, and some are not. Most companies start out small, then grow. Usually the more successful a company is, the larger it becomes. But not all companies turn out to be successful. Many businesses fail each year.

Becoming an Entrepreneur

People become entrepreneurs when they think they can "turn a profit" with a new idea. You see examples of this almost every day. Entrepreneurs are constantly introducing new restaurants, new stores, new services, and new products. Entrepreneurs are the people who "make it happen." They organize and manage the **start-up** of a company. This is a big, tough job. But many people do it every year.

Many famous companies started out as small-time operations in someone's basement, garage, or spare bedroom. Examples include the Coca-Cola Company, Kentucky Fried Chicken, and even Apple Computer. See Fig. 26-1. In each

Apple Computer, Inc.

Fig. 26-1 **Steven Jobs and Stephen Wozniak were the entrepreneurs who started Apple Computer. From its beginnings in a garage, the company grew into a billion-dollar industry.**

of these examples, one or two people took a good idea and developed it. Then they found someone willing to invest money in order to start the business. (To *invest* means to put money into something with the hope of making a **return** or profit on the money.)

Basic Stages of Development

The human body goes through several stages of growth and development. Companies also go through several stages of development. The start-up of a company generally involves the following steps (see also Fig. 26-2):

1. Discovering a product or service that people need and want
2. Getting together a group of people who are willing to help start a company
3. Finding a source of money so that the new company can start operating

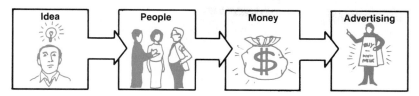

Fig. 26-2 **The four stages of business development.**

4. Letting consumers know about the new company so that they'll want to buy the new product or service.

Once people find out about the new product or service, they *may* start buying and using it. If the entrepreneur's idea was a good one, the company will probably grow.

Venture Capital

Every new business needs money to get started. There are many things to pay for even before you have your first paying customer. These things include the light bill, employees' wages, production materials, and many others. Money, or **capital**, is one of the *inputs* to a business. Since starting a business is a new *venture*, the money used is called **venture capital**.

Venture capital can come from different sources:

• The entrepreneur may put up his or her own money.
• The entrepreneur may try to convince other people to join in the business venture. In return for the money they invest, the other people are made **part-owners**. If the company becomes successful, the investors get their money back, with a profit. But if the company fails, the investors lose their money. As you can see, investing in a new company involves taking a *risk*.
• A third source of venture capital is a loan from a bank or another person.

Types of Company Ownership

There are three basic types of ownership. See Fig. 26-3. The type of ownership may depend on where the start-up money comes from. Or it could depend on how the company grows as it becomes more successful.

Proprietorship. A business that is owned and operated by one person is called a *proprietorship*. The **proprietor** (owner) puts up some or all of the money needed for the operation. He or she is also responsible for making all of the decisions. Finally, the proprietor gets to keep all the profit from the business.

Partnership. In some businesses, two or more people share the responsibility for starting and running the company. These people form a *partnership*. Each partner contributes money and helps run the business. The profit is split between the partners.

Corporation. Most businesses are corporations. A corporation has many owners. The ownership of the business is called **stock**. Each person who pays money into the company receives **shares** of stock. **Stock certificates** show that each investor owns part of the company.

A corporation is recognized as a legal **entity** (body) by the state. This means that the corporation has the same rights as a person. Also, corporate ownership has another big advantage: No single person is responsible for all of the decisions or debts.

Fig. 26-3 **Types of company ownership.**

Proprietorship   Partnership   Corporation

LEADERSHIP

A *leader* is a person who directs others or shows the way. We all have the opportunity to be leaders. For some, the opportunity will be for only a short time, with a small group. For others, the leadership opportunities will be greater. Some people like leadership roles; others don't. In all activities, however, it is important to have both leaders and followers.

The Styles of Leadership

There are three basic styles of leadership. These are autocratic, democratic, and laissez-faire ("lay-zay-fare").

Autocratic leaders determine what should be done. Then they give orders without asking the group members for their opinions.

Democratic leaders also determine what should be done. However, this time the process is different. A democratic leader discusses the

matter with the group. He or she asks for the group's advice and cooperation. The leader then assigns the work and sees that it is completed properly. See Fig. 26-4. The third style of leadership is the **laissez-faire** approach. *Laissez-faire* is a French expression meaning *to let people do as they please*. A leader who uses this approach just lets the group members do whatever they want. The group then goes on by itself.

Good leaders must often use different styles of leadership. The style to use depends on the situation. For example, the group members may not be able to agree on something. In this case, the leader may need to be more *autocratic*. Usually though, the best leaders use the democratic style.

People in groups react differently to the different leadership styles. Some followers may get angry at the leader for making too many decisions. Other people may get angry at the leader for not making *enough* decisions!

Jeff Stoops: Trucker and Entrepreneur

Stoops Express is a trucking firm that operates out of Daleville, Indiana. This company had a gross income of *30.5 million dollars* in 1983. That amount was a long way from the $8000 yearly salary that Jeff Stoops made as a junior high school teacher in 1971.

1971 was the year in which Jeff Stoops decided to go into business for himself. With the help of his father, Stoops bought a tractor-trailer. He then leased his truck to other trucking companies. By 1975, Stoops had a fleet of 12 trucks hauling materials for the auto and steel industries.

A major setback in the auto industry forced Stoops to re-think his own business operation. As a true entrepreneur, he was already on the look-out for ways to expand Stoops Express. He looked around the country and saw that there was a need for better shipping of produce. Stoops got the inspiration to buy some refrigerated trailers. With these, he set up shipping routes and started hauling fresh fruit and vegetables from California to the Midwest.

The produce shipping idea proved to be "right on the money." Stoops Express is now well-established in this profitable area.

The ability to "roll with the punches" and think of new ways to use your resources is the mark of a good entrepreneur. Jeff Stoops didn't limit himself to one way of thinking. He was flexible and open to change. With this ability, he made his business the nation's fastest-growing privately owned business.

Jeff Stoops didn't set out to create a transcontinental trucking empire. He was just looking for ways to increase his income. Stoops has had several good offers to buy his company. But he says he's having too much fun to quit!

Fig. 26-4 **Democratic leadership requires a sharing of problems and solutions.**

What Do Leaders Do?

What a leader does is determined by the needs of the group and the leader's own skills. For example, if the group has a job to do, the leader becomes a **starter**. In this role, the leader helps the group understand what's involved in the task. Then, as a **planner**, the leader helps the group develop a plan to accomplish the task.

The leader's next role is to **guide**. He or she guides the group toward accomplishing the task. While the group is working on the task, the leader also reviews the work and checks the progress. Therefore, he or she also acts as an **evaluator**.

Some Traits of the Successful Leader

One important leadership trait is **understanding**. Of course, this means an effort to understand the thoughts and feelings of other people. However, it also means understanding *yourself*. For example, do you honestly know why you do the things you do? Many people don't understand themselves, and actually, self-understanding is not an easy thing to achieve. But if you want to be a leader, you must make an effort to understand the things that motivate you.

A leader must also understand the **job** or **task** to be done. You must know the task thoroughly yourself before you can guide others through it. Remember, even leaders have leaders. Look for direction from your own leader. If you don't understand the task, get some advice from others. Don't be afraid to ask questions.

There are several other leadership traits. Not all leaders have these traits. However, good leaders display most of them. The "key" traits are:

- Honesty
- Fairness
- Initiative
- Tact (the ability to make and keep friends)
- Enthusiasm (the ability to motivate others)
- Judgment (the ability to evaluate)
- Thoroughness
- Resourcefulness

When you work with others, the most important thing to develop is a sincere relationship built on trust and respect. Here are six things that can help you as you work with people:

- Greet each person promptly and pleasantly.
- Use the person's name.
- Be interested in each person.
- Listen more than you talk.
- *Ask* rather than *tell*.
- If you must say "no," be tactful.

There are many other traits of a successful leader. Some people believe a great leader is just "born that way." However, many people believe that leadership qualities can be *developed*. The different skills can be *learned*. In fact, practicing leadership skills will increase your chances of becoming a leader.

Leadership Tasks

There seem to be two general types of leadership tasks. **Goal achievement** is a type of leadership task that involves setting goals and accomplishing them. Leaders who like this kind of work are called **builders**. Some builders lose interest when the task is accomplished.

Once the goals are achieved, the group needs **maintenance leadership**. A maintenance leader keeps the activities going. He or she makes sure that the goals are continually met. Some leaders work well as both builders and maintenance leaders. Others work best at one type of leadership task.

Skills or Circumstance?

The most important responsibility of a leader is *to give directions*. See Fig. 26-5. There is no other single task or duty that is more important. However, not all leaders are able to carry out this responsibility all of the time.

Sometimes a leader does his or her best, and still the group is not successful. It may not be the leader's fault; it could be because of *circumstances*. For example, a football team might play their best game ever and still lose because

Metropolitan Edison Co.

Fig. 26-5 **Good leadership includes giving instructions patiently and carefully. The engineer on the left is giving special instructions to a control-room operator at a nuclear power plant.**

the other team played better, or because the field was muddy. Maybe a correction could have been made and the game could have been won. Maybe not.

Yes, circumstances do play a role in the success of any leader. Some leaders seem lucky. They seem to be at the right place at the right time. But luck or not, the skills of a leader can be developed. Being *prepared* for leadership opportunities is the important thing.

STUDY QUESTIONS

1. What is an entrepreneur?
2. Identify the four stages of developing a new company.
3. What is venture capital?
4. Name the three types of company ownership.
5. What two advantages does a corporation have?
6. Name the three basic styles of leadership.
7. Name four leadership roles.
8. List five traits of the successful leader.
9. What are the two general types of leadership tasks?
10. What is the most important responsibility of a leader?

ACTIVITIES

1. Think of a business you might start now or in the future. Describe the product or service you will provide. Explain why you think the business is needed. List possible sources of venture capital for your business venture.
2. Interview a leader you know. He or she should be someone who is responsible for the work of others. Ask the leader to describe the traits he or she believes to be most important for leadership.

How a Good Group Works...

1 The Industrial Arts Club is holding its regular meeting tonight. The secretary reported that the annual conference of the American Industrial Arts Student Association is coming up soon. The members are discussing the possibility of attending the conference.

2 The president guides the members through an organized discussion. The members expressed their interest in attending the conference. Now the president is selecting a conference committee. Committee members will review conference information, find out the cost of attending, and consider other points about the trip. They will report their findings at the next regular meeting.

3 The committee and the club advisor meet with the school principal. They ask about the school policy on student travel, teacher responsibilities, and other matters.

A special thank you to the American Industrial Arts Association group at Carthage Junior High School, Carthage, Texas, for their participation in this photo essay.

4 The conference committee gathered a lot of information. Now they must organize the material and develop a presentation for the next club meeting.

5 At the regular meeting time, the president asks if there is any "old business" to discuss. The committee chairperson asked to be "recognized." Here she is presenting the results of the committee's investigations.

6 When the committee's report is finished, the club advisor gives her approval for the proposed adventure. However, she suggests that the students' parents be involved in making the final plans.

7 The president entertains (asks for) a motion from the floor. A *motion* is a formal proposal. One member moves (makes the motion) that the group vote on the issue of attending the conference. Another member agrees with, or "seconds," the motion. Making motions and seconding are both parts of an orderly process called parliamentary procedure. The president "tables," puts aside, the motion until the next meeting.

Tabling the motion gives students interested in the conference trip time to speak with their parents. They talk about the trip in general and go over the information presented by the conference committee.

Many parents support the conference idea. Along with the school principal, they attend a special club meeting to discuss the trip. The parents ask several more questions, and suggest ways that they might help the club in its efforts. By the end of the meeting, the president has appointed a fundraising committee and a planning committee.

The planning committee is in charge of developing the travel plans. Their main resource person is a local travel agent. At the travel agency, they discuss ways to transport the members to the conference by an enjoyable, yet economical, mode.

The club secretary has a job to do, too. The president has assigned him to complete the registration forms for the conference. The club advisor helps the secretary make room reservations with the conference housing office.

The fund-raising committee is hard at work. In the past, the club has been successful at raising money by selling items that they have manufactured in the tools and materials lab of the school. The members of the committee must think like entrepreneurs: "Let's see — what can we make that the people would want to buy?"

12

Meeting time again! The travel agent's assistant tells the club about the interesting travel package planned for them. The transportation will be by chartered luxury bus, with interesting side trips on the way and when they return.

13

The president now asks for a vote on the travel package. The vote is unanimous! Everyone responds with "yes!" The club is on its way!

14

Departure day finally arrives! Three other adults will be traveling with the group to assist the advisor. The president has done a fine job with the leadership responsibilities. With his guidance and direction, each member was able to contribute to the group's goal. And the adventure has just begun!

15

GLOSSARY

acceleration. An increase in the speed of an object.

acid rain. Rain containing an unusually high quantity of acid. Acid rain usually results from high concentrations of sulfur dioxide (SO_2) in the air.

actuator. A fluid power device that changes fluid power to either rotary, linear, or reciprocating motion.

alternating current (AC). Current that alternately changes polarity (direction). The electrons flow in one direction, then reverse.

alternator. An electrical device that uses stationary conductors and a rotating magnet to generate alternating current.

ammeter. An instrument used to measure the amount of current in a circuit.

ampere (A). A unit of measurement for the rate of electrical current flow.

artificial intelligence (AI). A developing computer system which will help solve complex problems through analysis of stored knowledge.

atmospheric pressure. The pressure produced by the weight of the blanket of air surrounding the earth: 14.7 pounds per square inch.

atom. The smallest particle of an element that can retain the properties of the element and take part in a chemical reaction.

autocratic. A leadership style in which decisions are made by one person, without involving others.

automated control. The continuous automatic operation done to perform a task such as operating industrial or home devices.

Automated Guideway Transit (AGT). Vehicles which operate automatically to move people from one location to another.

automation. A set of operations performed or controlled in a planned order.

battery. Two or more cells connected together.

bearings. Devices that reduce friction between moving surfaces.

bioconversion. The process of obtaining energy from society's waste products.

bit. A binary unit. It is the basic unit of information in a computer. It has two possible values, 0 and 1 or yes and no.

Boyle's Law. *The volume of a gas varies inversely with the pressure applied to it, provided the temperature remains constant.*

British thermal unit (BTU). The amount of heat needed to raise the temperature of one pound of water one degree Fahrenheit.

byte. A unit of information for a computer. Consists of eight bits of information.

cableways. Special cars or chairs hanging on a moving cable, used to move people, such as skiers.

capacitive discharge (CD) system. A solid-state ignition system that eliminates the use of breaker points.

capital. In business, the money used to start or operate a business.

carburetor. The part of a gasoline fuel system that provides the proper mixture of air and fuel to the engine.

career. A job or position. Usually a person's main source of income.

cargo. Solid, liquid, or gaseous materials being moved from one place to another.

carrier. A transportation company which moves people and/or cargo.

Celsius temperature scale (°C). The scale used in the metric system to measure temperature.

central processing unit (CPU). The part of the computer which interprets and executes programs.

centrifugal force. A force that causes a rotating object to move away from the center of rotation.

Charles' Law. *The volume of a gas varies directly with the temperature applied to it, provided the pressure remains constant.*

chemical energy. Energy produced by chemical changes. Chemical energy is the source of energy for all living things.

circuit. See **electrical circuit**.

clutch. A transmission device that controls the transfer of power from an input shaft to an output shaft.

cogeneration. Using a single energy source for two useful purposes. Example: Steam run through turbines for generating electricity can also be used for heating.

computer. An electronic device designed to store and manage information, and solve problems.

computerized ignition and fuel control. A computer-controlled ignition and fuel-injection system used on some modern automobile engines.

conduction, thermal. The movement of heat through a substance.

conductors. Materials that easily transmit (conduct) electricity.

consignee. The person or company receiving freight.

containerization. In transportation, shipping goods in a large box which can be filled at one location, transported via different modes, and delivered to its destination without rehandling the goods.

convection. The movement of heat through air, water, or other fluids. The heated fluid carries the heat from one place to another.

conveyors. Continuous devices that move constantly, such as belts or overhead trolleys. Conveyors move goods from one location to another.

corporation. A legal structure, licensed by a state or country. The corporation owns and operates the company or business. Many people share ownership of the corporation.

coulomb. The basic unit of measurement of electricity. It is 6.28×10^{18} electrons.

cracking. The process of further refining crude oil through the use of a catalyst, thereby producing additional gasoline.

cryogenics. The science that deals with the properties and behavior of materials at very low temperatures.

current, electric. The movement of electrons through a conductor.

democratic. A leadership style in which people share in the decision-making.

diesel engine. An internal-combustion engine that burns diesel oil to produce heat. Pressure inside the combustion chamber causes the fuel to ignite.

diode. A solid-state electrical device that permits electrons to flow in only one direction.

direct current (DC). Current that flows through a conductor in only one direction.

dynamometer. A device used to measure the power produced by an engine or motor.

efficiency. The percentage of available energy converted into usable energy.

electrical circuit. A system of conductors and electrical devices through which an electrical current moves.

electrical energy. The motion of electrons. Electrical energy is measured in joules.

electricity. The movement of electrons from one atom to another.

electrolyte. A current-carrying solution, such as is used in storage batteries and fuel cells.

electromagnet. A magnet consisting of an iron core wrapped with a current-carrying wire.

emission control. A system of devices on automobile engines to reduce the amount of pollution produced by the engine.

energy. The ability to do work, or the capacity to produce motion, heat, or light. Energy is measured in foot-pounds, British thermal units, and joules.

energy conversion. The process of changing energy from one of its six forms to another.

energy system. A system for controlling energy. It includes the original source of energy, all the conversions and transmissions the energy undergoes, and the eventual use of the energy.

engine. A device that converts any form of energy into mechanical energy.

entrepreneur. A person who starts a new business.

environment. Natural surroundings.

exhaustible energy sources. Sources of energy that cannot be replaced after use. These sources include fossil fuels and uranium.

external-combustion engine. A heat engine that uses heat and pressure produced outside of the engine.

Fahrenheit temperature scale (°F). The customary temperature scale.

fiber-optic system. A communication system that uses semiconductor lasers and strands of flexible glass for transmission.

fission. The splitting of atomic nuclei to release energy.

fluid. Any liquid or gas.

fluid power. The use of fluids under pressure to control and transmit power. Fluid power includes the use of both gases (pneumatics) and liquids (hydraulics).

foot-pound (ft.-lb.). Unit for measuring mechanical energy (work).

force (F). Energy exerted. Any push or pull on an object. Force is measured in pounds.

fossil fuels. Fuels that developed over millions of years from dead plants and animals. Fossil fuels include coal, oil (petroleum), and natural gas.

four-stroke cycle engine. An engine that requires four strokes of the piston to produce a single power stroke. The four strokes are intake, compression, power, and exhaust.

fractionating. Part of the refining process in which vapors rising in a tower condense at different levels.

freight. Cargo being transported to a new location.

friction. The resistance to motion produced when two objects rub against each other. Friction produces heat energy.

fuel. Any substance that can be burned, fissioned, or used in a fusion reaction to produce heat.

fuel injection. The supplying of fuel to an engine's combustion chamber through an injection device rather than with a carburetor.

fuse. An electrical device that protects a circuit from excessive current flow.

fusion. The fusing (combining) of atoms to release energy.

gas turbine. A jet-type engine used to provide rotary motion for vehicles and electrical generators.

gasohol. A mixture of nine-tenths unleaded gasoline and one-tenth ethyl alcohol.

gasoline piston engine. An internal-combustion engine that uses gasoline to produce heat. It is the most common type of automobile engine.

gear ratio. The ratio of the number of teeth on a driven gear to the number of teeth on the drive gear. This ratio determines the relationship between force and speed (distance per unit of time).

generator. A device that converts rotary motion into electrical energy.

geothermal energy. Heat energy generated within the earth. It is the result of the natural decay of radioactive materials.

governor. A device that automatically regulates the speed of an engine. It is usually used on small engines.

ground. An electrical connection that permits electricity to return to its source. In automobiles, the ground is the automobile frame.

hardware. The mechanical parts of a computer.

heat. The motion of atoms or molecules. Heat is present wherever there is motion.

heat engine. A device that converts heat energy into mechanical energy.

heat exchanger. A device that transfers heat from one fluid to another.

hertz. Electrical cycles per second. In the United States, alternating current is 60 hertz.

highway transportation. Moving passengers and cargo over roads and highways.

horsepower (hp). A unit of measurement of power. One horsepower is the energy needed to lift 33,000 pounds 1 foot in 1 minute.

hydraulics. The use of liquid under pressure to produce motion.

hydrocarbons. Compounds consisting of combinations of carbon and hydrogen. Hydrocarbons include the fossil fuels and such biomass materials as wood.

hydroelectric energy. Electrical energy produced by flowing water. Water is collected behind dams. When released, it is used to rotate a turbine.

induction. The process of transmitting electrical energy from one circuit to another through the building and collapsing of a magnetic field.

inertia. The tendency of an object at rest to remain at rest, and of an object in motion to continue in motion.

inexhaustible energy sources. Energy sources that will always be available. These sources include solar energy, hydroelectric energy, tides, ocean thermal energy, solar salt ponds, and geothermal energy.

instrument. A device to sense, measure, and display needed information.

insulation. Material used to reduce the flow of heat.

intermodal. In transportation, means moving passengers or cargo using more than one mode or method.

internal-combustion engine. A heat engine in which the heat and pressure are produced inside the engine.

isotopes. Different atoms of the same element. The difference is in the number of neutrons in the nucleus. All isotopes of a single element have the same number of electrons.

jet engine. An internal-combustion engine that produces linear motion through the principle of jet propulsion.

jet propulsion. The principle of Newton's third law of motion: *To every action there is an equal and opposite reaction.* Escaping gases reduce pressure at one end of a cylinder, producing thrust (force) at the other end.

kinetic energy. Energy in motion. All energy performing work or producing power is kinetic energy.

laissez-faire. A leadership style which permits members of the group to do as they wish.

laser. A concentrated beam of light that travels in a very narrow straight path.

Law of Conservation of Energy. Energy cannot be created or destroyed; the amount of energy in the universe is fixed. However, energy can be changed from one form to another.

leadership. A personal responsibility for the performance and accomplishments of other people.

light energy. The visible part of radiant energy. Light consists of electromagnetic waves traveling through space.

linear motion. Motion in a straight line, such as produced by a jet engine or rocket.

machine. A device that changes the relationship between force and speed (distance per unit of time). Simple machines include the lever, wheel and axle, pulley, inclined plane, wedge, and screw.

magnetic field. The area around a magnet or current-carrying wire in which magnetic attraction or repulsion takes place.

matter. Any substance that has weight and takes up space.

mechanical advantage. An increase in force or speed (distance per unit of time) gained through the use of a machine.

mechanical energy. The energy of motion, the most common and visible form of energy. It is measured in foot-pounds or joules.

memory. The computer unit which stores information.

methanol (methyl alcohol). A clean-burning liquid fuel made from wood or plants, including waste wood and vegetable products.

microcomputer. A computer constructed around a single integrated circuit. Also referred to as a personal computer.

microprocessor. A single-purpose computer that has integrated memory and controller units which control a transportation, home, or industrial device.

microwaves. High-energy electrical waves used to transmit electrical energy through the atmosphere and outer space.

minicomputer. A mid-sized computer used in business and industry for data management and mathematics.

molecule. The smallest particle of a chemical compound that retains the properties of the substance.

momentum. The measured force of a moving body. The faster a body moves, or the greater its weight, the greater its momentum.

monorail. A train that runs on a single track.

motor. An electrical- or fluid-operated device that produces rotary motion.

Newton's Third Law of Motion. See **jet propulsion.**

nuclear battery. A type of battery that produces electrical energy from radioactive materials.

nuclear energy. Energy produced by reactions in the nuclei of atoms.

nuclear reactor. A device in which a fission or fusion reaction is started, continued, and controlled.

nuclear wastes. The byproducts of a nuclear reaction, usually very radioactive and dangerous.

ocean thermal energy. Energy generated by the difference in water temperature between surface and deep ocean water.

ohm (Ω). The unit of measurement of electrical resistance.

Ohm's Law. *It takes one volt to force one ampere of current through a resistance of one ohm.*

parallel circuit. An electrical circuit in which current flows in more than one path.

particulates. Tiny particles of matter released into the air by burning fossil fuels.

partnership. Two or more people who share responsibility for starting and operating a business or company.

passengers. People who are being transported from one place to another.

personal computer. See **microcomputer.**

photovoltaic. Pertaining to the process by which light energy is converted directly into electrical energy, usually through the use of a photovoltaic cell (solar cell).

photosynthesis. The conversion by plants of sunlight, carbon dioxide, water, and nutrients into food and plant material.

pipeline transportation. Movement of cargo through a tube-like vessel, a pipe.

plutonium. A radioactive element found in small quantities naturally and produced from uranium-238 during nuclear fission. It can be used as a nuclear fuel.

pneumatics. The use of air under pressure to produce motion and perform work.

polarity, electric. The direction of current flow. Current always flows from negative (area of more electrons) to positive (area of less electrons).

polarity, magnetic. The direction of magnetic lines of force. Around a magnet, the lines of force move from north to south. Inside a magnet, they move from south to north.

pollution. Any undesirable change in the air, land, or water that harmfully affects living things.

port. A place where ships dock to load and unload passengers and cargo.

potential energy. Stored energy, or energy ready or available for use. When used, potential energy changes to kinetic energy.

power. Energy (work) per unit of time, or work accomplished in a given period of time. Power is measured in horsepower, BTUs per hour, and watts.

pressure. A measurement of force determined by the area on which the force is applied. Pressure is force per unit of area. Pressure is measured in pounds per square inch and inches of mercury.

primary cell. A device that stores chemical energy. The chemical energy is converted to electricity as needed. Primary cells cannot be recharged.

program. A planned set of instructions placed into the computer memory.

proprietorship. A business owned and operated by one person.

pump. A device that converts mechanical power into fluid power.

radiant energy. A form of energy produced by any warm or hot object, such as the sun. It is a combination of light and heat energy. Radiant energy changes to heat energy when it strikes a solid object.

radiation, atomic. Energy released during changes in the nuclei of atoms. Atomic radiation includes alpha particles, beta particles, and gamma rays.

radiation, thermal. The transfer of heat by electromagnetic waves through space and air.

radioactivity. The release of atomic radiation during the disintegration of the nuclei of certain atoms.

rail transportation. Moving passengers and cargo in vehicles that run on rails.

ramjet engine. A jet engine that uses the forward motion of the engine to bring air into the combustion chamber and to compress the air.

reciprocating motion. Back and forth movement, such as a piston moving inside a cylinder.

refining. The process of separating a crude substance into purer, useful substances. In crude oil refining, this process is called fractionating.

renewable energy sources. Sources of energy that, with proper management, will be available indefinitely. Renewable sources include wood, plants, and waste products.

resistance, electrical (R). The opposition to current flow through a conductor. Resistance is measured in ohms.

robot. A computer-controlled device which performs human-like operations.

robotics. The field of study dealing with the construction, maintenance, and use of robots.

rocket engine. An engine that operates on the principle of jet propulsion and which carries its own supply of fuel and oxygen.

rotary engine. An internal-combustion engine in which a triangular rotor rotates within a housing. The motion of the rotor rotates an output shaft.

sea-lane. An established shipping route across an ocean.

secondary cell. A chemical storage cell that can be electrically discharged and recharged repeatedly.

semiconductor. A material that has some of the properties of both insulators and conductors. Semiconductors are the basic materials for solid-state electrical devices.

series circuit. An electrical circuit that has only one path for current flow.

shipper. Person or company that sends cargo.

short circuit. An accidental bypassing of the normal resistance in an electrical circuit.

slurry. A mixture of solids and liquids. The resulting product (like a mud) is transported through a pipeline.

smog. A major form of air pollution produced from sunlight acting on hydrocarbons and nitrogen oxides in the air.

software. The programming components of a computer system, including the programs, languages, and procedures.

solar battery. Two or more solar cells grouped together to increase the electrical output.

solar (photovoltaic) cell. A device that converts light into electricity.

solar energy. Energy from the sun. Solar energy is our most basic source of energy.

solar heating, active. A solar heating system that collects solar energy and moves it by mechanical means to where it is needed.

solar heating, direct-gain. The heating of living space directly with sunlight.

solar heating, indirect-gain. The heating of living space by first heating a thermal mass such as water or masonry. The thermal mass then heats the living area by radiation and convection.

solar heating, passive. A solar heating system that collects, stores, and transfers heat by natural means. Passive heating systems include both direct-gain and indirect-gain systems.

solar panel (collector). A device used in an active solar heating system to absorb solar energy.

solenoid. A device that converts electrical energy to linear motion.

steam engine. An external-combustion engine that converts the heat and pressure of steam into mechanical energy.

steam turbine. An external-combustion engine used to produce rotary motion. Steam turbines are commonly used to produce electricity and power large ships.

stroke. The movement of a piston from one end of a cylinder to the other.

superconductor. An electrical conductor that has lost all detectable electrical resistance because its temperature has been reduced below $-418°$ F.

synfuels (synthetic fuels). Liquid or gaseous fuels made from existing solid fuels. Synfuels are produced from coal, tar sands, and oil shale.

system. An organized way of doing something. Also a set of things that are related.

technology. The knowledge of doing things that can make life easier. Also the technical method of accomplishing practical purposes.

terminal. A building used for loading and unloading passengers and cargo. Also, a part of a generator or battery to which the conductor attaches.

thermal energy. See **heat.**

thermal mass. A heat storage material, such as water, masonry, or concrete, used in passive solar heating systems.

thermal pollution. The release of waste heat into air or water by power plants. This waste heat upsets the delicate balance of nature.

thermoelectric coupling. A device used to generate electricity from a heat source. The coupling uses two different materials joined together. Heat applied at one end produces electricity at the other.

thermosiphoning. The circulation of water by natural convection. Thermosiphoning operates on the principle that as water is heated it expands and rises.

thermostat. A device that senses and regulates temperature.

thrust. A forward push or force produced by a jet or rocket engine. Thrust is measured in pounds.

tidal energy. Energy generated by using the flow of ocean tides.

torque. A twisting or turning effort; also, a measurement of force applied to a radius. Torque is measured in pound-feet.

transformer. A device that transmits electricity from one circuit to another by induction, changing amperage and voltage in the process.

transistor. A solid-state electrical device that is used as an on-off switch.

transportation. The movement of people or cargo from one location to another, using vehicles.

turbine. A rotating device driven by wind or water and used to drive an electrical generator.

two-stroke cycle engine. A heat engine that requires two piston strokes to produce a single power stroke. The two strokes are compression and power. Intake and exhaust take place at the bottom of the power stroke.

uranium. A radioactive element used as fuel in nuclear reactors.

vacuum. A pressure below atmospheric pressure. Zero pressure is a perfect vacuum. Vacuums are measured in inches of mercury.

velocity. The speed of an object.

viscosity. A measure of the thickness of an oil.

volt (V). A unit of measurement of electrical pressure. One coulomb of electricity exerts a pressure of one volt.

Wankel engine, See **rotary engine.**

water transportation. Moving passengers and cargo in vessels on water.

watt (W). The measurement of electrical power. One watt equals one joule of electrical energy per second. One watt is also equal to a flow of one ampere at a pressure of one volt. 746 watts equals one horsepower.

wind power. The use of wind to generate electricity or produce motion.

work. Motion that produces a desired outcome or accomplishment. Work is equal to force times distance and is measured in foot-pounds.